Beyond Natural Selection

Contents

Acknowledgments

I am deeply indebted to the many people with whom I was able to discuss the manuscript for this book. It has been read, in whole or part, at various stages of completion, by Francisco J. Ayala, Raymond Barnett, William Warren Bartley III, John H. Campbell, David Depew, Ronald Fox, Michael Ghiselin, Norman Goldstein, Robert Jurmain, Vida Kenk, Christopher Kitting, Mark Kirkpatrick, George L. G. Miklos, John Steiner, John Thomas, and Bruce H. Weber. As anyone familiar with writings in the area may guess, their opinions have been divided ideologically in much the way the biological profession is divided, between those who are firm supporters of the neo-Darwinist synthesis and those who are hospitable to criticism of it. All, however, have given useful suggestions. I am especially grateful to John H. Campbell and Bruce H. Weber for their wisdom and encouragement.

Loren Rusk's editing was extremely helpful and produced a more readable text. Several biology students assisted in checking facts; Carol Lee and Kara Takeuchi in particular showed themselves capable young scientists.

Introduction

"How have we come about?" is one of the most profound of questions. How we see our origins, whether by a higher will or by chance-generated changes interacting with the physical environment, lies, with the kindred question of the relation of mind to matter, at the heart of philosophy, religion, and human relations.

Religions usually attribute human existence to the will of a supreme being. Charles Darwin gave a plausible nonreligious explanation in *On the Origin of the Species by Means of Natural Selection,* published in 1859. Since then it has been scientifically accepted that all living creatures are related, having evolved and diverged from common ancestors, back to the distant genesis of life on earth. A vast array of evidence, such as fossils showing stages of the development of plants and animals and the resemblances of living creatures in structure and physiology, supports this idea.

The reality of evolution is not questioned in the scientific community. But we do not know what to make of the kinship of living beings unless we understand why different species came to be so very different. It is important for our self-estimate how the first little self-reproducing, that is, living globules gave rise to such admirable animals as ourselves.

Darwin's answer to this question was that the differences among species, not only between wolves and foxes but between elephants and spiders and bacteria, result from the accumulation of random variations in the heredity of organisms. Because some are born with traits enabling them better to survive and reproduce, over the generations there will be more and more of the superior stock, and the species will change, or evolve, accordingly. This process, going on through unimaginable ages, is called natural selection or survival of the fittest.

Although a large majority of biologists accept Darwin's theory with few qualifications, many were dubious of it from the time Darwin proposed it until well into this century, when it was systematized in the neo-Darwinist synthesis. The orthodoxy became very firm, especially in the 1960s. Recently, however, there have been increasing tendencies to doubt that the role of natural selection is as great as has been assumed, and a growing number of biologists believe that it is not a wholly satisfactory answer. Its inadequacy is a thesis of this book.

This position implies no denigration of Charles Darwin and his legacy. He was a great thinker and a dedicated researcher and writer in an age when most scientists were more or less dilettantes. He not only worked out and impressively argued for his theory of evolution but industriously carried out many detailed studies of animals and plants. He more than anyone else laid the foundations of the science of biology (Mayr 1982, 108). It was a remarkable achievement that he was able to build a theory that, as it has been developed, sheds much light on the organization of living nature. But it was impossible that he could come to a fully satisfactory explanation of evolution on the basis of the limited information available in his day.

In the light of the vast amount of knowledge of all aspects of living creatures piled up in the last century and especially in the last decades, this book seeks to present a soundly based and objective critique of Darwinism. There is room for such an effort, if only to call attention to matters needing closer scrutiny; it underlines many puzzling things with which exponents of conventional evolutionary theory have felt little need to deal. It is not intended as support for any religious teaching, although those who feel uneasy with a mechanistic interpretation may find comfort in it.

Unhappily, however, pointing out the need for a better explanation means attacking a theory that scientists find useful, if not always satisfying. They certainly do not want to surrender the accepted doctrine unless they have something better. A natural rejoinder to criticism is, What do you have better to put in its place? Natural selection is credited with seemingly miraculous feats because we want an answer and have no other.

There probably cannot be another general answer—certainly no equally broad and basically simple answer. Biologists, it seems, must do without a comprehensive theory of evolution, just as social scientists have to make do without a comprehensive theory of society. But this book ventures a number of ideas and shifts of emphasis that point broadly to the autonomy of life, its self-organizedness or self-directedness.

Reductionism, the effort to understand by reducing something to simpler basics, has limited value in evolutionary theory and may be misleading. Chaos, the recently much discussed domain of the turbulent and unpredictable, seems to have been important in evolutionary innovation. It also appears evident that evolutionary change, although chaotic, is by no means random. Patterns of development channel change. Evolutionary trends resemble neither cowpaths free to wander anywhere nor rivers fixed by the landscape but are like roads that conform to various needs and constraints.

Moreover, the interaction of organisms and environment can hardly be limited to chance and the selection of survivors that will propagate their kind. Adaptive innovations, especially of higher animals, may be commonly initiated by behavior or learning. That is, exploration, choice, and learning seem to play a key part. This idea has implications quite different from the view that random variation and selection are the sole movers of change. The emergence of the human species in particular is not to be comprehended simply as the outcome of accidental mutations and natural selection for intelligence.

In this view, evolution is not only adaptation, as Darwinism stresses, but also the realization of inherent potentials. What a species becomes depends on both its relationship to its environment and its heritage or constitution. Evolutionists have strongly emphasized the former, which is definite and subject to investigation. How the evolution of a species is conditioned by internal factors is difficult to evaluate, but there is much reason to believe that evolution, although conditioned by natural selection, is to a large extent innerdirected.

This book can only skim the disciplines involved and the knowledge available. Chapter 1 takes up the conventional Darwinian theory. Although it is ordinarily accepted without question by practicing biologists, there is persistent controversy over its application. But it is charmingly reductionist, offering an encompassing explanation on the basis of simple axioms and material relations, and its appeal is partly ideological.

The next six chapters examine reasons for regarding explanation by natural selection as insufficient. The most general of these is chapter 2, which contends that the reductionist emphasis is discordant with the nature of the cosmos as revealed by modern science. The effort to understand complex systems by breaking them down into simpler entities has limited usefulness, and new and unpredictable relations enter at each level of higher organization of matter. Living nature cannot be fully described by any simple law. Evolution seems to be infused with what is rather misleadingly called chaos.

Those who are more interested in verdant nature than gray evolutionary theory might skip much of the first two chapters and go directly to chapter 3, which takes up the tantalizingly incomplete fossil record of life. The crucial parts of the story— the transitions from one form to a significantly different one— are missing. The origins of such novelties as birds or whales raise difficult questions.

Creationist critics of Darwinism emphasize the incompleteness of the record, but many facts about living creatures are more difficult to account for within the accepted theory. Chapter 4 notes some of the countless ways of animals (and to some extent plants) in the "believe it or not" category. Among them are the classic problem of the eye, the sonar apparatus of bats and porpoises, the electric organs of some fish, the problem of migration, and seemingly intelligent instincts. Many traits require both elaborate instincts and structures, which can function only as part of the entire set.

Although it may be hard to envision how many incredible adaptations were developed, their utility is obvious. Other traits have no apparent utility or are negative, including infertility or seemingly premature death. Such matters are treated in chapter 5.

Chapter 6 takes up the often extravagant and apparently useless developments around sex and why it should be necessary, despite high costs, to mix genes from different individuals.

Chapter 7 deals with a problem that has received much attention: the apparent sacrifice of individuals in favor of the group (the extreme examples are such colonial insects as ants and honeybees). Sociality or altruism is difficult to reconcile with natural selection based on the individual, but it is important in nature and for our social species.

1
The Conventional Theory: "It Is Solved"

The Reductionist Approach

Today's evolutionary theorizing suffers a problem the opposite of Darwin's: a surfeit of information. Specialists today do well to keep up with their small corner of the immensity of life. Biochemists and physiologists have more than they can comprehend in its molecular properties. Paleontologists can only begin to decipher the book of rocks. Many volumes are filled with the genetics of the bacteria that all of us carry around by the trillions, *Escherichia coli*, or little fruit flies, *Drosophila*. Ethologists are confounded by the marvelous medley of animal behaviors. Ecologists can hardly begin to understand the interactions of species and environments. Taxonomists make only a brave beginning of sorting out the multiplicity of forms. Although what we know is an infinitely small part of what there is to know, it seems beyond human capacity to bring all of this knowledge together.

In this situation, a simplifying theory is greatly to be desired, and many biologists believe that they have such a theory, agreeing with a widely read book on evolution that "it is a mystery no longer, because it is solved." There remain only questions of detail (Dawkins 1986, ix). –selfish gene–

The accepted modern theory, essentially that worked out by Darwin a century and a half ago, rests on a few obvious and plausible propositions. Animals and plants have more offspring than can survive and reproduce in the long run. The young are not exact copies of their parents, and differences are frequently inheritable. If an inheritable variation gives some individuals a competitive advantage, they will leave more descendants. Differential reproduction with inheritable v

tion and the sieve of selective survival account for the development, or evolution, of all living things.

This idea is summed up as natural selection, although there is no selection in the sense of choice. More descriptive is "survival of the fittest," a phrase Darwin took from Herbert Spencer. This sounds tautological because the definition of fitness is the ability to survive (and reproduce). However, since the race is to the fastest (or fittest), the winners enter the next race and produce the next set of contestants. Just why the winners win we may not know, but they are enabled by their varying qualities to procreate others like themselves. This provides the framework for a complete theory of how life evolves.

The theory of natural selection is neat and appealing. Undeniably, offspring often differ from their parents, differences can be inherited, and inherited traits can enable some to leave more descendants than others. The logic seems so solid that, in the view of Dawkins, "even if there were no actual evidence in favor of the Darwinian theory, we would still be justified in preferring it over all other theories" (Dawkins 1986, 287).

Most biologists are not quite so sure, but they accept the conventional theory as their frame of reference. As zoologist John Alcock expresses it, "An animal's appearance, physiology, and behavior can be assumed to be perfectly adaptive, for the purposes of productive speculation, because less adaptive characteristics should have been replaced by superior ones as these arose by genetic mutation over the millennia" (Alcock 1988, 7–8). We may not have an answer to many enigmas of structure and behavior, but we know where to look.

Saying "it is solved" expresses a hope. Biologists, like ordinary folk, want to understand, or to feel that they understand, how to relate visible things to an idea or theory, to find some sense in things, as when the child asks, "Why is the sky blue?" Science, a synonym for systematic learning and objective knowledge, seeks to give explanations. Representing the work of millions of minds over centuries, it is the most successful human endeavor. Scientists naturally take pride in their ability not only to make useful discoveries but also to analyze and explain.

The principal method of science is to take things apart, reducing the complex to simpler components—reductionism, it is called. By studying the interactions of parts, one achieves

Chapter 8 considers aspects of the dynamics of evolution. This requires some discussion of the apparatus of heredity and the problem of conveying the vast amount of information needed to make an animal. Much of this information seems to be independent of environmental needs; that is, organisms are varied beyond need in an essentially chaotic manner. Feedback also has a large role, apparent in symbiosis, specialization, parasitism, and sexual selection. It is helpful also to think of the genome in terms of dynamic attractors.

Chapter 9 looks at the ways in which genetic patterns or attractors influence evolution. Evolutionary development is not only externally but internally constrained by propensities to certain directions of change.

Change of pattern is the subject of chapter 10: speciation, microevolution, modes of change, stability, diversification, and extinction.

Chapter 11 considers the relationship of heredity and environment and genetic response to external signals. Much of the adaptation of animals, however, is behavioral and the relation of instinct to learning is problematic.

Chapter 12 applies evolutionary theory to humans. The evolution of intelligence and the mind defies understanding in conventional natural selectionist terms. Efforts to interpret human behavior genetically have not been successful. Evolution has quite changed its character in its recent cultural dimension.

Chapter 13 summarizes the argument for the autonomy of the genome and notes the long-term trends of evolution. It applies the anthropic principle—that many apparently improbable conditions have been necessary to produce observers of evolution. Finally, it touches on evolution as the culmination of the cosmic order and on its meaning for our values.

Evolution clearly is much more interesting than would appear from the simple theory of change by random variation and selection, and it is much more awesome and meaningful for the comprehension of our existence.

understanding of more involved phenomena, and facts of one branch of science are made derivable from the results of a more basic science (Ayala 1985, 65–79). This usually implies quantification; to handle things with assurance, one should be able to express them in measurements, and scientists are happiest when writing equations.

The reductionist approach seeks to analyze complex matters into simple concrete entities in exact, measurable terms. So far as we can do this, we master phenomena. "Reductionism is without question the most successful analytical approach available to the experimental scientist" (John and Miklos 1988, vii). Scientific reductionism deserves much credit for countless wonders from antibiotics to spacecraft.

The successes of reductionism inspire the hope that fullness of understanding requires only more and better knowledge and analysis. That is, all natural phenomena should ultimately be made describable in terms of the elementary particles of matter, electrons, quarks, and the rest; their qualities should be knowable and are to be tied into a "Grand Unified Theory." Reductionist science would like to see everything from physics through chemistry and biology to psychology as potentially or theoretically explicable in purely material terms. Although many phenomena are admittedly too complex for concrete analysis, it is the scientific faith that everything is ultimately learnable except why the universe exists—a question that can be ignored as unanswerable.

Michael Ruse holds that "the whole is composed of nothing but its parts. . . . An organism is nothing but the molecules of which it is made" (Ruse 1988, 24). In this view, living organisms are nothing more than elaborate physicochemical systems, the product of genes, or nucleic acid sequences, reacting with their surroundings. Thoughts are the workings of the coordinative capacities of the organism; they are electrochemical phenomena, produced by neurons and their synapses with neurotransmitters and changes of potential across membranes. Everything should be mechanistically understandable as the behavior of material substance guided by the laws of physics. "The ultimate aim of the modern movement in biology," according to Francis Crick, "is to explain *all* biology [his emphasis] in terms of physics and chemistry" (Crick 1966, 10).

Most biologists would probably agree with Crick. The quantity of scientific knowledge, the successful analysis of many problems, and the ability to manipulate nature have increased so tremendously and so rapidly as to give scientists confidence that they can find intellectual order—a theory or set of theories bringing the whole of living nature under a discipline of reason. If all things result from knowable particles' behaving in predictable or at least describable fashion, we can hope in due course to learn reasons for everything. We have scored many victories; surely we can win the war.

The materialistic approach is also a useful working hypothesis and hence easy to take as truth. Scientists think in terms of experiments and verifiable results. The view of nature as essentially nonmysterious and knowable helps them frame hypotheses to test and encourages them to dissect their compartment of reality. Fanciful explanations, which amount to a renunciation of exact knowledge, are to be cast aside. A hard-headed approach admits no ghosts in the machine.

The phantoms, however, refuse to be banished. The faith that all things can be attributed to analyzable material causation is, in the end, only a faith like more candid faiths. The contention that reality consists of only material particles and their modes of interaction is not even a clear-cut theory. It implies a narrow definition of reality, making the thesis true by definition: if only material substance is real, then material substance contains the whole of reality. But are the laws of nature not real? Are mathematical theorems real? Are patterns real? Are thought and consciousness? It is paradoxical to deny their essentiality, for science could not exist without them. As idealists see it, reality is in the mind, and quantum theory, the glory of physics, although it seems mysteriously inherent in the particles it governs, logically leads to treating events as dependent on being observed (the thesis of Herbert 1985).

Great scientific advances and better ways of understanding nature, though based on facts, come from intellectual leaps. Einstein reached the ideas of relativity not from dissecting data but from seizing on a generalization: the invariance of the velocity of light (and of physical laws in general) under motion. Quantum mechanics and many other innovations of physics and chemistry have also come from conceptual integrations. Nature is not to be taken apart like a mechanical alarm clock, with gears, balance wheel, and spring interacting in a clearly comprehensible fashion. Many things are more like a quartz watch: relations between parts are nonmechanical, and disassembling it destroys it.

Science often reaches a point where further research and analysis seem unable to provide an answer without new concepts and paradigms. In biology, for example, though much has been learned about the role of sexual reproduction, we do not understand why sex exists or why such a scheme is so general in living nature. The fossil record is filled with minor or major enigmas that cannot be dismissed as due simply to lack of evidence. Countless traits of plants and animals defy the imaginative efforts of naturalists to devise plausible scenarios for their origination. To learn more is to confront more riddles. The enlargement of understanding is like making a bonfire in a forest at night: as the fire blazes up, more and more shapes loom in the darkness.

If many biologists feel confidence in the possibilities of ultimate explanation, physicists have much less. They once thought they had the world wrapped up in the laws of material particles, but new discoveries have shattered certitudes. Those who explore the frontiers of reality today are disposed to acknowledge the obscurity dwelling in the should-be simple little bits of matter that they study, admitting that hard data and intricate equations lead to bafflement as well as comprehension. Biologists, however, would explain in purely material, preferably biochemical, terms such a perplexing problem as how living forms took shape to make bacteria, amoebas, jellyfish, trees, lizards, and humans. They do not care to see anything inscrutable about it. Yet the theory, despite its logicality and deep appeal, leaves many questions unresolved.

Before taking up problems of evolution, however, we should look briefly at the theory as it has been developed.

The Neo-Darwinian Synthesis

A generation before Darwin, Jean-Baptiste Lamarck, who coined the word *biology*, related fossils to living organisms and proposed a consistent theory of evolution to explain their differences. Charles's grandfather Erasmus and others had put forward the idea that species grew out of one another. Several writers in the first part of the nineteenth century also proposed something like natural selection of those more capable of self-propagation (Darlington 1961, 14–24). But Darwin crystallized these ideas in the intellectual atmosphere and mustered a mass of evidence in their support. No one else treated the matter with comparable learning and persuasiveness and Darwin's passion for detail appealed to experimental scientists.

Darwin's claim to greatness rests not on his advocacy of the idea of common origins of species but on his simple and logical explanation by inheritable variation and natural selection. This idea was by no means as clear, however, as the thesis of evolution ("We are descended from apes"), and it raised many questions. It was not thoroughly accepted until much later when it was fortified with Mendelian genetics and, in this century, formed the neo-Darwinist synthesis regnant today.

Without Darwin, the doctrine of evolution would probably have been named Wallaceism, after Alfred Russell Wallace. Wallace came to conclusions very close to Darwin's about the same time but received much less credit, partly because he did not support his ideas with such a mass of observations. Moreover, Wallace's education was meager and his family impoverished, in sharp contrast to Darwin's aristocratic background and excellent connections with the British scientific elite. While Darwin studied and wrote, Wallace made a living by collecting specimens for nature fanciers. Wallace also lost scientific stature because he abandoned natural selection in dealing with human evolution and eventually became a spiritualist. Although Darwin grew up as a firm Anglican and a candidate for the clergy, he adhered strictly to material causation; Wallace found "causes of a higher order than those of the material universe" in evolution (Wallace 1901, 474–476).

Darwin—who was better situated, presented more evidence, and was more consistent in his scientific attitude—became the symbol of evolution personified. Acceptance or denial of the theory of evolution came to be and has remained nearly equivalent to loyalty or opposition to Darwin. Nonetheless, his theory of change by natural selection was based more on plausibility and analogy than solid evidence. It was fairly clear that variations like those observed in domestic animals brought about some changes in nature; Darwin extrapolated to assert that all differences between living creatures were thus caused, ultimately back to the separation of humans, fish, protozoa, and plants. In order to exclude anything savoring of divine intervention, Darwin also assumed that change had to be gradual and random.

A serious weakness of the theory was ignorance regarding how variations are transmitted. Darwin theorized that particles from all parts of the body gather in the reproductive organs to determine the inheritance of the offspring (pangenesis), but he had no evidence. The theory also had the defect that it led back to Lamarck's earlier idea, which Darwin wanted to supersede, that evolution progresses by the inheritance of acquired traits. In the conventional example, the giraffe came about because a gazellelike animal stretched to reach higher foliage, its offspring were born with longer necks, and, after many generations, voilà: a giraffe.

Without knowledge of genetics, it was difficult to understand how random variations could change a species. Any rare improvement would be diluted by half through mating with an animal without the improvement, and it would be lost in the general population, critics argued, in a few generations. Darwin was troubled by this difficulty; especially in later editions of his *Origin of Species*, he admitted the inheritance of acquired characteristics along with the chance variations that he stressed. For example, disuse of an organ would weaken it in descendants: ears of domestic animals drooped because they ceased to listen intently (F. Darwin 1897, 10).

There seemed to be no answer until the work of Gregor Mendel. Darwin, however, ignored this work when he published his findings in 1866 (he never even cut the pages of his copy of "Experiments with Plant Hybrids"). But in 1900 three investigators rediscovered Mendel's theory that traits are inherited in simple unitary fashion; hence they are not diluted by mating and reproduction but can reappear in subsequent generations. This meant that the features making up the plant or animal are mechanically transmitted and are more or less independent of the rest of the organism. It became difficult to conceive of acquired characteristics being somehow rather mysteriously incorporated into the genetic makeup, or genotype, of the individual and species. "It needed only the discovery of genetics . . . for the secret [of life as revealed by Darwin] to stand before us in all its naked glory" (Crick 1988a, 32).

Mendelian genetics initially seemed to contradict Darwin's gradualist variation. But in the first decades of this century, biologists melded Mendelism with Darwinism to make what came to be regarded as a definitive synthesis of evolutionary theory. In a research binge, geneticists working with cages of fruit flies discovered more and more of the complications of genotypes and phenotypes, heterozygotes and homozygotes, alleles, spontaneous and induced mutations, transpositions, gene frequencies, heritability, linkages, and the like, to construct what seemed to be a solid theory of evolution. The synthesis completed in the 1940s became an imposing theoretical edifice.

This success, however, was more imposing than solid. Genetics proved to be unexpectedly and confusingly complicated. Mendel's experiments had worked out well partly because he had extraordinarily good luck: the several genes that he traced through generations of peas happened to occur on different chromosomes and manifested themselves in simpler fashion than usually occurs. Genes were found to combine variously, to have multiple effects, and to change, or mutate, in ways difficult to explain. Moreover, the mountain of data produced little understanding of evolution. Fruit flies had mutations of eye color or even sprouted a leg from the head, but nothing was learned about how new organs originated and little about how different species of *Drosophila* may have arisen.

Consequently, many biologists looked for forces or directions in evolution beyond those postulated by the basic Darwinist theory. Some resorted to vitalism—the theory that living beings are alive by virtue of a special life force—or to some variant of Lamarckism, or to a teleology—the belief in a built-in purposiveness of evolution.

The variation-selection theory was refortified, however, by another big advance of genetics. The discovery in the 1950s that nucleic acid (DNA and RNA) is the carrier of heredity, demonstrating the physical materiality of the genes, restored confidence in the materialistic explanation of evolution. Because nucleic acid both replicated itself and transmitted information to proteins, molecular biologists concluded that information flows only from the nucleic acid to the body, never in the reverse direction. This implied that changes in the genetic apparatus, or genome, could come about only by errors of replication. Variation had to be accidental; there could be feedback only through the relative success or failure of variants in propagating themselves. This beautifully mechanical-materialistic idea was hardly to be questioned. As a biochemist put it, "Advances in understanding of the details of genetics [through discovery of DNA] confirmed Darwin's theory and documented the processes to the same extent that Newton's laws of physics had been validated" (Loomis 1988, xii).

The successes of the new molecular biology in analyzing and manipulating enzymes and nucleic acids made it easy to believe that unraveling the material substance of heredity would reveal the secrets of evolution. Again, however, answers remained elusive. Relations between genes and organism turned out to be excessively complicated, and the voluminous information coming out of the laboratories was of little help in interpreting the fossil record or understanding structures and behaviors. But the treatment of heredity in concrete molecular terms set the tone of the discussion for most practicing biologists.

The standard or purist evolutionary doctrine, as commonly expounded in biology textbooks, is simple and firmly reductionist. It would like to treat the organism pretty much as a bundle of traits, a genetic "beanbag." This term would seem pejorative, but it was adopted by Darwinists who liked to see one gene corresponding to one trait. Each trait would be established by selection because of its relative advantage, that is, its contribution to reproductive success. As stated by Edward O. Wilson, "Each phenomenon is weighed [by natural selection] for its adaptive significance and then related to the principles of population genetics" (E. Wilson 1980, 4). It is difficult to understand how anyone contemplating the infinite variety of life could believe this, but it has been favored as a working assumption.

An unconditional thesis of neo-Darwinism is that "Chance is the only source of true novelty" (Crick 1981, 58). This means that innovations spring from errors in the reproductive process. A large majority of the mistakes are insignificant or harmful, dislocations in a well-organized and tested apparatus, but these are weeded out by the stern culling of the environment. A few errors are improvements, enabling the organism to cope better and reproduce itself more abundantly, especially when environmental changes raise new demands. As variations accumulate, the interbreeding population changes indefinitely, eventually altering the species or making a new species. Darwin assumed that variation could tend in any direction without definite limits (C. Darwin 1964, 41), and the modern theory tends to favor this assumption, which leaves the organism subject to the roulette wheel of variation and the pressures of its world.

This idea was promoted strongly by August Weismann, who nearly a century ago postulated the complete separation of the reproductive cells—the germ line—from the functional body, or soma. According to Weismann, reproductive cells are already set apart in the early embryo, and they continue their segregated existence into maturity when they contribute to the formation of the next generation. This does not occur in plants and a large majority of animal phyla, but the idea was influential and attractive because it supported the nonheritability of acquired traits with no direct feedback from environment to heredity. The mostly nonexistent "Weismann barrier" between somatic and genetic tissues is still cited (Lewin 1981, 316). The important point is that there can be nothing purposive or teleological in evolution; any notion of inherent purpose would make nature less amenable to objective analysis. For a biologist to call another a teleologist is an insult. Even the idea of direction in evolution caused by internal factors, or orthogenesis, is disliked. The sole force for change must be adaptation.

Many biologists go on to refuse to recognize any overall direction in evolution. They even dislike the notion that some creatures are in any important way "higher" than others. In spite of the fact that natural selection implies improvement and that a mammal is much further from its presumed one-celled ancestor than is an amoeba, they sense a contradiction between the idea of "higher" forms and mechanistic means of change. As R. L. Trivers stated, "There exists no objective basis on which to elevate one species above another" (Trivers 1976, v). Is not an amoeba as perfect as a human in its small way?

Conventional theorists also prefer to think as much as possible in terms of the material particles of heredity, the genes. For Darwin, the unit of selection was the individual, which might be enabled by its qualities (plus good luck) to give rise to more than its share of the new generation. The discoveries of genetics led to a shift of emphasis, however, from the organism to its reproductive material. In the view of population genetics, as developed by Sewall Wright, Ronald Fisher, and others, a species is characterized simply by the frequency of various genes and their alternatives at any given site, or alleles at a locus. Evolutionary change thus was reduced to a matter of gene statistics (as in the classic formulation by Crow and Kimura 1970 and Feldman 1988). The product is a set of impressive and sometimes useful equations governing the probable alteration of a population (that is, evolutionary change), equations that are sometimes taken as the biological equivalent of Newton's laws.

Population genetics is less firm, however, than classical mechanics. Its chief variable, gene frequency, is seldom measurable in practice; its principal independent variable, fitness, can only be guessed because it is impossible to determine to what degree survival is a matter of special genes or accident or special circumstances. More broadly, an organism cannot be treated simply as the product of a number of proteins, each produced by the corresponding gene. Genes have multiple effects, and most traits depend on multiple genes. The selection of individual genes is most important in very simple organisms. That is, population genetics is best applicable to bacteria, and it does not tell much about the evolution of organs and higher animals.

Some biologists have gone far in exalting the gene over the organism and demoting the animal itself to being merely the means of replicating genes (Dawkins 1976). The essence of evolution is said to lie in the competition of genes and their (unconscious) struggle to survive and multiply. In a typical expression, "The individual bodies . . . throwaway 'survival machines' . . . are designed by genes simply as a means of enhancing gene survival and perpetuation" (Barnard 1983, 119). In other words, "The individual organism is only their [the genes'] vehicle, part of an elaborate device to preserve and spread them with the least possible biochemical perturbation" (E. Wilson 1980, 3).

The stark affirmation of the "selfish gene" appeals for its counterintuitive boldness. But to say that the genes are in some indefinable way primary is more of an ideological than a scientific statement. Genes are not independent entities but dependent parts of an entirety that gives them effect. All parts of the cell interact, and the combinations of genes are at least as important as their individual effects in the making of the organism. Selection operates not on genes but on organisms or perhaps groups (and possibly species).

In the cycle from sex cells or gametes to organism to gametes, all links are essential. There is no reason even to assume that the gene came first in the making of life; proteins and RNA are generally believed to have preceded DNA. To make the simplest and smallest part the reason for all the rest no doubt appeals as a token of sophistication, a claim to profundity by paradox. But it is odd to claim that the function of the elephant, a complex, seemingly purposeful, and responsive creature, or of a human is to copy sequences of nucleic acid bases, which can do nothing outside the body and are of no significance except as they contribute to the making of a new elephant or a new person. An organism interacts with the world and has a destiny; a gene only assists in making an organism.

Apart from such exaggerations, the selection-variation theory is a useful approximation. It doubtless accounts for, or helps to account for, very much. Natural selection serves to eliminate those less qualified in the competition; that is, it stabilizes forms. Darwin, however, thought of selection not so much as elimination of the unfit as the opportunity of the fitter, that is, adaptive change.

The best studied and most frequently cited example is the alteration of the British peppered moth (*Biston betularia*). In this species, as in many other species of moths and other insects, a dark-colored variety has taken over in areas where industrial smoke has blackened much of the landscape, replacing a mottled gray form that was camouflaged on lichen-covered surfaces. This adaptation is selected for by natural agents; birds have difficulty spotting dark-colored moths on a dark background. A clean air act was followed by a return of lighter-colored varieties (Clarke, Mani, and Wynne 1985). But things are not simple. The dark (carbonaria) gene was not fully dominant at first but became so (Wöhrmann 1990, 20). Melanic moths have invaded rural areas with lichen-covered tree trunks; it seems that they may have become more viable for reasons other than camouflage (Grant 1985, 107–108). The victory of the darker variety is interesting, but it does not prove that such a selective process can account for the ability of chameleons to camouflage themselves by reflex in a few minutes.

Variants of the Theory

To meet problems abundantly arising, many biologists, while adhering in principle to Darwinism, have suggested modifications to the basic theory, and evolutionary doctrine has become less coherent. In principle it is held correct, but when one tries to apply it—as in the question of human origins—controversy arises.

A mild challenge to the theory is the idea that populations change not only through adaptation but through mutations that are neutral or at least not seriously negative (Rothwell 1983, 596–599). Strict adaptationists claim that there "must be an evolutionary advantage to a trait if we only look hard enough" (Barash 1982, 51). But for each positive change, there must be thousands or tens of thousands that are not clearly useful, and unuseful traits (except for those harmful enough to be eliminated by selection) can lead to changes of the population as variants are increased or eliminated by the workings of pure chance. Suppose one has a barrel filled with a mixture of black beans and white beans. After stirring it, one removes half and throws them away, and then doubles the other half and returns it to the barrel, like a new generation being born, and repeats the operation. The proportions of black and white will change in each "generation." After enough times, one will have all black or all white beans. How many "generations" it will take depends on luck and the number of beans.

Such change, called "genetic drift" by Sewall Wright, takes part of the honor of driving evolution away from the principle of adaptation. That drift is a reality is indicated by the universality of molecular change in proteins and its greater frequency where selection is less important (Kimura 1985, 80). Resolute selectionists sometimes seem to find neutralist drift an ideological affront, but it is a fairly acceptable supplement to natural selection because it is nonpurposive. It is especially important in theorizing about the formation of new species.

A sharper challenge to mainstream theory is the rejection of gradualism in favor of "stasis-punctuation," the idea that change does not simply flow along but is sharply punctuated. Darwin insisted (despite an occasional contrary remark) on gradualism, although his theory did not really require it. The continuity of nature was a major discovery of his age; the corollary of present-day continuity was that the past was to be understood in terms of forces at work in the present. Attributing change of species to an accumulation of small variations made everything seem most natural and understandable. This idea also fitted Darwin's philosophical outlook: gradualism was the opposite of the special creationism that he combated.

The conviction that change must have been gradual harmonizes with the rejection of any idea of purpose in evolution. If organisms grope their way, so to speak, into new adaptations, they do it by little steps. Hence, biologists for the most part prefer the gradualism that Darwin preferred. But the fossil record fails to show continuous series linking different groups, and some biologists early in this century, led by Richard Goldschmidt, attempted to show that evolution must have proceeded by radical leaps—in Goldschmidt's phrase, "hopeful monsters." This idea was pretty much abandoned. It was argued that no such leaps have been observed, and when gross mutational changes do appear in the laboratory, such as a doubled thorax in a fruit fly or a leg coming out of the head, they consist of errors in the formation or placement of old parts, never the appearance of a new organ. *Drosophila* may mutate to have four wings instead of two, but the additional wings are just extra copies. It is also doubted that a major remodeling of the organism could be viable. And if, by some rare accident, a much improved animal appeared, it would supposedly find no mate to perpetuate its special endowment.

Recently, however, some theorists, led by Stephen J. Gould and Niles Eldredge, have forcefully argued on a more sophisticated level that evolution is by no means a mere accumulation of tiny, continually occurring changes. They contend that there is a dichotomy between stability of species and rapid change, or "stasis and punctuation" (Stanley 1979, 13–22). More conventional theorists dislike this non-Darwinian thesis because it seems to contradict the strictly mechanistic approach, at least in spirit.

Another qualification of the principle of selective adaptation arises from the fact that genes are pleiotropic, that is, have multiple effects, perhaps a dozen or more without any apparent connection, many of them probably useless or negative. This was quite familiar to Darwin, who wrote of "laws of correlation," linking traits, such as the deafness of white cats with blue eyes. Darwinists prefer to regard all or nearly all traits as brought about individually by selection, but traits otherwise difficult to explain can be accounted for by postulating that they go with useful or essential traits. On the positive side, a useless trait that comes as a by-product may become useful under changed circumstances. But this represents a dilution of selectionist theory. Traits are attributed not directly to knowable selective advantages but to postulated relations to other traits. If an insect has a spine, this may not be because a spine adds to its reproductive success but because it is tied to effective kidneys. This complication has far-reaching ramifications.

Evolutionists also differ in their emphasis on single genes or combinations, on genes that make proteins or on genes that turn other genes on or off. If most traits are the effect of multiple genes, regulatory genes, or gene-enzyme systems, large changes become more conceivable as results of recombinations, and mutations to make new proteins are correspondingly less important. The process of inheritance is evidently much more complicated than appeared from Mendel's tracing simple, unitary traits in peas—smooth or wrinkled, yellow or green, dwarf or tall. This implies shifting emphasis from relatively well-understood structural genes (which produce not structures but proteins that may or may not go into structures) to the poorly understood regulatory genes, which turn batteries of genes on and off. No simple theory can cope with the enormous complexity revealed by modern genetics.

A different problem is the existence of many traits that are evidently advantageous for the group but not for the individual. One prairie dog stands guard while others feed, forgoing a meal and exposing itself to predators. A honeybee not only rushes out to defend the hive but sacrifices its life by leaving its barbed stinger in the flesh of the intruder, along with the poison sac and part of the bee's guts. To resolve this contradiction, evolutionists have postulated that selection operates not only on the basis of individual but of group advantage: a group possessing traits useful for its collective success would prosper and reproduce, eventually perhaps replacing groups lacking the socially useful trait.

On the other hand, individuals lacking the altruistic gene would presumably have greater individual reproductive success, and the social trait would be wiped out. To overcome this contradiction, William Hamilton in the 1960s and 1970s elaborated the idea of indirect fitness, a significant addition to the Darwinist canon. An animal has "direct fitness" in its ability to propagate its lineage; it also has "indirect fitness" insofar as it helps other individuals with which it is related by sharing genes. This idea has proved handy in evolutionary theory despite its logical weaknesses.

Evolutionary theory may be modified to meet such difficulties, and evolutionists differ widely in their views regarding the pace, focus, and mechanics of change. They firmly maintain, however, the central ideas: there is nothing purposive, and organisms adapt genetically only by success or failure in leaving descendants. In the words of Ernst Mayr, "The one thing about which modern authors are unanimous is that adaptation is not teleological" (Mayr 1983, 324).

The Commitment to Darwinism

Darwin answered the intellectual need of the day, and the age recognized itself in him (Barzun 1941, 80, 85). He has been elevated as perhaps the greatest of scientists, and his name stands for a theory that has grown far beyond his work. What is commonly called the neo-Darwinian synthesis, or simply the modern synthesis, has taken on somewhat ideological overtones, especially in the United States. It becomes a little like a revelation by a prophet, whose every word in his major works is recorded in concordances. Darwinism is to be guarded against irreverent attack; in the words of Bruce R. Levin, "New ideas are liable to attack for no reason other than their real, or even apparent violation of orthodoxy" (Levin 1984, 452). Some biologists find reason to believe "that preoccupation with Darwinian orthodoxy shrouds the field to little benefit" (Miklos and John 1987, 279).

Mundane scientific work is predicated on shared assumptions and general understanding (Kuhn 1962, 5), and practicing biologists do not spend much time on profound questions for which there are no answers. Theoretical controversy is less rewarding than specific investigations, developing ideas not about evolution but about birds' nesting habits, finding how genes can be engineered, or breeding useful plants. The Darwinist model is a good working hypothesis and paradigm for research. Karl Popper, in fact, regarded it as more of a "metaphysical research program" than a scientific theory (Schlipp 1974, 134).

In a common view, the accepted evolutionary doctrine, rough hewn as it may be, has to be regarded as true unless it is proved false, even though the evidence for it is admittedly incomplete. Mark Ridley, for example, again and again makes the case for natural selection simply on the grounds that we

Adaptability = survivability

have no other plausible explanation (Ridley 1985). This perspective is understandable, perhaps persuasive. Theories in which many scientists have invested their careers are not set aside until they can be replaced by more satisfactory theories, usually brought forward by younger thinkers. Neo-Darwinism is an accepted "mode of cognition" as conceived by historians of science (such as Fleck 1979).

Science advances by testing, modifying, and so far as necessary replacing hypotheses, but standard evolutionary theory is not usually treated as a hypothesis to be investigated. A single counterexample refutes a mathematical theorem, but evolutionary theory is in practice not falsifiable (Saunders 1988, 279–282). Many very simple facts, such as that all the millions of species of insects and no species of noninsects have six legs, might well be considered to disprove natural selection as a generalization. But such broad problems are usually ignored, and it is assumed that any puzzle must be solvable in its terms if adequately studied. As an enthusiastic Darwinist puts it, "Given almost any such [behavioral] correlation, a competent evolutionary biologist can generally point out how it is, in fact, 'adaptive'" (Barash 1982, 41). He might well have said "always" instead of "generally," as does McFarland: "We can always invent a plausible adaptive advantage for an observed or supposed trait" (McFarland 1985, 528).

Despite the infrequency of any useful mutation, it can always be postulated that the appropriate mutations came along by accident and were selected, bringing about the adaptation in question. For example, it is hypothesized that natural selection has led the female sedge warbler to prefer full-throated males because they should make good foragers for the family. On the other hand, the female lyrebird supposedly has been selected to prefer the male who neglects his offspring and so avoids bringing the nest to the attention of predators (Alcock 1988, 80–81). The female spotted hyena, in the opinion of some, has a set of external genitals like those of the male in order the better to greet her friends (Kruuk 1972, 229). Some weaverbirds are monogamous because food is scarce, others because food is abundant (Crook 1972, 304). Marmot families stay together longer at high altitudes because there is less vegetation (Barash 1982, 59); if the young ones dispersed sooner at high altitudes, it would probably be because where food is scarce they have to seek new pastures.

Choices - "logical" thought ...

Instead of defecating on demand, like other tree dwellers, a sloth saves its feces for a week or more, not easy for an eater of coarse vegetable material. Then it descends to the ground it otherwise never touches, relieves itself, and buries the mass (Forsyth and Miyata 1984, 27–28). The evolutionary advantage of going to this trouble, involving no little danger, is supposedly to fertilize the home tree. That is, a series of random mutations led an ancestral sloth to engage in unslothlike behavior for toilet purposes and that this so improved the quality of foliage of its favorite tree as to cause it to have more numerous descendants than sloths that simply let their dung fall, and thus the trait prevailed.

In practice, however, biologists are usually rather realistic. They think of more evolved plants and animals as "advanced" or "higher," and they use those words. However scrupulously they avoid teleological language, they recognize certain directions in evolution. They are likely to speak of animals having purposes and acting to secure ends, if only because it is awkward to say that the animal has genes causing its limbic system to direct certain actions conducive to its reproductive success. Having devotedly spent thousands of hours observing lions or jays or damselflies doing nothing in particular, they faithfully report facts that do not accord with the standard theory.

It is comforting, however, to see nature as basically mechanical, and hence totally understandable. "To maximize fitness" is an offhand explanation for almost anything, usually persuasive until critically examined. The ordering principle of natural selection has an appeal of paradox and seeming profundity: "Look how all the superficially disorderly multifariousness of living nature can be reduced to a few simple principles!"

In the view of some researchers, "selection has always been and still is more a logical construct than a well-documented process" (Miklos and John 1987, 279). But it is useful as an experimental program, and scientists turn their method of work into a philosophical attitude (Peacocke 1986, 1). The principle of variation-selection represents a mental economy and suggests a way to seek answers, a key to unraveling the infinite variety and complexity of living nature. It gives a clearcut orientation, whether or not its explanations are adequate.

No other science has such a comfortable foundation. The physicists had something like it in the Newtonian synthesis, but they lost it because they learned too much. Sociologists and other social scientists (except for devotees of certain schools) have never been able to cherish even an illusion of total intellectual mastery. But the neo-Darwinian synthesis gives biologists a satisfying basis on which to work, most suitable for those who simply want to get on with their interesting investigations. The variation-selection theory even offers an uncomplicated view of the human condition and promises a means of coming to grips with the bafflements of human nature. The doctrine being axiomatically true, there must be a basis in natural selection for all human as well as other animal behavior; we have only to search for it. Sociobiologists find satisfaction in the bold assertion of a firmly scientific attitude in a controversial area where no one else can offer satisfactory answers.

A penchant for simplification is not, of course, a specialty of biologists. Historians, sociologists, and thinkers in other areas of great complexity where proof is difficult use theories lacking a sound factual basis. Intelligent social scientists have accepted in Marxism a dogma (that history amounts to economic class struggle) with less empirical support than the most extreme Darwinist position. Many psychoanalysts have subscribed to a Freudian thesis counter to common experience and entirely without a scientific basis—that all dreams are wish fulfillments. Even physicists play with odd theories without factual basis because they fit an intellectual need, such as that a near infinity of universes are continually being created in order to satisfy quantum conditions (Herbert 1985).

There are strong reasons for reluctance to admit any modification of evolutionary theory that might lead away from its mechanistic essence. Humans have adhered much more blindly to many less rational beliefs.

The Religious Challenge

Biologists are the more reluctant to admit weaknesses of neo-Darwinism because alternatives deal in intangible or mysterious forces. Critics of evolution commonly postulate influences unknown and perhaps inaccessible to scientific investigation, such as a vitalist essence or an inherent purposiveness in development toward higher creatures, ultimately ourselves. They may, like Arthur Koestler, have a penchant for the paranormal, parapsychology, and the like.

There is, however, an even sharper attack: the orthodoxy of special creation. Biologists, more than any other scientists, are subject to an organized assault that varies in intensity but never ceases, insulting and even injurious to their professional worth. No lay groups try to check or abolish the teaching of chemistry in the schools, but biologists see their science cramped and put on a level with ideas lacking in empirical foundation. They naturally assume an indignant defensive posture. Biblical fundamentalism in the United States may be the chief reason that Darwinist fundamentalism is especially strong in this country.

This is an old fight. Darwin made himself the champion of natural science when its intellectual prestige was rising sharply and the intellectual community of Britain, then the most advanced country in the world, was seeking to liberate itself from theological traditions. In an area of the utmost philosophical, ethical, and religious significance, Darwinism became the banner of those who would overthrow what they saw as an irrational, superstitious view of human origins.

Darwin was much more destructive of old faiths and ideas of divine guidance than was Newton two centuries earlier or Copernicus before him. The theory of evolution became the focus of the confrontation of science and religion. The debate was emotional, and decades elapsed before the fires of controversy burned low and most churches came to terms with evolution by a qualified surrender. The temperature is raised from time to time, however, especially as the advocates of Creation science press political authorities to impose their views on the public schools, or at least to check teaching of the naturalistic approach to the problem of human origins. Biologists, under attack, do not want to admit doubts that might undermine their central theory.

This defensiveness should not be necessary. The fact of evolution can hardly be doubted, unless one supposes that God so constructed the universe, with fossils in good order and receding galaxies, as to deceive His rational creatures into doubting the biblical account. There is confusion, however, between acceptance of common ancestries, implying the community of life on earth, and the analysis of how species diverged. One can and should question how a dinosaur gave rise to a bird without doubting that birds had dinosaur ancestors.

The antievolutionists are much more concerned with denying the reality of evolution than with the way in which it is theorized to have occurred, to which they do not usually pay much attention. But they welcome any uncertainties about it. And if they retreat from the dogma that all species were individually created in their present forms, they would at least like to see the evolutionary process as purposeful, perhaps divinely guided. Their position would, of course, be much stronger if they accepted the reality of common ancestries and concentrated their fire on the vulnerable issue of how natural selection can account for many seeming miracles of nature, including thinking beings. Evolutionists, in counterpoint, often seem to take the very strong evidence for the reality of common ancestry as proof of the complete correctness of the mechanism they postulate.

The conflict of views regarding human origins represents a profound clash of philosophies. The anti-Darwinist camp finds being placed in the company of animals a gross indignity. Surely something so glorious in moral and intellectual qualities ought to be the special work of divine will. Specifically to combat this pride, Darwin strove to sweep away illusions, and biologists reject what they regard as a narrow anthropocentrism, denying us any special status. Their religious opponents see the other side of the picture. If we are the work of a chanceled process, the result of the interaction of random variations with a totally impersonal environment, how are we really different from apes, except for being more successful at the moment? In the opinion of a neurologist, "There is an eternal war between what might be called the Darwinist side of our nature and the civilized side" (Restak 1988, 15). If we are ethically equivalent to animals, what becomes of moral values? (Concerning creationism, see Bowler 1989, 354–364.)

If conventional evolutionary theory did not have socially disruptive connotations, the creationists would have much less standing. The determination not to be demeaned fuels the emotional intensity of their attack, which raises the evolutionists' defensive insistence on the completeness of the mechanistic explanation. Antievolutionists feel that the doctrine of material origins and drives ("the gene's the thing") would be fatal for the social order if applied to human society. Evolutionists, on the other hand, are resolved to defend science, the search for the truth, against what they see as self-serving mythology. Objectivity suffers in the heat of controversy.

2

The Universe of Complexity

Limits of Reductionism

The theory of evolution by natural selection of randomly occurring variations is presupposed to be true because it is logical and simple. For this very reason, however, it should be regarded with suspicion; this inscrutable universe does not lend itself to facile explanations. A mechanistic approach to evolution oversimplifies thinking on an immense subject of the greatest intrinsic complexity.

The hope of many biologists theoretically to base their discipline on physics, the model science, is delusive for two reasons. One is that the complexity of organisms makes it impossible to learn much biology from facts of physics; biologists leave physics far behind when they consider adaptation, behavior, and evolutionary change. Living beings operate on a very different level from atoms, and evolution is not a mechanical but a historical process. More fundamental, understanding evolution in strictly material terms is vitiated by the fact that physics itself is riddled with conceptual difficulties and contradictions. The material particles that should theoretically form a solid foundation for biology turn out to be not solid building blocks of reality but enigmatic, if not incomprehensible, entities. And in all but the simplest and most constrained interactions of bodies and forces, new relations enter.

A profound principle of the universe is something that may be called self-ordering: the tendency to complexity, the growth of arrangements neither random nor merely regular crystal-like, especially those capable of generating new order. From its inception in the fireball of 10 to 15 billion years ago, the universe has been growing ever more complex and more structured. On the large scale, complexification has meant formation of gas clouds, galaxies, stars, and their planets. On a small scale, different kinds of atoms have come together to make ever more elaborate compounds, with large molecules joining in more intricate systems and combinations, up to the evolution of life on at least one planet.

In the study of such processes, physics has learned much. But if physicists read Dawkins' effort to reduce evolution to mechanical events, *The Blind Watchmaker,* they must smile to see their subject described as "the world of simplicity" (Dawkins 1986, 15). True, the world of physicists is basically simpler than that of biologists, or at least the questions they confront are more exact; one may suppose that the universe at the level of atoms is tolerably orderly, though by no means simple. But the world of simplicity exists only in the imagination, a nirvana where science, like desire, would cease.

The simplest components of the material universe—quarks, electrons, and interaction particles (chiefly photons and gluons)—are all puzzling. The proton, the core of solid matter, is a little zoo of quarks and gluons skipping about, perhaps orbiting one another. Most of its mass seems to be due to the gluons, whose task it is to hold it together; its spin is yet to be explained.

The electron is the best understood particle, and it must be the simplest thing in the world, or the simplest thing that (unlike a photon or a neutrino) can stand still. Yet physicists continually learn new facts about it. At once wave and particle, it has no definite location until it interacts with something. Its charge repels other electrons and attracts protons of opposite charge, thanks to a little cloud of virtual or intangible half-real particles around it. It has magnetic moment. Despite its lack of shape or size, it has something called spin; the spin has a definite axis and direction, and it has to turn over twice in order to turn over once. When an electron joins with a nucleus to make an atom, it becomes even more interesting, going through modulated gyrations and figuratively joining hands with electrons in adjoining atoms.

The behavior of the electron is described by quantum mechanics, a mathematical method entirely different from the equations of classical mechanics used for macroscopic objects. But only the simplest equations governing the electron are solvable, and there are no equations for much of its behavior. Nothing about the solitary electron suggests superconductivity.

The ability of electrons to move without friction in many substances at temperatures near absolute zero is attributed to each electron's forming a bond with another at a considerable distance (by particle standards) even though there may be millions of other electrons between. How or why they pair up is obscure. Recently discovered higher-temperature superconductivity is still more puzzling. New superconducting materials have been discovered almost entirely by accident (Cava 1990, 42). Electrons show many other oddities. For example, at a given distance from the nucleus of an atom, the spin of an electron is aligned parallel to that of the atom; a little farther away, it becomes antiparallel; still farther, it reverses again to parallel, and so forth (Stein 1989, 54). WARP

Matter is composed of only a few stable particles, electrons, protons, and neutrons, but in atom smashers, more and more kinds of exotic particles are created. Empty space, the vacuum, has a complex structure, according to Einstein's general relativity, described by ten mostly unsolvable tensor equations. In order for theories of physics to cohere, space should be governed by a cosmic constant, but apparently it is not (Abbot 1988). Space is curved on the largest scale, however, and it is expanding and carrying the galaxies with it.

Space is bubbling with an infinity of virtual particles that borrow mass from nowhere to enjoy a fleeting existence in conformity with the uncertainty principle of quantum mechanics. It is crammed with energy, and the emission of light is influenced by the interaction of atoms and the vacuum. Electrons have a minimum energy level because they absorb energy from space. The vacuum exerts pressure in all directions, and it tends (on a minute scale) to draw surfaces together. Space has handedness, and some particle reactions proceed in a left-handed or right-handed direction. Nothing can go faster than light, but quantum events are believed to occur instantaneously over an indefinitely extended distance; they must be in a degree independent of space. It is theorized that at the very tiniest scale, space has a large number of curled-up dimensions.

There are many fundamental theoretical problems. There is no known relation between gravity, described by general relativity, and electromagnetic and other forces, treated by quantum mechanics. The mathematics of the two are entirely different; both are essential to the modern understanding of the nature of things, but they are contradictory and cannot both be correct (Hawking 1988, 12). In quantum mechanics, specific causation is replaced by probabilities or statistical regularities. At least seventeen arbitrary constants have to be inserted in the equations, none of them related, so far as is known, to any other (Leggett 1987, 146). Such basics as the rules of quantum mechanics and relativity seem to be as arbitrary as the existence of the cosmos itself.

The problem of reality in quantum events is far too abstruse for nonphysicists; physicists cannot agree on a convincing interpretation (Herbert 1985, 16–27, gives eight "quantum realities"). It is not possible to separate observer from observation. The medieval question of whether a tree falling in a forest makes a noise if no one hears it returns to trouble modern physicists. There are good reasons to regard events as becoming real only as recorded; observation "collapses" the quantum function. This idea defies common sense, which has the world going its way whether anyone is watching or not, but many things in physics are at variance with common sense.

The universe cannot be made understandable in the sense of deriving its manifestations from a small set of intuitively compelling principles or axioms and logical procedures, in the way that Euclid constructed his geometry. It may be possible to bring certain phenomena together into a common framework or a law of nature, and so far as this can be done, it is cause for scientific rejoicing. It is always at the cost, however, of postulating particular relations or mathematical procedures that have to be accepted because they work. For example, it is a recent triumph of physics to have united the electromagnetic force with the "weak" force, but the way in which they are joined requires profound and detailed study. To bring everything somehow together in a Grand Unified Theory (GUT) would be grand, but the GUT could only be a sort of generalization of the limits of the knowable. Super String

The science of life, like the science of matter and energy, gropes toward the frontiers of the knowable. Biology, however, is fundamentally different from physics (Mayr 1985, 43–62). It has no laws of nature and makes no exact predictions. Understanding primarily means learning why species are as they are.

In biology even more than in physics, the whole has prior significance over parts and their specific interactions (Bohm 1987, 37). Biologists are often averse to holism because it is suggestive of vitalism or a more or less mystical approach. But the organism is an entity, like a cathedral. By taking it apart—reductionism—we can learn how the stones are carved and put together, but they cannot give us the meaning of the stately edifice.

Higher Laws

Nature is an immense orderly jumble of what Paul Davies calls "spontaneous self-organization" (Davies 1988, 6). New and different relations enter as elements are joined into larger units; these interact to form still larger entities, with more and more varied elements, interconnections, and interactions. The ladder of natural systems consists of elementary particles, atoms, molecules, macromolecules, cell parts, cells, organs, organisms, populations, species, and ecosystems.

Quarks make protons and neutrons; these, forming nuclei, join with electrons to make atoms, which link up to make molecules. Small molecules come together to make big ones, including long chains, polymers and proteins, in an infinite variety of coiled and twisted interlocking shapes. Large molecules somehow formed bacteria and other cells. Cells made higher plants and animals. Certain animals became able to process information and respond to their environment. Having grouped themselves into societies, they built culture.

Each level has its own reality, as real as any other; and more complex systems require appropriate levels of explanation and display regularities, or laws, that cannot be deduced from simpler components (Mayr 1988, 34). Systems are composed of subsystems, but the interactions are decisive, and taking a system apart to analyze it destroys it. In wholes, new properties "emerge," and in living creatures more than anywhere else the whole is superior to its parts. "Emergent evolution" has been a popular concept (Mayr 1982, 63).

It may be confidently asserted that properties of compounds depend on forces and bonds as described by quantum mechanics, but in practice chemists try to predict the characteristics of a new compound by analogy with similar known compounds. Even the form in which it crystallizes may be decided, so to speak, by the caprice of the compound. As noted by Mario Bunge, "We still do not have an adequate understanding of a single organic molecule" (Bunge 1989, 206).

Gases are fairly understandable because their molecules are not bound in fixed relations, but solids are tolerably analyzable only in the regular array of a crystal. The flow of electrons in a conductor is incompletely understood, not to speak of the semiconductors of the computer industry. Aggregate matter behaves very differently from single atoms or molecules and is essentially unpredictable (Dyson 1988a, 8). It cannot be treated atomistically; in dealing with it, one applies an appropriate set of concepts, such as voltage, resistance, induction, waves, hydraulics, temperature, specific heat, or elasticity (Leggett 1987, 112, 115, 124). To get more than an approximate idea of the properties of a new alloy, one must make it and test it. Changes of state, as between water and ice, raise still unsolved puzzles; the shapes that icicles fantastically assume are problematic.

The famous second law of thermodynamics is not derivable from knowledge of particles; it rather contradicts their nature: elementary particle reactions are time reversible, but the essence of the second law is irreversibility. It is based on the simple proposition that energy flows from warmer to cooler regions, and from this fact develops a nonintuitive concept, entropy, which is roughly equivalent to disorder. Entropy is not easily measured or exactly defined, but scientists have found the concept useful, have generalized it, and have built theories on or around it, down to the ultimate (happily very distant) "warm death" of the universe. Thermodynamic entropy is related to information theory and thereby to evolution; some theorists see organisms primarily as systems of dissipation of energy into entropy.

In the growth of complexity, new relations provide the basis for new integrations. Many simple switches appropriately joined make computers with indefinite capacities. For practical understanding of a computer, it is unnecessary to know how the elements function, how intersections of doped semiconductors make gates and switches; at the higher level, lower-level laws become irrelevant. Computer scientists deal with chips holding thousands or hundreds of thousands of switches organized for certain purposes, and they have their own vocabulary and "laws" of computer behavior. Outcomes become unforeseeable as the organizational principles become more qualified and difficult to formulate in step with the capacities of the machine.

Animals far exceed computers in complexity and are more difficult to study. There are countless subtle interactions continually going on within any organism, as there are among individuals of a species, different species, and their environment. Biologists deal with many irreducible concepts, such as instinct, fertility, adaptation, behavior, speciation, dependence, and specialization, in relation to which lower-level knowledge—as of genes or physiology—has limited utility.

If we turn to our own interesting species, the effort to make generalizations on a materialistic basis breaks down entirely. There are far too many factors and interrelations, mostly more or less unknowable. Many things are, of course, partially explainable in terms of simple material facts, such as the need to eat and the sex drive, but such broad facts lead to few interesting conclusions. Their results in terms of institutions are unpredictably dependent on an intangible, human choice. The social order is especially unfathomable because it is supremely will dependent.

There is thus a sequence from the inscrutable electron to the most complex systems of which we have knowledge, chaotically inscrutable civilized communities. In this sequence, biology ranks high. It involves laws of its own, profound laws not to be understood without much study and not fully derivable from simpler entities. As a result, in the words of possibly the most respected living evolutionary biologist, Ernst Mayr, "Reduction is at best a vacuous, but more often a thoroughly misleading and futile, approach" (Mayr 1982, 63).

This is no secret to biologists. Unfortunately, however, to recognize such theoretical limitations implies an intellectual retreat. As early as the seventeenth century, the hope of mechanistic understanding of nature took strong root with the systematics of René Descartes and the triumphs of Newton's mechanics. In the glory days of classical physics a century and more ago, scientists and philosophers exulted in the discoveries of gravitation and mechanics, giving rational explanations for a host of puzzling phenomena, from the ocean's tides to the orbits of the planets. Many thought all the basic questions had been answered: the universe had been found to be a machine. In his classic boast, Laplace proclaimed that it was only necessary to know all the positions and motions of everything in the universe in order to predict the whole of the future.

This was an intoxicating perspective. We can see in retrospect that it was ridiculous, as Laplace might have realized if he had pondered how he could be sure that his great thoughts were simply equivalent to predictable motions of material particles. But the mechanistic philosophy was believable because thinkers were awed by the flood of discoveries changing the intellectual landscape. Those who investigated nature badly wanted to believe that the key had been found to unlock the treasure chest of her secrets. "It is solved!" they proclaimed.

Confidence in science grew through the nineteenth century with victories on many fronts, including the laws of electromagnetism, the flowering of chemistry, and Darwin's rationale of evolution. By the end of the century, as Darwin's theory was gaining general acceptance in the scientific community, many believed that basic discovery was coming to an end; knowledge of electrons and atomic nuclei promised that all natural phenomena would be reducible to a few basics. Questions enough remained, but it seemed that science had only to keep on the well-proved trails to find the detailed answers.

Yet nature is not so amenable, and the seemingly clear trails became confused in the first decades of this century. Relativity and quantum mechanics brought an intellectual revolution, as physicists found that they had to sacrifice simple certainties in order to overcome contradictions. But at the very time that Max Planck, Niels Bohr, Albert Einstein, Erwin Schrödinger, and their brilliant colleagues were revising the Newtonian view of the physical universe, biology was becoming more reductionist with the application of Mendelism to Darwinism. A little later, molecular biology came to reinforce the materialistic approach.

Biology remains laggard. Despite awareness of the inadequacy of reductionism, it generally insists on a reductionist approach to its primordial problem, evolution, accounting for everything by random variation (mutation) and selection, with unessential qualifications and allowance for various unpredictable influences. Many or most of its practitioners would treat organisms in the fashion of classical physics, like objects subject to forces of the environment. During the past decade or so, there has been something of a ferment as more questions are being asked and the certitudes of mid-century are questioned, but evolutionary theory "persists in adhering to the Cartesian and Newtonian mechanical paradigm" (Ho 1988, 87).

The New Outlook of Science

oh yeah

"Physicists have lost their grip on reality" (Herbert 1985, 15). This grip at the beginning of this century was confident and outwardly firm. The universe consisted of a few particles and fields, in reversible interactions, and everything was theoretically predictable on the basis of forces and motions (Nicolis and Prigogine 1989, 3). The rapid progress of thermodynamics, electronics, chemistry, and astronomy nourished the hope that remaining problems in a fully knowable universe were not of principle but of complexity and should be solvable by diligent investigation.

The solidity of physical science began to crack with the discovery that radiation behaved quite differently from the expectations of classical physical laws. In order to describe how light is emitted from a hot solid, Max Planck found it necessary to invent something entirely new—infinitesimal packets that came to be called quanta. Then it was learned that the emission of light by excited electrons was governed by a whole new set of mathematical laws, quantum mechanics, which were found to describe all interactions of elementary particles. Its hallmark is the notorious uncertainty principle and the inherent unpredictability, or only statistical predictability, of almost everything. Einstein's relativity led in the same direction by finding enormous complexity in what had seemed the simplest of things, space and motion.

The relativistic-quantum mechanical revolution was about consummated in the first three decades of this century. Since then, however, the problems of physics have continued to deepen, and the once-stable world of elementary particles has become a confusion of instability. Recently, especially in the last decade, science has been caught up in a much broader ferment (Pagels 1989, 52). Concepts that had been marginal, such as entropy and complexity, have acquired new importance in subtle relationships that defy exact analysis. Mathematically, a world of nonlinear relationships without exact solutions has replaced the classical, Newtonian world of simpler linear relations and theoretically exact solutions.

Attention increasingly has turned to what may be called sciences of complexity: study of larger-scale and higher-level questions, such as ecology, behavior, economics, information theory, cognitive science, artificial intelligence, and computer design. Much of science has turned from reductionism, which reaches a dead end in study of evanescent and elusive elementary particles, to more holistic approaches.

The strongest force in this upheaval has been the computer. Making all manner of problems more accessible invited attention to them, and the ability of computers to solve previously unsolvable equations invited researchers to formulate new equations and attack questions that would hardly occur to the unaided human intelligence. Complex systems became scientifically more important because they could be simulated on the computer screen. Symbolic of the old science was the calculation of planetary motions by arithmetic; of the new, the modeling of weather systems by computer.

New concepts come to the fore. Entropy has been extended from its original thermodynamic meaning of unusability of energy to take on broad significance as part of or key to the irreversibility of events. Increase of entropy implies energy flow, which makes possible the self-organization of systems. Unpredictability derives not only from the behavior of elementary particles but from deterministic chaos. Symmetry breaking—the tendency of uniformity to yield to differentiation—is a critical factor in the development of complexity. Autocatalysis is crucial for change and the creation of novelty (Nicolis and Prigogine 1989).

The paradigm of the new mode is the broad principle rather confusingly called chaos. Science discovered (because it was prepared to discover) a dynamic new realm, which had always been there. Chaos seems to reign over the universe; in the view of physicist Joseph Ford, "The whole way we see nature will be changed" (cited by Pool 1989b, 26). In another opinion, "Chaos has opened new horizons in science, and it is already considered by many the third most important discovery in the twentieth century, after relativity and quantum mechanics" (Tsonis and Tsonis 1989, 31).

A meteorologist, Edward Lorenz, opened the field in the 1960s by calling attention to the inherent unpredictability of the weather. In the 1980s, in large part because of the high-powered computers that can work out the mathematics of dynamic systems, the study of chaos became a fashion and something of a science in itself (Gleick 1988; Briggs and Peat 1989; A. V. Holden 1988).

Chaos

Chaos is not easily or exactly definable—the concept itself is a bit chaotic—but is fairly understandable. It may be called the uncertainty principle of the macroscopic world, a broad tendency to irregularity and unpredictability within regularity and determinism. It represents the tendency of nonlinear systems with positive feedback to behavior that is deterministic but cannot be predicted. In practice, it is not easily distinguishable from random or stochastic behavior, but it is theoretically quite different. The concept of chaos often makes better sense of disorder.

The first principle of chaos theory is sensitive dependence on initial conditions; indefinitely small differences may lead to indefinitely large differences of outcome. Not only do differences result from the inevitable inexactitude of measurements; deviations increase exponentially because of feedback. A disturbance cascades as interactions change conditions of further interactions. Lorenz was originally drawn to the study of chaos by the realization that a tiny turbulence in the atmosphere could expand indefinitely. For example, the fluttering of a butterfly's wings on the Amazon could trigger a disturbance leading to a thunderstorm in Kansas. To model the weather accurately would require an apparatus as large as the weather system; even such a model would soon begin to diverge indefinitely from reality.

The orbit of Pluto is very much simpler than the weather, but it is considered chaotic although it is assumed to be fully determined by the gravitational attraction of other bodies. A difference of a few inches in Pluto's position could grow in a few million years to billions of miles because each approach to one of its neighbors magnifies the deviation. In the very long run, the solar system itself may be unstable as planets chaotically transfer momentum from one to another. Many resonant gravitational interactions of both inner and outer planets indicate chaotic behavior (Kerr 1989, 144–145). But the sun will swell to swallow up the inner planets long before any planet can escape its grip.

Chaotic systems never exactly repeat a previous state; if they did so, there would be not chaos but cycling. Chaos is much more, however, than a statement of the impossibility of exact prediction of outcomes and the tendency of systems to develop irregularities. It has mathematical structure. Many iterations lead to 2-cycles, 4-cycles, 8-cycles, and so on. Moreover, at a definite value of a parameter, periodicity ceases and chaos enters. The simplest example is the classic equation (May 1974, 645–647),

$$X_{n+1} = X_n \, k(1 - X_n),$$

in which X is population density (treated as less than 1) and k is a parameter. For low values of k, X tends to a steady figure. X begins to oscillate as k passes 3; things get complicated around 3.5, and chaos begins at 3.57. Along the way, values split, or bifurcate, for different initial values of X. The ratio of the intervals between bifurcations is constant, the Feigenbaum number, 4.669202 . . . , one of the very few pure numbers in mathematics, like pi and e, the base of natural logarithms, and this number holds for a great variety of functions. There is a profound generality in chaos.

Fractals—patterns repeated on a smaller and smaller scale, with nonintegral dimensions and unpredictable variations within the general motif—are akin to chaos. The conventional example is a typical coastline. A map with a scale of an inch to 1,000 miles shows a wavy line. At an inch to a 100 miles, or to 10 miles, one sees a generally similar wavy line with ever greater detail. Fractals are beautifully demonstrated by computers iterating simple functions, that is, starting with arbitrary values and using them to derive new values, which are inserted in the functions to give new values, and so on until they rise to infinity or shrink to zero. If points of the complex plane on the computer screen are colored to correspond to initial values that become infinite and those that lead to zero, the resultant patterns (as shown in Mandelbrot or Julian sets) make all manner of strange, often very ornate displays, complexly and infinitely varied within general resemblances, repeating motifs on an ever smaller scale. This mathematical artistry, which can be produced only by computers making millions of calculations, may look like the feathery branches of frost on a wintry windowpane.

Chaos is fractal in that values are subject to repeated subdivision on a smaller scale; for example, trajectories may be divided into a number of lines, which, when multiplied, can be subdivided, and so on indefinitely. A look at a fractal Mandelbrot set shows graphically that one is dealing with the same broad phenomenon: indefinitely detailed and unpredictable disorder that is at the same time deterministic and held within a general pattern, repeated with infinite variety. In chaos, variation is irregular but bounded, as it is in living creatures.

While indefinitely varying, chaotic behavior remains regularly within certain limits, "attractors" (or as they are often called, "strange attractors"), with number of dimensions corresponding to their complexity (Pool 1989a, 310). An attractor describes the limits to which a system tends—a point, a cycle, or a space. Functions may circle erratically in a certain vicinity, like a moth flying about a light, or they may course around more than one attractor, shifting irregularly between them. Instability enters when positive feedback, or negative feedback that takes the system farther from equilibrium, causes a trajectory to escape the attractor. Attractors are commonly chaotic; the more closely they are observed, the more complicated they are (Crutchfield, Farmer, and Packard 1986, 53).

Mathematicians have made chaos a new specialty, but it includes much more than can be described mathematically. To observe it, one needs only to open a faucet enough that laminar flow gives way to a stream taking never-repeated forms or to observe a flag that cannot remain still in a breeze. Chaos can be seen or surmised on every hand. The breakers on the beach are chaotic, like highway traffic pileups and the fluctuations of the stock market; especially a crash resulting not from economic conditions but from the interplay of perceptions. Sunspot cycles, varying in period and intensity, are apparently chaotic, as are El Niño, which brings rain to the Peruvian desert, and the red spot on Jupiter. So are cyclic networks in autocatalytic systems. If one arranges to give a circular pendulum a little kick with a period not commensurate with the period of the pendulum, it soon becomes erratic in amplitude and direction. Today's yen-dollar rate depends on yesterday's, as traders project trends, but also on the trade balance, interest rates, and capital flows, which are circularly influenced by the exchange rate, a complexly chaotic fluctuation (*Economist* 1990, 93). Chaos is important for electronics, chemistry, hydrodynamics, optics, and other disciplines. It is "a basic mode of motion underlying almost all natural phenomena" (Tomita 1988, 213).

Where there is an input of energy into an unsettled condition, chaos generates new and unpredictable order. This is the way of entropy-defying life, and there must be large elements of chaos in something so dynamic and governed by feedback as evolution.

Modernizing Darwinism

Biologists can maintain an essentially Newtonian, statistical-mechanist outlook because the many anomalies and unanswered questions in biology do not present such clearly defined challenges to accepted doctrine as those that brought the downfall of classical physics. The invariance of the velocity of light or the spectral lines of hydrogen were facts that could not be ignored. Evolutionary theory, on the other hand, is elastic and can be stretched to cover many things. It is always possible to assume that there must have been appropriate mutations. So much is unknown or unknowable that it can be supposed that facts would fit the theory if they could only be learned. Evolution is history, history is subject to interpretation, and not much can be proved or disproved about it.

Yet traditional evolutionary thinking does not escape corrosion from the modern intellectual climate. A scientific theory is not an autonomous entity. Scientific theories are shaped by the attitudes and presuppositions that scientists bring to their handling of facts, which are selected according to the presuppositions prevalent in the scientific community and the society at large.

The Darwinian theory of evolution was the heart of a revolutionary change of outlook in life sciences in the mid-nineteenth century comparable to the rationalistic Newtonian-Cartesian revolution in exact sciences two centuries earlier. It made intelligible the process of change in the living world by the law of natural selection, much as Newton had made the movements of the planets and much else intelligible by his laws of motion and gravitation. Change proceeded by fixed and mechanistic processes in a closed universe. Evolution by the natural selection of "fitter" individuals was analogous in its concreteness and simplicity to classical mechanics—and to the economic system of liberal England, in which the competition of production units brought progress to the whole.

The union of Darwinist natural selection with Mendelian genetics carried forward the tradition of classical mechanics married to the statistical dynamics that became prominent in the latter part of the nineteenth century. Population genetics—

making the gene the atom of evolution—grew into the neo-Darwinist synthesis of the middle of this century, following the dominant fashion of physical science with a lag of about half a century.

This synthesis seemed satisfactory. It well suited the image of most biologists of their science and their intellectual role. Now it seems outmoded. Biology increasingly must deal with process and pattern, with self-organizing and self-regulating systems. Like physics, it is beset by uncertainty, instability, and complexity in a universe that is open and unbounded. Symmetry breaking is even more relevant for morphogenesis—it is the essence of morphogenesis—than it is for physical systems under stress. By extension, it is crucial for evolution, and complexity can be defined as the ability to make transitions, that is, to evolve (Nicolis and Prigogine 1989, 36).

To advance understanding of living things and evolution, we must look to new principles. One promising idea is that the more complex a dynamic structure is, the more endogenously it is driven. Its change depends not only on its external compulsions but on its internal conditions. Animals, the most complex of structures, are the most self-governed and able to respond to external conditions according to their internal dynamics. Self-organization is the essence of the origin of life and its complexification, that is, evolution. It lies at the heart of morphogenesis, ecology, and the aggregation of human culture. In this mode of evolutionary thinking, organisms may be considered "autocatalytic energy-processing systems stabilized far from equilibrium," which are selected for their energy-processing capacities (Depew and Weber 1989, 256, 260). Their time-directed evolution is tied to the inevitable increase of entropy in the universe, as an energy flow enables matter to organize itself (Wicken 1988, 140).

Such views are abstract, but they have important implications for our understanding of evolution. According to such theorists as Depew and Weber, we must see organisms not as closed systems subject to outside forces and constraints but as open systems undergoing continual flux, internally as well as externally generated. Complexly interactive, they evolve within an ecological context; change is irreversible because there is no stable equilibrium to which they can return (Depew and Weber 1988, 333).

The search for new approaches does not mean that natural selection is to be overthrown. Science does not advance by rejecting major achievements but by refining and redefining them; the deeper change is not of results but of approaches. Einsteinian mechanics did not invalidate Newton's well-reasoned principles but modified them for motion at an important fraction of speed of light. For all ordinary calculations, classical mechanics is amply accurate. Einsteinian gravitation left Newtonian gravitation valid for all practical purposes; physicists have difficulty finding ways of demonstrating the difference. But Einstein's gravitational force, resulting from the geometry of space, is wholly different from and far more sophisticated than Newton's simply postulated attractive force.

The core of the neo-Darwinist synthesis will remain valid. No one doubts that there are small, random mutations, that mutations affect the ability of organisms to survive and propagate, and that gene frequencies in a population vary. But the meaning and centrality of these Darwinian propositions will surely be reassessed. The new mode of scientific thinking calls for a broadened agenda for evolutionary thinking, asking different questions and expecting different kinds of answers, and it is certain to be more sophisticated in its reasoning.

Contemporary theorizing about fundamentals of evolution offers more interesting perspectives than specific answers. To apply the ideas and methods of the new science of complexity to biological processes and evolution is difficult. This book seeks to present evidence of the need for new thinking about evolution and to suggest some helpful ideas.

3

The Cryptic Record

Problems of Origins

The remains of extinct creatures are probably the most convincing proof of the reality of evolutionary descent of living creatures, but they cast doubt on the theory that random variation and natural selection suffice to account for it.

The study of fossils was already fairly advanced in Darwin's day; since then, it has produced a huge mass of information about the life of the past. There are many obviously ancestral or near-ancestral forms, yet many pages of the history of life are conspicuously missing—generally the most interesting pages.

Darwin insisted on gradualism as the essence of naturalism and the repudiation of divine intervention. His theory implied, and he quite reasonably believed, that there should be most evolution in large populations, which would produce a large number of variations, and hence that there should be much evidence of evolutionary change. Consequently he was much concerned with the incompleteness of the fossil record, to which he devoted 28 pages of *On the Origin of Species* (C. Darwin 1964, 279–311). He attributed it to the accidental absence or erasure of parts of the record and the inadequacy of exploration, and he was confident that in time the gaps would be filled.

This was not implausible in his day. But since then the hundredfold multiplication of the number of known fossils has not much improved the continuity of the record. The most impressive intermediate—the reptile-bird *Archaeopteryx*, the most famous of all fossils—was aptly discovered in 1861 when debate over the new theory was most heated, encouraging the hope that more digging would uncover many more such discoveries. But no equally admirable bridging form has been found.

The problem cannot lie merely in the scantiness of fossilization. True, it is a rare event for an animal, especially a land animal, to leave its skeleton to be dug up millions of years later. It is always possible to say that a transitional form must have existed but has not yet been found. Nevertheless, an enormous amount of information is available. A large part of the earth's crust consists of compacted sediments, which contain countless remnants of creatures that lived in the waters or were washed down and buried under sand or mud. The bulk of these are shells of marine animals, from diatoms to large molluscs, but here and there bones, and occasionally traces of other parts, of land animals have been preserved.

Remains of some 250,000 extinct species have been recovered and classified, and they ought to provide a reasonably good picture of the life of the past (so far as fossilizable). They probably do so, at least for marine forms, deficiencies owing less to failure of fossilization than to destruction or erosion of deposits (Valentine 1989, 90). Antiquity does not seem to matter very much; early Cambrian species are about as well known as those of much later periods (Valentine and Erwin 1987, 86).

But the fossil record does not tell us what theory promises. We expect to find a great tree, with many forks sending branches in different directions. The tree of life is frequently so drawn, with groups diverging from common origins, dividing and redividing, giving rise to the different phyla, classes, orders, families, genera, and species. Some large branches and countless smaller ones come to an end at extinction; others lead to the twigs of contemporary species.

The tree of life as it appears in the rocks is strangely different from this ideal. The beginnings of new limbs are seldom even close to the part of the tree from which they supposedly sprang, and a number of branches usually appear close together without any connection. Charts depicting ancestries through the ages are sometimes fudged by drawing connections where they are assumed; the more honest ones have dotted lines. By corollary, there is little indication of actual change. Stability or stasis is normal. Gradual change appears mostly in dimensions, as increases of size or enlargements of parts (Eldredge 1985, 23, 75). When one species slowly changes into another (chronospecies), there is ordinarily no more difference between species than among the members of a single contemporary species

(Stanley 1979, 14). It is as though life goes behind the bushes and emerges in new clothes.

A few gaps would be expected in a haphazard record but not the absence of documented transitions. Not only are relationships between the great groups, the phyla, obscure; lesser divisions are also undocumented. Logic suggests that there should be many intermediate forms between widely differing groups, such as the bat and the four-footed insectivorelike animal from which it must have arisen. One is more likely to find transitional forms where change has been less drastic, as between modern carnivores and those of 50 million years ago. The width of gaps tends to lessen, in a taxonomic sense, as one approaches the present because structural change has slowed as organisms become more complex and ecological spaces are filled. But Ernst Mayr goes so far as to assert that there is "no clear evidence for any change of a species into a different genus or for the gradual emergence of any evolutionary novelty" (Mayr 1988, 529–530).

In a few cases there is fair continuity. The lineage of the mammals is relatively well documented, leading back 230 million years to mammallike reptiles. But there are blanks separating amphibians, early mammallike reptiles, later mammallike reptiles, and mammals. Each new family appeared suddenly and thereafter changed slowly (Kemp 1982, 327). Gaps also appear between mammals of dinosaur times and those after the great extinction, when the modern orders seem to have arisen independently about the same time (Bonner 1988, 11; Carroll 1988, 361–362, 415).

The gradual emergence of the horse has been frequently cited as proof of evolutionary theory. Nearly a century ago, T. H. Huxley, "Darwin's bulldog," made much of the series leading from the little four-toed *Hyracotherium* (formerly known as *Eohippus*) of 55 million years ago to modern one-toed horses. But even this lineage proceeds by jumps between many branches, and modern horses appeared rather suddenly about 3 million years ago (S. Gould 1987a, 20; Stanley 1981, 6).

The human sequence from *Australopithecus* through *Homo habilis* and *H. erectus* to *H. sapiens* is among the most nearly continuous. Authorities differ regarding the number of coexisting species, the extent to which change was within species or between a species and its offshoot, and whether hominid species emerged or were replaced, but gaps are not wide. However, the earliest hominid fossil ("Lucy") of 4.5 million years ago is fully bipedal, with legs and pelvis very different from those of apes and fully of hominid type.

Australopithecus is only one of very many unheralded appearances of markedly new forms. Among the reptiles, the oldest known turtle had a complete shell like turtles today. The snakes appeared in practically modern form and rapidly became abundant. Dinosaur genera were periodically replaced throughout the long reign of that order, but no genus is clearly ancestral to any other (Carroll 1988, 180, 207, 234, 328).

The origin of reptiles is obscure, and it becomes more puzzling with the recent discovery of a small reptile dating back 338 million years, nearly as old as the oldest amphibians (Smithson 1989, 676). There is no record of a transitional form between bony-finned fish and amphibians. Modern amphibians are very different from those that first colonized the land, and there is a blank in the record of amphibians for some hundred million years after the beginning of the age of reptiles. When it resumes, all the older types are gone. The three modern orders—frogs (anurans), salamanders (urodelans) and caecilians—appear on the record in approximately modern form, and it is disputed whether they constitute branches of a single line (Pough, Heiser, and McFarland 1989, 374).

Further back, there are no fossils leading to primitive chordates or linking them with the vertebrates to which they must have given rise. The latter showed up possessing such advances as a brain case, specialized sense organs, and calcified bones. The earliest known members of major vertebrate groups are very different from each other. A major step was the development of a jaw, but no known jawless fish seems to qualify as ancestor (Carroll 1988, 17, 21, 44).

Much the same is true of invertebrates. The first known insect looked much like a modern bug, and the interrelationships of different orders of insects (dragonflies, cockroaches, beetles, flies, and so forth) are unknown. Each makes its appearance independently, although some 15 orders go back about 300 million years (Cloudsley-Thompson 1988, 44–45).

In the more distant past, multicellular animals of modern phyla appeared abruptly about 570 million years ago in the spectacular Burgess shale formations. About 50 phyla (compared with half that number in today's world) and a large number of classes appeared—about 300 new major body plans

AGE IN M.y.	EON	ERA / ERATHEM (SUB ERA)	PERIOD / SYSTEM	EPOCH / SERIES	ESTIMATED EUSTATIC CHANGES IN SEA LEVEL (Meters above and below present-day sea level) +400 0 −300	MAGNETIC POLARITY (Normal / Reversal / Mixed)	GLACIAL PERIODS	ORGANIC REMAINS	GEOLOGICAL & BIOLOGICAL EVENTS
			QUAT — Q	Pleistogene Ptg	Pleistocene Ple	Td		❄	Northern Hemisphere continental glaciation
5					Pliocene Pli E				Desiccation of Mediterranean Sea
10 15			Neogene	Miocene L / M / E	Tc				FA of Hipparion horses in Mediterranean Sea
20									FA of hominids
25		Cenozoic Tertiary	Ng Mio						Red Sea opens
30 35				Oligo-cene L / E					
40			Paleogene	Oli	Tb				
45				Eocene L / M				🐀	FA of rodents
50	Phanerozoic			Eoc E					Australia and Antarctica begin separating
55				Paleo-cene L	Ta			🌾	FA of grasses / FA of Equidae / Diversification of mammals / FA of primates Primates
60			Pg Pal	Paleo-cene E					
65 70		Cz TT			Kb			🐚	LA of globotrucanid foraminifera, rudists, ammonites, and dinosaurs
				Late Senonian					Tasman Sea opens
100			Cretaceous						FA of marsupials, placentals
				Early Neocomian	Ka				FA of diatoms / FA of angiosperms / Beginning of Labrador Sea opening
150		Mesozoic	K	Late Malm		M1 M16 M10 M22 M29			South Atlantic opens / FA of birds / India, Madagascar, Antarctica, all separate
			Jurassic	Middle Dogger	J	?			FA of Globigerinacea foraminifera / Gondwana and Laurasia separate / Central Atlantic opens
200			J	Early Lias					LA of conodonts
				Late	Tr	?			Rapid diversification of mammal-like reptiles
248	Ph Mz		Triassic Tr	Middle Scythian Scy	Rising ← / → Falling			🦎	FA of hexacorals, dinosaurs, LA of rugose corals

The geological time scale, showing more recent times in more detail, including first appearance (FA) and last appearance (LA) of major groups; also changes of sea level, magnetic polarity, and glacial periods (indicated by snowflakes). (Illustration from *The Fossil Book* by Carroll Lane Fenton and M. A. Fenton, © 1958 by Carroll Lane Fenton and Mildred Adams Fenton. Used by permission of Doubleday, a division of Bantam Doubleday Dell Publishing Group, Inc.)

AGE IN M.y.	EON	ERA / ERATHEM	SUB ERA	PERIOD / SYSTEM	EPOCH / SERIES	ESTIMATED EUSTATIC CHANGES IN SEA LEVEL (Meters above and below present-day sea level) Rising — Falling	MAGNETIC POLARITY (Normal / Reversal / Mixed)	GLACIAL PERIODS	ORGANIC REMAINS	GEOLOGICAL & BIOLOGICAL EVENTS
250	Phanerozoic	Paleozoic		Permian	Late / Early — P	P		✳		LA of trilobites, tabulate corals, orthid brachiopods
										Proto-Atlantic Ocean finally closed
300				Carboniferous — C (Pennsylvanian — Penn; Mississippian — Mis)	Gzelian, Kasimovian, Moscovian, Bashkirian — P; Serpukhovian	PP				FA of winged insects
										FA of pelycosaurs
350					Visean; Tournaisian — Mis	D–M		✳		FA of cotylosaurs
										LA of graptolites
				Devonian — D	Late / Middle / Early		?			FA of amphibians (labyrinthodonts)
400				Silurian — S	Pridolian, Ludlovian, Wenlockian, Llandoverian	O–S				FA of ammonoids
										FA of land plants
450				Ordovician — O	Ashgillian, Caradocian, Llandeilian, Llanvirnian, Arenigian, Tremadocian	(2nd-order cycles (Supercycles))	?	✳ ✳		
500										FA of echinoids, bryozoans
				Cambrian — €	Merioneth, St. David's, Caerfai	C–O				FA of vertebrates (jawless fish, graptolites)
550										FA of many invertebrate phyla
590	Ph	Pz		€						FA of exoskeletal material
600	Proterozoic	Pt₃	Sinian (Z)	Vendian — V	Ediacaran / Varangian			✳ ✳		
		–?–		Sturtian — U						
1000		Pt₂	Riphean (R)	Yurmatin — Y						FA of eukaryotes (organisms with a nucleus)
		–?–		Burzyan — B						FA of common red beds
										LA of banded ironstones
2000		Pt₁		Huronian — H				✳		
	Pt	–?–						✳		
3000	Archean	Ar₃	Swazian (Sw)	Randian — Ran						
		–?–								
		Ar₂								FA of stromatolites, ? first microorganisms
		–?–								Oldest sedimentary rocks
		Ar₁		Isuan — I						Oldest dated rocks
4000	Ar			Hadean						
	Priscoan				Hd					
5000	Pr									

developing in a few million years. Many of these were quite odd looking to our eyes, and they were extremely varied. There is no indication of ancestry; no invertebrate class is connected by intermediates with any other. There is very little continuity between the more complex Burgess Shale animals, with hard parts, and the preceding Vendian-Edicaran soft-bodied animals (Morris 1990, 33; Valentine 1985, 263–267).

Animals were simpler when they had just learned how to join cells into organs and multicellular organisms; nonetheless, it is not clear how species-level changes could have made new phyla in the time available (Valentine 1985, 270). The great radiation of new forms may have taken place in as little as 10 million years, roughly the time since the separation of apes and humans (Clarkson 1986, 48). These facts were contrary to the implications of Darwin's theory. He wrote, "If my theory be true, it is indisputable that before the lowest Cambrium [fossile-bearing] stratum, long periods elapsed, as long as or probably longer than the whole interval from the Cambrium age to this day, and that during these vast, yet quite unknown periods of time, the world swarmed with living [multicellular] creatures" (Darwin 1960, 331).

The record of plants is even more discontinuous than that of animals. When fossils of land plants appeared, without recorded ancestry, about 450 million years ago, major lines had already been formed, with no evident linkage among them. Many types arose in about 30 million years in the Silurian period (Thomas and Spicer 1987, 21). Some plant families, such as horsetails, club moss, selaginella, ginkgoes, and cycads, have been almost unmodified for tens or hundreds of millions of years. Flowering plants (angiosperms) appeared about 120 million years ago; for many millions of years, their rise was slow (Stebbins 1974, 318). However, "as soon as angiosperms became well represented in the fossil floras of the Cretaceous, they are largely referable to modern families and even genera" (Bell and Woodcock 1983, 318). Abundant fossils give little evidence of gradual change (Thomas and Spicer 1987, 61–67).

The impression that many groups arise suddenly at about the same time may be exaggerated by the system of classification. As one traces different orders, such as carnivores or ungulates, back to their earliest appearance, one naturally finds that the ancestral forms differ less than do their modern descendants. Similarly, it was possible for the principal animal types, the phyla, to diverge very rapidly, leaving no traces of intermediates, because they were much simpler and less deeply separated than their distant descendants. The differences, although basic, were not yet deeply embedded.

The gaps in the record are real, however. The absence of a record of any important branching is quite phenomenal. Species are usually static, or nearly so, for long periods, species seldom and genera never show evolution into new species or genera but replacement of one by another, and change is more or less abrupt (John and Miklos 1988, 307).

This contradicts the Darwinian approach. Natural selection—and Lamarckian evolution by use and disuse—would imply gradual, progressive change, with randomly diverging lines of descent. This would make a great irregular bush, not the branching ideal tree of life, much less the record that we have, with big and little branches suspended without junctions.

Those who study the fossil record, dealing not with equations of population genetics but with hard facts of the past, have been most inclined to be skeptical of Darwin's insistence on slow, more or less steady change. Such paleontologists as Stephen J. Gould, Niles Eldredge, and Steven M. Stanley have recently been in the vanguard of the critics.

The Leap into the Air

Along with the blanks in the record, evolutionists face the problem of how important changes could have come about; the origin of no innovation of large evolutionary significance is known (Langridge 1987, 248). Perhaps the most discussed transition is that from reptiles to birds. The appearance of mammals is passably understandable. One can fairly easily imagine reptiles' becoming mammals by degrees: standing more erect, improving the heart, stabilizing body temperature and acquiring hair to keep warm, producing a nutrient secretion from epidermal glands to nourish their offspring, and so forth. But the leap into the air is a theoretical crux.

Six skeletons of a primitive pigeon-sized near-bird, *Archaeopteryx,* about 150 million years old, have been discovered in a German limestone deposit (Wellnhofer 1990, 70–77). It is classified as a bird primarily because outlines of feathers were preserved in some specimens. Its skeleton is reptilian: long tail, no sternum to attach flight muscles, fingers not fused to make

Archaeopteryx, the size of a large pigeon, was a small dinosaur in bird's clothing. Its feathers were fully developed, but otherwise it had made hardly any of the skeletal changes distinguishing birds from reptiles. (Illustration from *The Fossil Book* by Carroll Lane Fenton and M. A. Fenton, © 1958 by Carroll Lane Fenton and Mildred Adams Fenton. Used by permission of Doubleday, a division of Bantam Doubleday Dell Publishing Group, Inc.)

the wing, and claws on the wing. The dinosaurian ancestry is obvious, and evolutionists point to it triumphantly as an excellent example of an intermediary between classes. It must have been close to the ancestral line of the birds because a modern Venezuelan bird, the hoatzin (*Opisthocomus hoazin*), by a remarkable surfacing of repressed genes, has almost identical wing claws as a fledgling. But the reptile-bird does not tell how a land animal became a flying animal. The question of the origin of birds has been muddied, moreover, by the discovery in 1986 of some crow-sized skeletons dubbed protoavis. This was a birdlike animal 65 million years older than *Archaeopteryx;* it was capable of limited flight at best but in some ways—fewer teeth, more of a sternum—was more birdlike than *Archaeopteryx* (Beardsley 1986, 677; Paul 1988, 251).

Specialists find good reasons to reject both of the theories of the origin of birds: that they developed flight as a supplement to running (because half-wings would slow the animal down) (Ostrom 1986) and that they started as gliders (because they have legs like ground-living animals) (Bock 1986). Many animals have acquired some kind of sail to extend their ability to jump, from frogs with expanded feet to gliding lizards to flying squirrels. Several modern lizards glide, mostly by flattening the body. One (*Draco*) can sail as far as 60 feet, thanks to a membrane supported by extensions of ribs (Bellairs 1970, 85). But *Draco* could not possibly advance to flight. Flying requires a surface to beat the air at some distance from the body (Paul 1988, 214). The gliding lizards of dinosaurian days were not ancestral to birds.

The first requirement for flight is a greatly enlarged surface area of the forelegs; it is useless to flap limbs without good airfoils. This problem was solved with feathers. The harder question of birds' origins is how feathers came about. Strangely, evolution gave *Archaeopteryx* feathers almost indistinguishable from those of a modern bird, such as a pigeon, even having series of primaries and secondaries, while leaving the skeleton so little modified that some specialists have doubted that it could fly. Feathers are complexly structured organs, with delicate interlocking details, barbules, and hooklets. One would suppose them to be very difficult to evolve, more difficult than wings (wings have arisen five times, feathers only once). It is speculated that feathers served originally for temperature control and were readapted for flight—a guess without evidence that is frequently asserted as though it were a fact (as by McFarland et al. 1985, 415).

Although they appear fully developed in an animal poorly designed for flight, feathers seem designed for this purpose. For warmth, something much simpler, like hair, would serve well, and some flightless birds, such as the kiwi, have hairlike feathers (Grant 1985, 323). The down of baby birds is better insulation than the plumes that make flight possible, but down would be rather an impediment to flight. No nonbird has anything like feathers, although many animals—not only mammals but arthropods like moths and spiders—have something like hair. If ground-living reptiles had found feathers useful for thermoregulation, it would seem likely that some would have kept the trait, or at least left some fossil trace of it, as found in *Archaeopteryx* and subsequent bird remains (Welty 1982, 594). A covering for warmth, moreover, would presumably be least developed on the limbs, but for flight, feathers are needed only on the forelimbs.

A possible explanation might be that feathers developed as a sexual display, scales being elaborated by the males for color and shininess. Feathers not only retain warmth and make an excellent aerodynamic surface but also serve for show, often being brilliantly colorful and iridescent. Many birds—from the gorgeous quetzal and birds of paradise to gaudily colored ducks—have spectacular displays.

One might fantasize that a small dinosaur attracted and stimulated its mate by waving its otherwise rather useless forelegs (bird ancestors were bipedal), somewhat as many lizards today display their charms. Such a mode of sexual selection might arise randomly, like innumerable others. In the competition for the attention of the female, showiness could be increased most economically by making brightly colored scales light and feathery. The gaudy display might be turned to advantage if the decorated forelimbs were set flapping as the animal took short leaps.

The rise of birds is the more remarkable because the air was already occupied by numerous and apparently efficient flying lizards, pterosaurs, which became extinct only at the end of the age of dinosaurs, nearly 100 million years after *Archaeopteryx*. The question of pterosaur origin is as unaccountable as that of birds. The earliest known pterosaurs were even more specialized for flight than *Archaeopteryx*. The wing was mostly an enormously extended finger; the sternum was developed for the attachment of flight muscles; and the main bones were thin-walled tubes, making the creature lighter for its size than birds. Like bats, pterosaurs had a membrane-wing. One species, *Quetzalcoatlus,* was by far the largest flier ever known, with a wingspread of 11 or 12 meters, three times that of a condor (Radinsky 1987, 132). The pterosaurs competed with the birds for some 70 million years, only to succumb in the extinction that removed the dinosaurs.

Soon after the great extinction, the bats made a spectacular leap into the air, where many kinds of birds were well established. The appearance of bats, the first mammals to reach modern shape, was as abrupt as that of pterosaurs. Their skeleton is very unlike that of a running animal, and the earliest known bat was almost indistinguishable from modern bats. It even seems to have had an advanced apparatus for echolocation (Novacek 1988, 70). Extraordinarily, it appears that bats evolved twice.

To attain flight, bats and pterosaurs modified the forelimb much more drastically than birds did. To achieve what the birds did with feathers, the bats and pterosaurs stretched fingers out to double the length of the rest of the body, a modification that would seem useless in its initial stages and difficult for gliders. Bats, however, are less aerially adapted than birds; they are capable of only slow, jerky flight, and they lack the birds' efficient cooling and respiratory system with air sacs and hollow bones. Yet they are generally much clumsier on the ground than birds.

The problem of the evolution of bats' flight is like that of birds. Modern gliding mammals, such as the flying squirrel or the flying lemur (colugo), have no tendency to prolong their leap by flapping the membranes stretched between fore and hind limbs, and it would seem difficult to do so. That bats somehow did so is suggested by the fact that their wings are attached to their hind legs, which, unlike those of birds, are poorly adapted for running. A difficulty is that if a glider like a flying squirrel began flapping its membrane to control its glide, the obvious course was to lengthen the forelimbs, not the toes. But the earliest known bats are finger flyers, almost indistinguishable from modern bats.

Insects are the only other animal to have achieved flight, which they did in the Carboniferous period, about 100 million years before flying lizards. Insects also have the distinction of sprouting wings, so far as appears, de novo. Insect wings are extensions of the integument of the thorax, and their genesis required the concurrent development of a light but stiffened membrane, a joint to the body, and suitable muscles and innervation, along with the controls necessary for flying.

There is speculation that insect wings originated as an outgrowth of larval gills or as thermoregulatory devices. It is postulated that the sails of some dinosaurs served this function, but no insects are known to have such an organ, and they hardly need it because they can easily regulate temperature by moving into or out of the sunshine. It is believed, in any event, that their flight arose from gliding (Kukalova-Peck 1987, 2342). It should be easier for a small animal to develop flight because of the relationship between surface and weight. But no other invertebrate class has done so.

Between Land and Ocean

It is an equal transition to shift between the ocean where life began and the land where it has attained its greatest complexity. Few groups have been able to do this. Although sundry worms and other small animals get along in a moist terrestrial environment, only three phyla (molluscs, arthropods, and vertebrates) have managed to equip themselves for life on dry land.

The stages by which a fish gave rise to an amphibian are unknown. There are resemblances between the first amphibians and certain (rhipidistian) fish with bony fins, but the earliest land animals appear with four good limbs, shoulder and pelvic girdles, ribs, and distinct heads. Certain bones in the fish fins correspond to bones in the tetrapod limbs, but the fin is very far from being a leg. In a few million years, over 320 million years ago, a dozen orders of amphibians suddenly appear in the record, none apparently ancestral to any other (Carroll 1988, 157).

The reverse transition, from land to water, has been easier. A land animal can more readily take to swimming than an aquatic animal fit itself for life on land. There are many semi-aquatic animals, from crocodiles through the muskrat and hippopotamus. Yet the whale is puzzling. The earliest whale fossils (named *Basilosaurus,* or king-lizard in Greek because the first skeleton was thought to be reptile) are about as specialized as modern whales. In fact, it was more eel-shaped than modern whales. A detail is that whales propel themselves by moving the tail, with horizontal flukes, up and down. Almost all fish, large and small, swim by flexing horizontally with vertical fins. Land animals move their tails mostly horizontally, as dogs do. There can be no intermediary between vertical and horizontal motion; muscles and fins or flukes have to concentrate on one or the other for efficient swimming.

Semiaquatic mammals, such as beavers, muskrats, hippopotamuses, polar bears, and otters, make good use of their hind limbs; none propels itself with its tail. The beaver comes nearest, steering with its tail while swimming mostly with its webbed hind feet (Hanney 1975, 47–48). In the first stages of a land animal's taking to the water, the legs have to be used for swimming, and they can hardly be eliminated before the tail becomes strong enough to propel the animal. It is hard to find

Basilosaurus, the earliest known whale, was about the size of modern gray whales. In general aspect, it was even more specialized for marine life, being quite elongated, almost eel-shaped. However, it had tiny hind limbs of unknown function (not shown here). (Illustration from *The Fossil Book* by Carroll Lane Fenton and M. A. Fenton, © 1958 by Carroll Lane Fenton and Mildred Adams Fenton. Used by permission of Doubleday, a division of Bantam Doubleday Dell Publishing Group, Inc.)

any reason that they should be lost; they are useful for maneuverability even if not needed for speed. A large majority of fish have fins on the rear as well as the fore part of the body.

Reptiles, beginning with strong tails, readily adapt them to swimming, as crocodiles have done. But the various families of marine reptiles (which also appear suddenly in the fossil record) kept hind limbs as paddles. Only in ichthyosaurs were they much reduced (Fenton and Fenton 1989, 425–432).

The straightforward adaptation is to develop webbing on the feet and turn them into something like flippers. Seals and walruses—whose transition from land to sea is also undocumented—have done this. To the contrary, the *Sirenia*, dugongs and manatees, like the whale, have lost hind limbs and move the tail vertically. They too appear in the record fully adapted to marine life (Carroll 1988, 544). It would appear that the loss of hind limbs was linked to other aspects of the whale (and dugong) pattern as part of the general development for aquatic life.

Whales have diverged more than any other mammals from the basic pattern of the mammal class. How long they (or seals, dugongs, ichthyosaurus, birds, and bats) may have taken to develop from quadruped ancestors is not known, but their extraordinary specialization (like that of the bats) must have been complete in about 10 million years (Eldredge 1989, 23). It could have been less because whales may have been around long before the first known bones show their presence. But 10 million years is less than a fifth of the time taken by *Hyracotherium* to become a not extremely different animal, the modern horse.

During this period, whales, besides converting forelimbs to flippers and growing a long and powerful tail, moved the nostril to the top of the head, modified their respiratory system, and made other adaptations for feeding in the depths. They remarkably developed new organs, dorsal fins and flukes, from skin and connective tissue (Young 1981, 498). In addition, before losing the hind limbs necessary to clamber onto the shore, they had to become able to give birth in the water, a process that must have involved new instincts for both mother and calf, including suckling the calf by pumping milk into its mouth, having surrounded the nipple with a cap to keep out seawater. It is difficult to imagine how all of this could have come about without a remarkable series of highly coordinated changes.

Genetic considerations also point up the difficulty of the whale's rapid evolution. By Mayr's calculation, in a rapidly evolving line an organ may enlarge about 1 to 10 percent per million years, but organs of the whale-in-becoming must have grown about ten times more rapidly over 10 million years. Perhaps 300 generations are required for a gene substitution (Mayr 1963, 238, 259). Moreover, mutations need to occur many times, even with considerable selective advantage, in order to have a good chance of becoming fixed. Considering the length of whale generations, the rarity with which the needed mutations are likely to appear, and the multitude of mutations needed to convert a land animal into a whale, it is easy to conclude that gradualist natural selection of random variations cannot account for this animal. After their perplexing rapid development, both whales and bats have for many million years evolved slowly, supposedly because their populations mingle widely, with no territoriality and much dispersal (Carl et al. 1977, 3945).

Perhaps we should not expect to understand major evolutionary innovations. None has ever been observed; indeed, no one has ever observed a mutation's making even the beginnings of a new organ. Innovation is the central problem that has troubled evolutionists ever since Darwin, and it is no less mysterious today than when he published his great book.

4

Inventive Nature

The Wonder of Life

In the miracle of life, material substance takes on complex, self-organizing order. Life is not merely the product of the past but a program to make a future, a novelty in the universe, structure shaped for needs.

The fundamental problem of life was how a biochemical system could multiply itself, in the long term improving its capacity to do so. Life uses energy (almost entirely from sunlight) to defeat the near-universal principle of increase of entropy, which means degradation or loss of faculties. In the short term, this requires growth; in the long term, it entails reproduction to surmount the decadent individual.

When molecules link together to make a crystal, their order serves as a template to which other atoms can adhere and enlarge the structure. But the distance from the most elaborate crystal to the simplest living organism is enormous. Organisms are self-regulating, or homeostatic, maintaining internal conditions despite fluctuations of the external medium. All animate beings selectively exchange substances with their environment, permitting certain materials to pass in and others to go out. Almost at their inception, living things had to become able to process materials absorbed or ingested, using them to carry out vital processes, to grow and reproduce. Such an exchange is the essence of animation. A minimum of about 300 biochemical processes are necessary; in the simplest known self-sustaining organisms, there are about 550 (Morowitz 1985, 248).

Much is theorized but little is known about how life originated. It is believed that the primeval ocean was a broth of large organic molecules, such as are found in meteorites and are known to exist in space, which could accumulate because the atmosphere was free of destructive oxygen. The initial sequence may have been that a compound had the capacity to catalyze another compound, this a third, and so forth until the chain reached a substance able to catalyze the original compound—a circle that would continue as long as energy-giving nutrients were available (Waldrop 1990, 1543). In some such fashion, there must have been formed ever more elaborate combinations, leading to a prebiotic competition for the capture of dissolved substances. One difficulty, however, was that the autocatalysis had to find a delicate middle way, typical of living processes: the catalyzer had to draw the appropriate substances to itself but not to hold them (Pool 1990, 1609).

The first—and very difficult—achievement of life would have been to reshuffle energy-containing compounds in such a manner as to manufacture more of the materials needed for its growth. This may have occurred in pools where sunlight helped energize combinations, perhaps with the help of mineral surfaces that catalyze reactions. Modern bacteria that live on iron or sulfur compounds produced by the earth may be continuing the next-oldest of all ways of life. Energy-containing compounds produced by electric discharges in the atmosphere could also have nourished protolife. But photosynthesis, doubtless crude at first, must have come near the genesis. Without using solar energy, life could make little progress.

Certain aspects of the conjectured beginning of life are fairly comprehensible. Amino acids, such as are easily formed from the probable components of the prebiotic atmosphere, combine to make proteins when heated, and the proteins spontaneously form tiny spheres (Fox 1988, 22). There must have been a near-infinity of molecules or spherules interacting over many million years, sometimes merging, sometimes growing overlarge and dividing ("reproducing"). A lucky invention, making copies of itself, would have to occur only once in order to become indefinitely numerous.

Yet the hurdles in the way of life's making itself were formidable. Organic molecules had to grow large enough for biological reactions yet not so large as to be immobilized. Intense ultraviolet light would have made it impossible for anything lifelike to exist at or near the surface of the ocean. It is believed that RNA must have been very close to the origin of life because it is chemically more active than DNA and can uniquely act as both self-reproducer and catalyst. But RNA is difficult to make and could not have come into existence by a chance combination; unless there is a guidance mechanism, it does not reproduce itself accurately (Waldrop 1990, 1544). There had to be a set of protein structures to permit nucleic acid to replicate, yet nucleic acid was necessary to make needed proteins. A membrane was needed to contain interacting proteins and nucleic acid, but proteins and nucleic acid were necessary to make the membrane. Moreover, it had to be semipermeable from the outset to admit useful materials and permit waste to diffuse out.

A minor problem is that although amino acids made nonbiologically are randomly optically left or right rotating, biological amino acids are always left rotating. All the amino acids in an enzyme must have the same orientation for it to be functional. The same is true of the sugars that form part of the nucleic acid chain. It seems that for life to begin, there had to be long chains with many units of the same rotational (isomeric) class, but the only known way to produce such a chain is by biological process (Hegstrom and Kondespudi 1990, 109).

In the simplest bacterium, reproduction is complex. The strands of nucleic acid must be replicated accurately; then strands and corresponding structures must be pulled apart in such a way as to make two complete sets, and a new wall has to be built to divide the new cells. This process requires hundreds of enzymes and proteins. It is subject to a high rate of errors, resulting partly from the never absolute stability of the intracellular environment, and errors have to be corrected in order to maintain the viability of the organism. Only a very short DNA sequence could replicate itself with sufficient reliability. But a fairly long sequence—the simplest modern genome has about 3 million bases—is necessary to produce appropriate enzymes to check errors. If a cell had a hundred bases or so in its DNA, there would be too many errors to maintain structures—certainly more than 1 percent going wrong—yet the bases would be far too few to code for the enzymes needed to correct mistakes of transcription (Maynard Smith 1986, 118). To surmount such barriers, life had to devise, through some process of self-organization, an interlocking structure of many essential components, none of which would seem possible without the others.

Life must have begun on a single track (or else only one track left descendants) because all creatures in their infinite diversity have the same basic chemistry, with similar metabolic processes. Most remarkable, the genetic code, which as far as known is arbitrary (there is no apparent reason that any particular set of bases codes for any particular amino acid except that is the way it started), is universal (with trivial exceptions). The code is believed to be as old as life itself (Eigen et al. 1989, 673). Once fixed, it could not be changed. It is also possible that the basic chemical reactions shared by all life are the only, or at least the best, attainable way to carry out many of its processes.

Despite its seeming near impossibility, life seems to have arisen relatively rapidly—within a few hundred million years of the formation of the planet. The earliest remnants believed to be fossil bacteria are about 3.5 billion years old. It may be assumed that these ancient bacteria were simpler than their modern descendants, but they must have solved the big problems, having developed most of the enzymes and proteins that enable organisms to function. Photosynthesis, a rather complicated process, developed at about this same time. For over 2 billion years—well over half the entire history of life—bacteria had the world to themselves. Apparently life rapidly came to or near a sort of plateau and continued for a long age with little apparent change, a pattern repeated countless times in evolution.

The more complicated nucleated (eucaryotic) cell appeared about 1.2 billion years ago. It was so long in coming that it must have been extremely unlikely, requiring several times longer than the genesis of life itself, and its advent marks the greatest known discontinuity in the sequence of living things (Glaessner 1984, 15). Between such very different organisms as bacteria and protists (protozoa and algae) there is no intermediary. The crucial advance was a membrane separating the directive nucleus from the supporting cytoplasm. Nucleic acid was divided into chromosomes instead of simply forming a ring, as in bacteria, and the amount of nucleic acid was multiplied manyfold, much of it seemingly being placed in reserve. There were also developed organelles (principally mitochondria and chloroplasts), which cooperate in the housekeeping but reproduce independently of the remainder of the cell.

Eucaryotic cell division, or mitosis, implied a more intricate apparatus for handling the much larger amount of DNA. When the cell on some signal undertakes to split, the genetic material strung out through the nucleus is gathered systematically into the correct chromosomes, and these replicate, forming paired sets. The body to which the chromosomes are to be drawn also replicates, apparently on its own, to make two poles. The nuclear membrane breaks down. Bundles of fibers extend from pole to pole, forming what is called the spindle. Some of these attach to central points of pairs of chromosomes. The two sets of chromosomes are brought into alignment along the middle plane of the cell. They are then pulled toward the poles, corresponding chromosomes always going to opposite poles. Next, each set surrounds itself with a new membrane. Meanwhile, in animal cells, a ring of microfilaments has formed around the equator of the cell (as determined by the poles); it constricts, and a new cell wall separates the halves and consummates the division. The process, which is much more complex than this description, is something of a ballet; how enzymes floating in the protoplasm dictate the complicated and dynamic structures is almost entirely unknown (Margulis and Sagan 1986, 146).

During the approximately 600 million years of the reign of nucleated or eucaryotic cells, there must have been many refinements and improvements, making possible such elaborate organisms as modern protozoa. However, what a single interacting system—the medley of enzymes and structures inside a single envelope—can do is inherently limited. More could be achieved by partitioning different processes in different compartments, that is, by instructing cells to differentiate for special functions.

The organization of many cells into different structures was the third great advance. Some bacteria (especially *Myxococcus*) act almost like multicellular animals, moving as a clump toward food (Shapiro 1988, 82–89). But it became possible for cells to take on different characteristics for the common benefit, forming a multicellular organism, only when the more complex eucaryotic cell was well developed. Multicellularity appeared by the pre-Cambrian, about 600 million years ago. It seems to have arisen several times, giving rise to several different lines, including plants, fungi, and animals. However, the fact that multicellular animals (with some primitive exceptions) begin life as a blastula, or hollow sphere, indicates a common genesis (Patterson 1990, 199–200; McMenamin and McMenamin 1990, 14–30).

Multicellularity seems inconsistent with the principle of natural selection of the individual. Some offspring of a mother cell have to surrender their capacity for reproduction in order to help a few sister cells make more reproductive cells in the long run. That is, the organism has to remodel its genome in such a fashion that most of its descendants are programmed to reproduce only for a limited number of generations. Cells, moreover, must turn off parts of the genome and activate other parts in order to make specialized tissues and organs.

The single-celled organism is marvel enough: the tiny quantity of nucleic acid has to contain instructions for both duplicating itself and making a complex cell to protect, nourish, and propagate the material of heredity. In the many-celled organism, the tiny part carrying its heredity has to contain instructions not only for replicating itself and its cell but also for making secondary classes of cells and organizing them into a body. The originally tiny element gives rise to creatures containing billions or trillions of cells, comprising a hundred or so different kinds of tissues. The means by which this ontogeny is accomplished remain obscure; geneticists know practically nothing about genes that make organs.

Multicellularity, like the previous great developments of life, was highly successful. Its potentialities were (and are) infinite because it could experiment indefinitely with changes and reproduce them to build upon them, and it led rapidly to a unique proliferation of forms. For about 570 million years, evolution has only been adding details.

Remarkable Structures

Although everything about life is in a sense miraculous, its achievements usually excite little wonder. One does not much admire the architecture of the weeds that are so well designed as to defy efforts to eradicate them; much less do we consider how ably the submicroscopic cold viruses entering our throats keep mutating to stay ahead of the immune system. But many seemingly simple or hardly perceived faculties are quite noteworthy. For example, some plants living in salty ground extract water against an osmotic pressure of as much as 200 atmospheres; others secrete salt against a similar pressure (Salisbury and Ross 1985, 696–697).

The number of admirable, more or less inexplicable traits that one might cite is limited not by the inventiveness of nature but by the ability of investigators to study and describe them—and those that are known would fill many large volumes. The ants, for example, are cited many times in this book, and few other families have received so much attention as these fascinating social insects. Yet the wealth of potentially intriguing information has hardly been tapped. There has been intensive study of only about 100 species of approximately 8,800 that have been described and probably 20,000 existent (Hölldobler and Wilson 1990, 2–4), and researchers continually uncover new facts about some of the best-known forms.

Not only do countless complicated instincts coupled with special organs challenge explanation in mechanistic terms; some organs are bafflingly elaborate in themselves. One is the brain, which processes words like these. Its complexity is largely unknown, but it may be appreciated by comparison with the best computers engineers can put together. Many million words would surely be needed to give complete instructions on how to build a supercomputer basically inferior in capacities to a good brain. The human computer not only has vastly more switches than a silicon-based machine but has many more different kinds of elements—at least 30 different sorts of neurons using as many neurotransmitters and up to a hundred neuropeptides (Guyton 1986, 567). One cell may receive inputs from 100,000 others; the brain must contain hundreds of trillions of interconnections.

The liver, a much simpler organ, has about 500 functions, including manufacture of bile and other digestive fluids; storage, conversion, and release of carbohydrates; the storage of iron and vitamins; the regulation of fat, cholesterol, and protein metabolism; the manufacture of materials used in the coagulation of blood (some 30 substances); the removal of bacteria from the blood; and the destruction of excess hormones and many toxic substances. And almost all of these functions are performed by cells of a single type (Guyton 1986, 875–876).

The classic example of cunning biological design is the organ transmitting these words to the brain. The human eye is much better than any camera of comparable size, with excellent resolution of detail. The retina continually produces images in full color, and its cells begin the processing of images for the brain. It is so geared that the periphery is especially sensitive to potentially threatening movement. In faint light, the eye registers an image instantly, whereas a camera with fast film requires several minutes to do so. A black pigment lining the retina eliminates reflected light, and the inner layer of the retina is pulled back around the center of vision, or fovea, to permit maximum acuity (Guyton 1986, 736–742). At a distance from the fovea, the retina is protected so that one can look at an object not far from the sun without burning it. The sensitivity of the retina can increase a millionfold, making vision possible over a range of brightness of a billion to one. Three kinds of cones make it possible to distinguish millions of hues. Retinoids, which are related to vitamin A, are stored in rods and cones as long as they retain a certain shape; when altered by light, they change shape and flow across the subretinal space into the black lining, where cells restore their shape and send them back to the rods and cones. This is only a beginning of the chemistry of vision (Stryer 1987, 42–51). The eye's diaphragm, the iris, is automatically controlled. The cornea has just the right curvature, and the lens, which is made achromatic by the varying refractivity of its layers, is focused by tiny muscles, controlled automatically by feedback from the brain. Paradoxically, the muscle controlling the lens stretches it by relaxing.

The brain melds the images from the two orbs into a single perception, coordinating for direction and distance. The eyes have two kinds of muscles and engage in several motions: they drift slowly, they follow a moving object smoothly, and they make little jumps, called saccades, to scan a scene; during the saccades vision is suppressed. They also jiggle 30 to 70 times per second to keep retina cells sensitive to margins.

The human eye is not outstanding. Some birds see several times as keenly as we do. Nocturnal animals such as owls and cats can see well with only a tenth to a hundredth of the light that we require (P. Burton 1973, 35). Some fish (*Anableps*) have bifocals for vision under and over the water surface. The archerfish (*Toxotes*), aiming at bugs above its pond, compensates precisely for the refraction of light at the surface of the water (Wheeler 1974, 353).

From different kinds of eyes in contemporary animals, one may guess how the organ evolved. Many primitive animals, even a few protists, have light-sensitive spots. In some flatworms (planaria) the pigmented spot becomes a cavity; if the opening is narrowed, it can form a crude image. Covering it with transparent skin could lead to the making of a lens, and so forth.

Darwin, troubled by the perfection of the eye, pointed out such gradations (C. Darwin 1964, 186–190), yet the existence of viable stages on the way does not explain how it was possible that many very unlikely genes came along in the right order to direct all the details, while at the same time an immensely larger number of continually occurring deleterious mutations were continually being eliminated. Nonetheless, complex single-unit eyes have originated at least five times (Pearse et al. 1987, 754). Some jellyfish (*Cubomedusa*) have as many as 24 eyes with retina and lens, apparently with means of accommodating for light intensity and distance (Pearse and Pearse 1978, 458). Some worms (nereids) also have eyes with lenses (Pearse et al. 1987, 394). Various molluscs, from scallops and snails up, have eyes. As Darwin noted, those of the octopus and squid are similar in pattern to vertebrate eyes, although they have evolved independently since the most elemental beginnings and have quite different embryology (C. Darwin 1964, 185). The compound eyes of arthropods have a completely different architecture, with thousands or tens of thousands of little receptors (ommatidia), but some of them are quite efficient. Certain deep-sea shrimp have lensless secondary eyes. Living in the dark and lacking the eyes that other shrimp have on their heads, they have substituted, for uses unclear, simple eyes on their backs with connections to the brain (Monastersky 1989, 90–93).

Pit vipers (rattlesnakes and relatives) are named thus because they have cavities on their heads with a small opening like a pinhole camera. Some 150,000 receptors sensitive to infrared light line the pit; they are connected to the optic tectum of the snake's brain. Sensitive to within .003 degree C, they form a rough image of a mammal, enabling the snake to strike accurately in the dark (J. Gould 1982, 156). The infrared sensor is in a way more remarkable than the eye because the eye was useful at every stage to discern radiation of the wavelengths most strongly emitted by the sun. Perception of the far infrared radiation coming from an animal can hardly have developed over a very long period because it is not widespread among snakes, and it would not be useful unless fully developed. In daytime it would be inoperative because of confusion with surfaces warmed by the sun, yet without it the snake could hardly hunt at night.

Hearing is a simpler sense, but some species have developed it quite remarkably. Owls can use a hundredth of a millisecond difference in the time a sound reaches one ear or the other to fix the direction of a mouse (Konishi 1983, 57). One ear of the barn owl is tuned to lower frequencies to locate a sound in the horizontal plane, the other to higher frequencies for the vertical plane (McFarland 1985, 227).

The echolocation system of bats is more elaborate. The bat not only registers the infinitesimal echo of its squeak from a mosquito but can also determine accurately its distance and direction. This requires a very high frequency—up to 220 kilohertz. The bat must overcome the fact that the signal it sends out, lasting as little as 1/2,000 second, is millions to billions of times stronger than the returning echoes. To prevent the echo from being totally swamped, there is an insulating pad behind the bat's inner ear, and muscles stiffen the eardrum up to 100 times per second synchronously with the squeaks. The brain is geared to ignore strong auditory messages while registering weak ones. Bats are also sensitive to a change of interval between click and echo indicating movement of the target, and they pick out echoes timed to their own signals to distinguish them from those of other bats. The membrane of the inner ear is thickened to impede perception of the clicks but to permit perception of a shifted echo. There is even an offset system between the two ears to increase contrast and permit more accurate direction finding. The brain of the mustached bat (*Myotis lucifugus*) can distinguish a difference of 1/700 of a note. Many species have a constant-frequency signal, to tell the direction of a prey, and a sharp falling signal, to tell distance (McFarland et al. 1985, 510; Suga 1990, 60–68). This apparatus enables the bat to fix range within a centimeter or two at a distance of several meters, to detect the shape of a tiny target, and to flit among a network of threads a tenth of a millimeter in diameter.

Yet this wonderful adaptation is less useful than might seem because prey may perceive the chirps much farther away than the bat can hear the echo. Moths, lacewings, crickets, and other insects have cells tuned to bats' wavelengths and take evasive action when one approaches (Alcock 1989, 114–121). Some moths, hearing the signals, produce ultrasonic clicks of their own, apparently to jam the bat's sonar. Bats have sometimes reacted by turning off their echolocation and hunting, as owls do, by sight and sound (Vaughan 1986, 505–506).

Various other animals, including cave-dwelling birds, shrews, some seals, perhaps some rodents, and probably all toothed whales and porpoises, also use echolocation (Vaughan 1986, 510). The porpoise apparatus is more difficult to study than that of bats and consequently is less well known, but it is quite as remarkable. Porpoises can find objects as small as a centimeter in diameter on the floor of a tank, even distinguishing a copper square from an aluminum one of the same size. At a distance of 10 feet, they can detect an object 0.1 inch in diameter. This accuracy is the more amazing because the porpoise has problems the bat does not have. Because the speed of sound in water is five times greater than its speed in the air, the wavelength is five times longer, and hence the sharpness of resolution is five times less. For good resolution, the signal must be of very high frequency; the large animal can produce very high intensity sound (said to be enough to stun fish) at frequencies even higher than those of tiny bats. The porpoise also has to overcome the great loss of energy of sound passing from air to water and vice versa.

Consequently the porpoise, even more than the bat, has had to develop new organs. Its signal is produced by a set of sacs and valves in the head without expelling air, and it is focused through a sort of lens in the front part of the head, which changes shape to direct the beam in the desired direction. The echo is picked up by the lower jaw, with a canal to conduct sound to the ear, which is shielded from extraneous noise. A wide, low-frequency beam is used for navigation and a narrow, high-frequency beam for hunting (McFarland et al. 1985, 510–511; Pryor 1975, 87; A. Popper 1980, 1–44).

The electrolocation system of some fishes seems even harder to achieve than the echolocation system of bats and porpoises. Darwin, who could propose an explanation for almost anything, admitted that he could not conceive how the electric organs of fish could have been evolved (C. Darwin 1964, 192). Modern science has not come much closer.

Certain sharks have developed such sensitivity to the discharges from the muscles of fish (like the impulses that make an electrocardiogram) that they can perceive as little as .01 microvolt per meter, equivalent to about 10 volts at a distance of a kilometer (Budker 1971, 80). Other fish, combining the abilities to generate electricity and to perceive it, use an electric field up to several volts as a sort of radar, sensing disturbances caused in the field by other fish or solid objects. The electrosensing fish generally live in turbid water or are nocturnal, and they have weak eyes. They may also court electrically.

Other electric fish use their discharge as a weapon. Electric eels (*Electrophorus*), with some 6,000 generating plaques, can produce about one ampere at 500 volts; they do well not to electrocute themselves. Some fish use electricity as both a weapon and a probe (Bennett 1971, 385). But one fish, the stargazer (*Astroscopus*), seems to have an electric organ for neither purpose. It is too weak (5 volts) to stun prey, and the fish apparently has no receptors to use it for location (Bass 1986, 21).

For electrolocation, many things must work together: an apparatus to generate fairly strong electric pulses at a rate of as many as 1,700 per second, consisting of a large number of plates stacked up like batteries in series; effective insulation of the electric generator from the body to make it possible to pile up voltage without allowing it to leak backward; special fins to swim without flexing the body and thus disturbing the field; a means of controlling the pulses; incredibly sensitive receptors capable of registering minute changes in the strong primary gradient of the field; means of filtering out the electric discharges of other fish, which are immensely stronger than the echoes of its own field; and a special structure in the brain to process and use the information received (Grundfest 1967, 399–428).

The various parts of the piscine radar system would seem hard to achieve even quite separately. For example, the electric generating apparatus is derived from muscle tissue, but how a muscle could turn into a generator while remaining always useful is hard to imagine. Muscles have minute electric discharges, but for any probable utility, the discharge of many plates must be strengthened, coordinated, and made part of an intricate response system.

Not least of the difficulties is insulation. Living tissues contain dissolved salts, that is, positive and negative ions capable of conducting electricity; it is difficult for any tissue to be non-conductive unless it is cut off from the ordinary processes of the body. The blood supply and the nerve fibers activating the organ have to be conductive. Moreover, fresh water is a much poorer conductor than animal flesh; when the fish makes a voltage difference between head and tail, the easiest path for the current would normally be through its body. The fish consequently has thin, conductive skin where the current exits and reenters the body and thick skin elsewhere, composed mostly of nonconductive connective tissue; it also has an extremely elaborate system of membranes to permit the transport of nutrients and nerve signals while preventing ions from travelling out of the electric organ despite the powerful discharge (Stuart Thompson pers. com.).

A detail is that the electric generating plaques have to be stimulated at exactly the same time. Since they are strung out along the body, the nervous impulse must travel more rapidly to cells farther from the brain; hence nerve fibers (axons) to farther parts of the electric organ are proportionally thicker or those to nearer parts take a circuitous route. The fish also has the same problem of interference as the bat: its receptors must be instanteously rendered insensitive to its pulses of current (Pough, Heiser, and McFarland 1989, 284–289).

Despite their apparent improbability, electric organs have been perfected more or less independently in most of the ten different families in which they occur. But a predisposition must have appeared very early because they are common to the shark family (Elasmobranchs) and bony fish (Osteichthyes), which diverged over 450 million years ago (Bennett 1971, 347). The question arises, of course, how such wonderful powers could have been developed unless they were extremely advantageous and, if so, why they have not prevailed more widely.

To mention other extraordinary traits may seem anticlimactic, but the capacities of evolution are extremely multifarious. For example, a simple mold causes the leaf of a blueberry plant it infects to look like a flower in the ultraviolet light to which insects are sensitive, which causes them to land on it and pick up spores to be spread to real flowers (Batra 1987, 57–59). Bacteria have decidedly elaborate structures, including rotatory flagella—wheel-like motion that is potentially useful but so complex to manage mechanically that it is almost nonexistent in living creatures. The base of a bacterium's flagellum has a set of discs fitted into membranes; they are somehow made to rotate continuously to propel the bacterium. The motor can shift into reverse and rotate in the opposite direction (Glagolev 1984, 37). Why the bacteria should have invented such a difficult mechanism is puzzling. Evolution seems to have gone to great lengths to do the unnecessary: cells have much simpler filaments swishing back and forth quite effectively.

Mimicry is also striking. Although it seems a quite understandable strategy for survival, how it could have been brought about is often incomprehensible. It is common enough in sundry orders; even some plants are mimics. Higher animals do not much engage in mimicry, but there is a wolf in sheep's clothing: the zone-tailed hawk (*Buteo*) of the American Southwest has become an almost perfect copy of the harmless turkey vulture (Alcock 1990, 144). Australian mistletoes have leaves like the host plant, for reasons unknown (Bernhardt 1989, 35). Some caterpillars mass together to imitate a pile of bird droppings (Cloudsley-Thompson 1988, 116). A still more inventive caterpillar (*Nemoria arizonaria*) lives on oak trees. When it hatches in the spring, it feeds on the oak flowers, or catkins, and looks just like another catkin, even bearing imitation pollen sacs on its back. After the catkins are gone and the caterpillar has to eat leaves, it changes its appearance entirely to look like a twig (Greene 1989, 643–645).

Hundreds of species of butterflies are near copies of different species, often not closely related; edible species especially mimic noxious ones. Most famous is the viceroy butterfly, so named because it is a slightly smaller copy of the monarch butterfly, which contains a toxic substance it gathered as a caterpillar from milkweed. The African milkweed butterfly is mimicked by 33 species of butterflies and moths (Cloudsley-Thompson 1988, 119). Some species of butterflies have multiple variants, or morphs, that mimic different models. *Pseudoacraea eurytus* has over 30, the male and female sometimes imitating different species (Mayr 1963, 248–249).

The phenomenal ability of some butterflies to shift wing patterns must be dictated by gene families working as units. Most butterfly wings are variants of a single basic pattern (Saunders 1988, 254; Murray 1981, 473–496). Since there may be more mimics than poisonous butterflies, it must be easier for an edible butterfly to mimic an inedible one than to develop its own poison. Inedible butterflies sometimes mimic one another—perhaps as much from convergence as selective adaptation. Butterflies from distant regions may have almost identical wings, although there can be neither mimicry nor environmental constraint (Ho 1986, 42).

The ability of the chameleon to adapt its color to its surroundings (and emotions) is renowned. The physiological mechanisms are incompletely understood; pigment cells react not only to what the animal sees but also to light falling on the skin. The chameleon seems to have a special capacity for innovation: the tail is prehensile, almost uniquely among lizards; the feet have an unlizardlike grip; the eyes are like independently moving telescopes; the tongue is as much as one and a half times as long as the body; and many chameleons have horns of eccentric shapes with no observable function (Bellairs 1970, 300–301).

Some bottom-living fish, especially plaice, sole, and flounder, outdo the chameleon. They take on not only the color of the seabed, from white through a variety of colors to black, but copy the pattern of the gravel, sand, or rocks on which they rest. If they are on a checkerboard pattern, they make a fair continuation of its pattern on their backs. How the fish brain can command the pigment cells so skillfully challenges comprehension; how the genes could come properly to program the brain is of a higher order of incredibility (von Frisch 1973, 70).

Fantastic Behaviors

By what devices the genes direct the formation of patterns of neurons that constitute innate behavioral patterns is entirely enigmatic. Yet not only do animals respond appropriately to manifold needs; they often do so in ways that would seem to require something like forethought.

A very small animal, the aphid *Pseudoregma bambusicola,* a millimeter long or less, leads a complicated existence. Being soft-bodied and sugar-filled, it is the delight of many predators. So *Pseudoregma* (and some 20 other aphid species of several genera) produces specialized soldiers. These, for no apparent adaptive reason, are sterile, do not eat, and die young, but they have sharp horns in their hardened little heads and an irritating secretion. Their numbers compensating for their puniness, the armed guards keep most hungry insects at bay (Dixon 1985, 37).

The flowerfly, however, contrives to thwart the defenders. The female fly, finding a colony of aphids, lays eggs among them if the weather is cool and the aphids are sluggish. In warm weather, she goes a little way off to find something like a strand of spider silk on which to lay. If soldier aphids can reach the eggs, they destroy them. If not, the eggs hatch, and the minute flowerfly larvae crawl to the aphid colony. There, protected from the soldiers by a tough skin and apparently a poison, they feast on aphids.

Another aphid eater, a miniature moth, solves the problem of safeguarding its eggs by making them so pancake flat that the aphid soldiers cannot get a grip on them. On hatching, the moth larva makes a silken tunnel to cover itself. When hungry, it lunges out to grab an aphid. The victim probably releases an alarm pheromone to warn its fellows and summon rescue. If the moth larva is attacked by a soldier, it drops down with it on a silk line and lets the soldier fall by cutting it off (Moffett 1989, 406–422). These must be only a few of the facts to be discovered about a little-studied, practically brainless insect.

One may admire the Amazon fish, which lays its eggs above the water and splashes them every 10 minutes or so for three days or more until they hatch (Kevles 1986, 64), and the little wasp that provokes an insect-eating ant lion into seizing it in order to lay an egg on the predator (Askew 1971, 167). The catalog of marvels has also to include the repertoire of the male marsh wren, the one hundred to two hundred songs (in California; eastern wrens know only about a third as many) with which, in conventional sequence, he regales his love; a substantial part of his small brain is devoted to music (Kroodsma 1983, 43–46). Brown thrashers, which also have as many as 200 songs, learn different ones for different uses, primarily to attract females or repel males (Kroodsma 1989, 51–58).

Many other adaptations are more spectacular. Birds, turtles, fish, and sundry other animals find their way over great distances. Birds fly without previous experience as far as from the Arctic to the Antarctic, often without landmarks. The 3-gram ruby-throated hummingbird goes 600 miles nonstop across the Gulf of Mexico. The young of the bronzed cuckoo, a month after being left behind by their parents in New Zealand, fly 1,200 miles over water to Australia and then 1,000 miles north to the Solomon and Bismarck Islands (F. Gill 1990, 243–245; Welty 1982, 585). The journey seems unnecessary, since there is no great need to escape the mild winter in New Zealand, and it is unclear why the parent birds require their offspring to navigate by themselves.

Many birds use the stars as a compass or navigate by the constellations, which young birds have to learn prior to the autumnal migration (Barnard 1983, 168–170; Emlen 1972, 283). Most seem to be able to perceive the earth's magnetic field—apparently not north and south but the inclination of magnetic lines of force from the horizontal (at the magnetic pole, they point straight down). But no one has any idea how the fliers hit small targets after long journeys, often in the dark, during which they may have been carried far off course by winds. They would seem to need not only an accurate compass but a good map. Yet they keep on course even when clouds hide the stars above and the earth below.

Not only strong fliers like the plover but also snipes, curlew, and sandpipers migrate between Alaska or Siberia and the Hawaiian Islands; and missing the islands in the featureless ocean would be fatal for nonoceanic birds. The golden plover, having made the trip to Hawaii, goes 2,500 miles farther on to the still smaller Marquesas Islands (Emlen 1972, 278). The Hawaiian Islands were never connected with the mainland, and they are only a few million years old; it is not easy to imagine how the instinct to migrate seasonally 2,500 to 4,000 miles from Alaska or Siberia could have started. Storms occasionally carry birds accidentally to Hawaii, but this good fortune would not make a migratory instinct.

The navigational ability of pigeons has been studied intensively for many years. Researchers have learned that they orient themselves by the earth's magnetic field and the position of the sun. They have to learn to use the sun on the basis of magnetic orientation but, having learned solar navigation, preferentially use it. They may follow olfactory clues. Their ability is only partly inborn, and different birds seem to use different strategies (Walcott 1989, 40–46). But their faculty for knowing the direction of home from a distant region where they have never before been still defies understanding. Even nonmigratory birds sometimes find their way home after being transported thousands of miles away. The ability, however, is irregular. Some species of sparrows have such homing abilities; others are lost if removed a few miles from their territory (Welty 1982, 580).

Sea turtles, after feeding for several years off the coast of Brazil, swim 2,500 miles to Ascension Island, which is only about 8 miles across, a speck compared with Hawaii. One may imagine the turtle breaking the surface occasionally to take bearings by sun or stars, but how it navigates so accurately without a modern chronometer and a good sextant remains to be discovered. The turtle cannot be guided by the taste of the island runoff in the sea, because the arid land has practically no runoff. The journey seems worth the trouble, however, because Ascension Island has no predators. On mainland beaches, turtle eggs are dinner for many animals, and the newly-hatched turtles are tender and easily picked off on their journey to the sea (Lewin 1989a, 1009).

Salmon can hardly guide themselves by the stars, but they find their way from hundreds and even thousands of miles offshore to the river and the same small stream from which they set out years earlier. No one has suggested a plausible means of navigation, although fish are known to perceive the earth's magnetism, the position of the sun, the polarity of light, and electric fields (Barnard 1983, 174). Salmon respond to chemicals to which they are exposed at a certain stage of life, but they range far beyond the possibility of detecting the taste of the native brook dissolved in thousands of cubic miles of seawater. The ocean contains vastly larger amounts of dissolved minerals of all kinds than the streams, which must be much alike in the dissolved salts and organic material they carry in minute and seasonably variable amounts. It is difficult at best to follow a dilute scent; the salmon must swim far enough to sense the odor's getting weaker or stronger, and the gradient will be infinitesimal in the ocean. It is some but not much help to suppose that they head for the coast and follow along it until they sense their river (McFarland 1985, 253). It would seem impossible for salmon to do what millions of them do.

European eels are hatched in the Sargasso Sea not far from North America. Doing the opposite of what the salmon do, larval eels cross the ocean to Europe, dawdle in the coastal waters, change shape, then head up rivers to mature. Eventually they head downstream and find their way back to the breeding grounds from which they set out many years before (Moriarty 1978, 60–89).

A migratory faculty must be contained in the brain, and the brain of the salmon is small compared with a turtle's, not to speak of a bird's. But the monarch butterfly (*Danaus plexippus*) does wonders with a brain hardly visible to the naked eye. The monarch winters at a few sites, especially in central Mexico, where hordes festoon the trees. In the spring, they migrate north, each generation going some hundreds of miles, as far as Canada. In the fall, the five-times-great-grandchildren return as much as 1,800 miles over lands they have never seen to the very grove, perhaps the very tree, from which the ancestors set forth. This, too, could reasonably be called impossible. One might also ask why they undertake the difficult and hazardous return journey. It would seem easier and safer to hibernate in a crevice (Cloudsley-Thompson 1988, 101). But the monarch is a numerous species; clearly it is successful.

Many other humble creatures have sophisticated means of navigation. Various butterflies migrate considerable distances, using the sun as a compass. Solar orientation, taking into account the movement of the sun across the sky, has been observed also in locusts, some beetles, and pond skaters. Sandhoppers, miniature crustaceans, seem to orient by the moon (Cloudsley-Thompson 1988, 108–111). Bees not only orient themselves by the time-compensated position of the sun but also perceive the polarization of light on cloudy days and guide themselves thereby. Ants do the same, finding their way over as much as a hundred yards of windswept featureless terrain without trails (E. Wilson 1971, 211).

There are countless problematic adaptations of parasites, of which there are millions of species; there are probably more parasitic than nonparasitic animal species (May 1988, 1447; Noble and Noble 1982, 3). In many cases, especially of internal parasites, there is no readily imaginable halfway house between free living and parasitic dependence. It is also difficult to adapt to two or more hosts as different as snails, cockroaches, and mammals. But among arthropods alone, some 1,000 species, mosquitoes, fleas, ticks, and so forth, are disease vectors (Nolan 1983, 182). Many parasites, especially worms, go through bewildering metamorphoses through a sequence of as many as four hosts (Inglis 1965, 85–86).

The brainworm (*Dicrocoelium dendriticum*) that reproduces in sheep uses ants to get back into a sheep. The worms get into ants by infecting snails that eat sheep feces. The snails expel tiny worm larvae in a mucus that ants enjoy, and some dozens of worms take up residence in an ant. But this would do them no good if the ant behaved normally; too few ants would be eaten by sheep. Consequently, while most of the worms make themselves at home in the ant's abdomen, one finds its way to the ant's brain and causes the ant to climb up a grass stem and wait to be eaten by a sheep. Ironically, the worm that programs the ant is cheated of happiness in the sheep's intestine; it becomes encysted and dies (D. Wilson 1977, 421–425).

The whole procedure seems unnecessary. Why do the worm eggs defecated by the sheep not simply hatch and climb up a grass stem to await being eaten by a sheep instead of making the hazardous trip through snail and ant? How could they become adapted to being carried by the ant unless the ant were already programmed to make itself available to be eaten by a sheep?

In many cases, one kind of animal takes advantage of another in raising its young. At least 53 species of birds, especially cuckoos and cowbirds, lay their eggs in other birds' nests. This requires or evokes a variety of strategies. The parasitic female may sneak around a prospective host nest, quickly laying in it while the owner is foraging. The male sometimes cooperates by challenging the host birds and luring them away. The cuckoo can insert eggs through holes it cannot enter, and it lays in a few seconds; its eggs resemble those of the principal host in the region. Some species mimic hawks to alarm prospective hosts (F. Gill 1990, 390). The parasite egg hatches first, and the chick pecks the foster siblings to death or pushes them out of the nest.

Various fish hatch their young in the security of their mouth and keep them there during the first stages of growth. A catfish, however, manages to get its eggs into the mouth of a brooding cichlid, where the catfish fry eat the baby cichlids (Alcock 1989, 254). The bitterling (*Rhodeus*), a freshwater fish, lays its eggs in a mussel. Doing this requires a specially designed ovipositor, a device for quick expulsion of the egg instantly before the mussel can close its shell, and appropriate instincts. The male, having taken a mussel as his territory from which he repels other males and to which he leads females, releases sperm for the mussel to suck in to where they can fertilize the eggs (Wiepkema 1961, 113–119). The mussel has its vengeance by a fantastic procedure for dispersing its young. It draws fish with a fishlike lure; when one comes by, baby mussels lying in wait clamp onto a fin or gill with a special tooth and become temporary parasites on the fish, dropping off a few weeks later, hopefully in a good location to start their sedentary life (Isom and Hudson 1984, 319–320).

Intelligent Instincts

Complex behavior can be directed by a simple central nervous system or none at all. An amoeba can perceive an edible smaller protozoan, surround it with pseudopods without touching it, and then engulf it. A sea anemone (*Calliactis*) can sense a shell inhabited by a hermit crab and climb onto it or note the approach of a hostile starfish and leap or wriggle away (Pearse et al. 1987, 29, 163; Burton 169, 62). Yet the anemone has not even the beginning of a brain to coordinate its nerve network.

The nervous systems of insects are more respectable, with up to 100,000 neurons, but it is difficult to comprehend how this number of elements, small for a computer, can mediate much plastic and functional behavior. The female of a parasitic ant (*Epimyrma*), on finding a colony of the host ant (*Leptothorax*), seizes a worker, rubs it with brushes on her legs, and then rubs herself to transfer its scent and make her acceptable to enter the host colony. There she goes straight to the *Leptothorax* queen and throttles her (Dumpert 1981, 172). Bumblebees somehow came upon the strategy of immobilizing an invading insect by smearing it with honey (Mosse 1981, 261). Assassin bugs attacking a termite colony fasten pieces of nests or husks of dead termites to themselves to confuse the termite soldiers (Griffin 1984, 123). Ants rob wasp nests, so the wasps smear ant repellent on the stalk holding the comb (Evans and Eberhard 1970, 165).

Social insects have more spectacularly developed intelligent-seeming capacities than their solitary relatives, doubtless because of the greater range of behavior opened by group interaction. The best studied are the honeybees, whose dances may be the most sophisticated communication among invertebrates. A lengthy book is necessary to describe in detail the life of the hive.

Weaver ants (*Oecophylla*) use the ability of the larvae to spin cocoons to build their home. Making chains up to 12 workers long, the ants pull leaves together or fold over a leaf and stitch the edges by causing the larva to exude silk. Then neighboring apartments are joined to make a many-chambered mansion. They also make barns for aphids. Incidentally, the weaver ants have 5 different recruitment signals to communicate the need for help in various tasks, such as bringing back food. For example, when fighters are needed, an ant makes movements equivalent to shaking a fist at a nestmate, and the latter transmits this message to others (Hölldobler and Wilson 1990, 618–629).

Many ants tend aphids for their sweet secretions, which are the aphids' thanks for many services. Ants vigorously protect the "cows" from insect predators instead of eating them and build shelters for them. Some ants bring the aphids into their nest overnight or for the winter, placing them on roots to feed. Others collect aphid eggs to allow them to hatch. A virgin ant queen going forth to found a new colony may carry a few aphids with her to start a new herd (J. Gould 1982, 420; Perry 1983, 28–29).

Fungus ants are the outstanding gardeners of the animal world. Their huge colonies send columns out for hundreds of yards to cut pieces of leaves, which are brought back, masticated, and used to nourish species of fungus not known to live elsewhere. The ants care for their crops, clearing out "weed" fungi, and harvest from their gardens their sole nourishment. Half a dozen different castes perform different tasks. The queen on her nuptial flight carries a packet of fungus with her, along with a few relatively tiny workers, to form a new colony. In order to nourish the fungus in the new home, she lays nonfertile (trophic) eggs. It may be guessed that this form of agriculture began with the ants' discovery that a mold growing on refuse or feces was edible (Hölldobler and Wilson 1990, 596–608), but it required something of a creative leap to put the casual discovery into practice.

Other tropical ants cooperate with certain bushes or small trees. An acacia provides an aggressive species (*Pseudomyrmex*) with nesting places in hollow thorns and with a nourishing secretion; in return, the ants keep off other insects and sting animals that may try to browse the leaves. They even kill competing vegetation for a few feet around the base of the acacia

(Dumpert 1981, 227; Hölldobler and Wilson 1990, 531–534), yet they take no notice of birds that nest in the branches.

In the perspective of the abilities of very small animals with tiny central nervous systems, the complex ways of larger animals may be less impressive. But it seems well calculated that the African shovel-nosed frog (*Hemisus*) makes a burrow near a stream to lay her eggs; when they hatch, she opens up a tunnel for the tadpoles to swim out into the water (Duellman and Trueb 1986, 137).

One parent or the other of numerous kinds of frogs holds the eggs on its back; one species has a chamber for each egg with specialized membranes and an apparatus to provide the tadpole with oxygen—with little apparent utility, because the young are released after as long as 4 months without having grown appreciably (del Pino 1989, 112–113).

A maternal tropical frog, the super-poisonous *Dendrobates*, lays her eggs in a moist place and covers them with a gel. The father comes back from time to time to aerate the eggs and moisten them with water from his bladder. When the tadpoles begin to hatch, the mother returns and allows one at a time to climb onto her back, where it is held by a glue. Then she seeks out an epiphyte in the branches of a tree with a little pool of water at its base. She checks whether it is already inhabited; if not, she leaves her tadpole. Having deposited perhaps a dozen young in as many private pools, she does not forget them; after a few days, she returns to check on them and, if they are well, leaves a few infertile eggs to nourish them in the protein-poor environment (Forsyth and Miyata 1984, 181–182).

As a variant of such behavior, the male of a South African frog, *Rhinoderma darwini*, broods the eggs in its vocal sac (von Frisch 1974, 168–169). Even more solicitous of its young is a small frog, *Rheobatrachus silus*, discovered in streams of southeast Australia. The female swallows its eggs and broods the tadpoles, about 20 in number, 6 or 7 weeks in the stomach, finally burping froglets, which she may reswallow. As M. T. Tyler, a specialist in gastric brooding frogs, remarks, it is not easy to contemplate how this custom could have started; there could be no progressive stages in ingestion of eggs or the reduction of digestion in the stomach, which is rather radically altered. The tadpole also had to change from a feeding and motile to a nonfeeding, nonmotile organism. In Tyler's opinion, the habit must have come "by a single, huge quantum step" (Tyler 1983, 129, 133). (Unhappily, *R. silus* has not been seen since 1980, apparently having shared the extinction that, for reasons unknown, has struck many amphibians worldwide [Barinaga 1990, 1033]. A related species survives, however.) A few species of frogs have found the obvious and, one might suppose, more practical way of protecting their eggs: retaining them until they are matured to fully formed young, in the manner of many reptiles.

The clever ways of birds are legendary, and it is difficult to pick out any few especially worthy of admiration. But humans would need much learning to weave nests of strands of grass like the weaverbird, some 150 species of which suspend their residences from branches. The birds crossweave and tie stems, not an easy task without hands (von Frisch 1974, 207–210). Some varieties tie overhand knots or slipknots; others braid a sort of rope to hang the nest as much as a meter below the branch (F. Gill 1990, 345–348). The Indian tailorbird sews broad leaves together with strands of fiber (Forsyth and Miyata 1984, 211). Chimpanzees, with hands and hundreds of times more brain mass, make nothing so well crafted.

Many less cunning instincts are difficult to explain. For example, how did some bats become bloodsuckers? It is wise for bats to stay clear of large animals unless fully prepared to bleed them. For this purpose, the vampires had to be able to maneuver on their legs much better than their less offensive cousins, and they had to have very sharp teeth and anticoagulant saliva, plus a modified digestive tract and suitable instincts.

Such an instinct required the animal to go against its ordinary nature. This is also true of the many fish that brood their young in their mouths. The normal reason for taking a hatchling into the mouth would be to eat it, and many varieties eat their own young on occasion. There would be no advantage in mouth brooding unless the mother fish not only repressed the urge to swallow but arranged to board the young (while herself going hungry) for a sufficient time to enable them better to face the world. The young would also have to be prepared for existence inside the mother's mouth. Yet mouth brooding seems to have come about independently many times in various groups of fish.

Fireflies have an admirable signaling system for finding mates: it requires that both sexes have a lantern, nervous control of it, an agreed signal code, and appropriate instincts. Unless all the elements are in place, luminescence is a waste of energy. One kind of firefly female (*Photuris*) mimics the signal of several different kinds of smaller fireflies and eats males that answer the siren call. It is harder to understand how this habit was established because the normal diet of adult fireflies is nectar and pollen (Klots and Klots 1971, 111–112).

Hedgehogs (*Erinaceus*) have certainly had unpleasant experiences with toads, which look edible but have poison glands in their skin, and they should easily learn to forgo the meal. But how or why would a hedgehog undertake to munch the repugnant skin in order to spit the poison on its own quills? This instinct is the more remarkable because the hedgehog is already rather well protected. Yet it is well ingrained; young hedgehogs do it automatically (Young 1981, 437).

Sea slugs (nudibranchs, such as *Aeolidia*) also make use of the defenses of other species. They eat sea anemones and jellyfish, loaded as these are with stinging cells, or nematocysts. The nudibranch somehow inactivates the nematocysts and keeps them from being digested, passes them through the body, lodges them in the skin, and reactivates them with the business end pointed outward or stores them in special sacs ready to be spewed into the mouth of an attacker (Greenwood and Mariscal 1984, 719–730).

The human botfly (*Dermatobia*) wants to lay its egg on the skin of a person (or a monkey), but it waits until a mosquito or a bloodsucking fly approaches, catches it, and sticks its egg to the mosquito's or fly's belly. When the mosquito alights on warm flesh, the botfly larva emerges rapidly from the pre-hatched egg, drops off, and burrows into its new host (Sancho 1988, 242–246). The botfly must have developed (1) an instinct to seek out a suitable host and, instead of laying its egg on its intended victim, to hover around; (2) the ability to recognize a potential vector, probably a mosquito; (3) the instinct to pursue the mosquito and catch it; (4) legs adapted to grasp and hold it; (5) a modified ovipositor to lay its egg on the underside of the mosquito; (6) a glue to hold the egg to the mosquito; (7) the ability of the egg to respond to the warmth of the mammal (only slightly higher than air temperature in the tropics); (8) a reflex of the larva to release itself, drop onto the human flesh, and bore in. Closer study no doubt would reveal more details that have to be in place. Yet it is not even certain that the behavior is very advantageous. Dropping an egg in a quick pass over the skin of the victim would not seem excessively dangerous; biting flies regularly drill through the skin for a meal. And it is disadvantageous for the egg to land where a mosquito bites and the victim is likely to slap or scratch. Botflies that afflict cattle engage in no such roundabout procedure.

We are so familiar with the achievements and eccentricities of the animal that calls itself "knowing" or "wise" (*sapiens*) that we think little of them. But evolutionary theory must apply to *Homo sapiens* as well as other forms of life, and it is very difficult to account for all the oddities and capabilities of humans, from storytelling to leadership to religion, in biological terms.

Problems of Creativity

Natural selection is a plausible explanation for many things. Thus one may assume that many little changes have led the koala to its noncompetitive life, munching eucalyptus leaves and sleeping in crotches about 20 hours a day. But how did the koala find its comfortable niche? It is descended not from arboreal animals but from the ground-dwelling wombat, and it could not utilize the indigestible and toxic eucalyptus leaves without an extraordinary digestive system and a special detoxifying liver (Lee and Martin 1990). It must have followed a circuitous route, graduating from more easily utilized plants to the repugnant eucalyptus, changing anatomy and habits in step.

Sometimes it is possible to figure out a sequence of adaptations leading to an observed behavior, each step supposedly improving the animal's reproductive success or at least not significantly injuring it. One may consider the habit of nest parasitism, which has arisen not only in cuckoos and cowbirds but also honeyguides, widowbirds, and ducks, which have nothing obvious in common. It is surmised that a shortage of nesting sites may have caused two females to lay in one nest; if one withdrew, leaving eggs behind, and went on to lay in another nest, this behavior would be favored. Alternatively, a female, having lost her nest before finishing her clutch, might easily lay in another nest of the same species (Anderson 1983, 220). Such a habit could be extended to laying regularly in nests of other species. Most cowbirds have ceased to build nests of their own, as have all cuckoos.

This assumes that mutations occurred appropriately to reshape instincts toward parasitic behavior, but it can at least be said there are steps that could lead from normal nesting to the cowbirds' and cuckoos' well-calculated laying in other birds' nests. But no gradual transition could lead up to a cliff swallow's deciding to take her egg in her bill and carry it to a neighbor's nest. Colonial swallows not only lay in neighbors' nests but often ferry eggs, even after days of incubation, to unattended nests. Male swallows help by removing eggs of the foster parents (Brown and Brown 1990, 34–40).

Some simple adaptations are hard to rationalize. For example, Stephen J. Gould wonders at the external cheek pouches of pocket gophers (*Thomomys*). Related species have developed large internal cheek pouches for suitcases, but an external pouch is of no use until it is fully formed, and it cannot be formed by a single mutation (S. Gould 1980, 190). Did a gopher, noting how its relatives transported things, decide to avoid dirtying its mouth by putting its pouch on the outside? If this was a bright idea, it cannot have been extremely difficult to reach. Other not closely related rodents, pocket mice and kangaroo rats, have similar external tote bags, with muscular arrangements for cleaning them (Voelker 1986, 114).

If something is not possible, it cannot occur no matter how many trials, and it is unclear whether random mutations can rearrange tissues to make new organs or alter the wiring of the central nervous system to make new instincts. The only information we have in this regard is negative. Thousands of mutations of *Drosophila* have been observed, but they are generally trivial or pathological (Mayr 1963, 159). Some give resistance to injurious conditions, such as heat, but none has been suggestive of a new organ.

The likelihood of deleterious change would seem greatly to outweigh the likelihood of a random change bringing improvement in the finely tuned organ. Hundreds or thousands of genes must be replicated with complete accuracy in order to keep an organ such as the eye functional. Defective genes must appear from time to time, and many of them would be insufficiently harmful to be eliminated immediately. Moreover, most deleterious genes would be recessive and consequently would accumulate until they constituted a large fraction of the genome. They do not. An incredibly strong selection would seem necessary to maintain the organ in good condition.

The more perfect a structure is, the more certain that any random tinkering will be harmful. Even if a piano factory had a very large output of pianos, it could not tune them by giving a monkey a wrench and letting it play with the pegs that hold the strings tight, discarding any instrument untuned. The more exact is the tuning needed, the larger is the proportion that would have to be junked; rarely indeed would the monkey improve a well-tuned instrument. Moreover, junking pianos made nonfunctional would not suffice. If the monkeys jiggered the posts a bit but not enough for the instruments to be discarded, all pianos would become slightly or eventually seriously untuned. That is, the organ would degenerate.

This analogy falls short of reality because the well-fashioned eye, for example, is more than a single instrument. It has coordinated systems for tracking, focusing, light control, light registration, color perception, and incipient image formation, all of which can and do sometimes go wrong. An additional complication is that genes have multiple effects, and it is always possible that a gene needed elsewhere may have negative consequences for the eye. It would seem difficult for simple selection to maintain the detailed structure of large gene control networks, such as must be responsible for such an intricately integrated organ as the eye, unless the pattern is somehow self-regulatory. Random changes would overwhelm a rather large selection rate, limiting the number of genes that can be maintained as a group (Winstatt and Schank 1988, 239–250). A combination of many genes would be vulnerable to deleterious mutations disturbing the delicate harmony if the structure did not have great powers of repairing mistakes.

This logic applies, with greater force, to instincts. An instinct of any complexity, linking a sequence of perceptions and actions, must involve a very large number of connections within the brain or principal ganglia of the animal. If it is comparable to a computer program, it must have the equivalent of thousands of lines. In such a program, not merely would the chance of improvement by accidental change be tiny at best. It is problematic how the program can be maintained without degradation over a long period despite the occurrence from time to time of errors of replication.

To theorize that highly evolved organs and instincts can come about by selection of mistakes of the genes requires quite as much faith as to suppose that the organism must have some means of biasing mistakes in useful directions and minimizing losses in refined structures. Perhaps the variation-selection theory requires the greater faith.

Countless instincts seem to require the genome to put together a pattern that would not be easy for even a well-developed central nervous system. Many observers have been amazed by the architecture of tropical termite nests, which may be 10 feet high in Africa and Australia. Many species have well-designed features that would seem to require something like a plan. The blind little insects, with no known means of detailed communication, build channels for cooling and ventilation with special porous walls in appropriate places, and they orient their towers to maximize exposure to morning and afternoon sun, thereby minimizing noonday heating. They seem to be assisted by sensing the earth's magnetic field, but how they coordinate the efforts of hundreds of thousands of builders is cause for wonder (von Frisch 1974, 139–142).

The improbability of a combination of adaptive changes, none of which is useful of itself, is equal to the product of the improbabilities of each necessary element. Such an unlikely combination is illustrated by the bombardier beetle (*Brachinus*), which makes a boiling hot defensive spray by the reaction of quinones and hydrogen peroxide with oxidative enzymes, somewhat in the way of binary nerve gases. The beetle stores hydroquinones and hydrogen peroxide in a reservoir, from which the mixture can be pumped into a reaction chamber containing enzymes. The valve is closed, and the explosive reaction at 100 degrees C forces the spray out through a turretlike orifice in the beetle's rear end, which sends it in any desired direction. The jet is not continuous but a series of about 500 pulses per second (Dean et al. 1990, 1219–1221). The chief defensive value is not from the quinones but from the heat, which greatly increases the otherwise mild effect of the quinones. There is also a nice question how the beetle avoids scalding its insides, (Aneshansley et al. 1969, 63) as the electric eel manages not to electrocute itself. The complex structures along with instincts would be of little utility unless nearly complete.

The development of new proteins by errors of replication, as selected by the great sieve of what is vaguely called the environment, is reasonably plausible. But to shape organs is vastly more complicated than to make an enzyme, and how it is accomplished probably cannot be understood until the processes of morphogenesis are clarified. Even relatively simple changes present difficulties. To alter the running forelimb of an insectivore to the digging arm of a mole required a large number of changes proceeding concurrently in bones, muscles, tendons, feet, and so forth. Genes, having to operate as an integrated system or a team, have to change as a team, and no one has any idea how this is managed. It must be that the instructions to make a tendon, for example, are so framed that the tendon properly coordinates with muscles and bones.

To develop elaborate behavioral patterns is far more complex than sculpturing an organ. It requires not only making exceedingly delicately organized structures, the branching neurons, but also dictating connections among them and the organization of their operations, as though composing computer software. We are nearly as far from understanding the origin of life's complexities as the origin of the universe.

5

Inconsistent Nature

Adaptation

Evolutionary theory should account for the perfection of nature—the countless admirable ways plants and animals have found to promote their kind. It is equally necessary, however, to take into account the many traits that seem to be without adaptive value or positively maladaptive and disadvantageous.

Adaptation is the central concept of evolutionary theory, and it often seems obvious. For example, a barrel cactus is fitted for existence in a hot, dry climate. It has converted leaves into spines, which reduce water loss and protect its flesh from hungry or thirsty animals. Its waxy cuticle conserves moisture. Its root system spreads out to profit from any shower, and it can store water for long droughts.

But if a species such as a cactus is adapted to a particular place in the ecosystem, or niche, numerous kinds of cactus, with varied shapes, types of thorns, and flowers, inhabit the same environment. Moreover, many quite different plants compete directly with the cactus; the mesquite and ocotilla, for example, are apparently equally well adapted to desert life. Can they all have near-optimal patterns? Nothing could be less cactuslike in appearance than *Welwitschia,* a strange-looking inhabitant of the southwest African desert that has only two straplike leaves, which have more numerous stomata per unit area than plants native to moister regions and generally lacks obvious adaptations to desert life (Bornman 1978, 27).

More striking is the superabundance of species in tropical forests. An endless variety of insects coexist in the canopy of the rain forest. Counting those on a few sample trees caused entomologists to raise their guesses of the number of insect species from under a million to 10 million or so. And why should there be eighty-one species of frogs in a square mile of jungle in Ecuador (del Pino 1989, 110), most of them living in about the same way in about the same environment? There are probably two or three in a square mile of Michigan forest.

Perhaps the more benign conditions of the tropics permit more specialization. Tropical seas, especially coral reefs, also have diversity far richer than mid-latitude oceans. The difference is not so great as that between temperate and tropical forests, but it exists despite the fact that cooler waters, holding more oxygen, produce a larger biomass. Stability seems to permit a self-compounding proliferation of types. It is not that the rain forest is intrinsically blessed with so many niches; life itself creates them.

The huge number of species in similar conditions in the rain forest contrasts with single species that live in quite diverse conditions. Some kinds of cockroach find a home throughout the temperate regions (*Blatta orientalis*) or almost worldwide (*Blatta germanica*). A small burrowing wasp inhabits all Australia, from steamy rain forests in the north to central deserts and the mild climates of the south (Alcock 1988, 54). Whereas there are hundreds of kinds of trees in a hectare of tropical rain forest (E. Wilson 1989, 110), in some parts of the Andes, the exotic Australian blue gum (*Eucalyptus globulus globulus*) is almost the only tree. This eucalyptus, sometimes called a weed tree because of its rapid growth and aggressiveness, competes successfully with native trees from northern California to the highlands of Ecuador, on to Brazilian uplands, and down to Uruguay and southern Chile, from semiarid to humid climates, in the tropics from warm uplands to chill highlands. It also grows vigorously in southern Europe, South Africa, and India, with rainfall from 20 to 80 inches per year and at altitudes from sea level to 3,000 meters. In central Portugal it has largely replaced native trees.

Like the eucalyptus, such unesteemed animals as rats, mice, houseflies, and pigeons show phenomenal adaptability, especially to fit into human-modified environments. English sparrows, after being introduced into the United States in 1850, had become the commonest North American bird within a half-century. They were outdone by later-arriving starlings, which have similarly prospered much more in North America than their native Europe. The common garden snail (*Helix aspera*) has made itself at home from Malaysia to Arizona. The Argentine ant (*Iridomyrmex humilis*) has spread widely in tropical and subtropical regions, as have a few other species, including the fire ant (*Solenopsis invicta*), in many places displacing much of the native ant fauna and becoming a serious pest.

Species that prosper across great areas in diverse environments cannot be selected for fitness in all of them. As a general rule, species extending over a broad range are remarkably uniform (Mayr 1988, 482). More striking is the fact that a species may be at home in restricted but quite different environments. For example, the savanna sparrow (*Passerculus sandwichiensis*) lives in both coastal salt marshes of the East and dry interior uplands of the southwestern United States. Swainson's warbler likes lowland canebrakes and mountain rhododendron thickets (Mayr 1963, 356). Supposedly it finds something in common in the two environments, but whatever it may be escapes naturalists.

A species that has lived for a long time in a certain environment, perhaps originated in it, might be thought best adapted to it. For example, in East African lakes, cichlid fish have diversified into many different species, surely adapting thereby to their environments. But a species (*Tilapia*) from half a world away has devastated many of them.

Invaders are often successful not by displacing natives but by going into unoccupied niches; evolution, despite the multiplicity of species, is very far from filling all potential spaces. Seasnakes, common and varied in the Pacific and Indian Oceans, are absent from the Atlantic. There are 27 species of woodpeckers in Borneo and Sumatra; in the similar forests of neighboring New Guinea, there are none (Mayr 1963, 87). By contrast, in the Argentine pampas, where no trees grow, there is a woodpecker with beak, tongue, toes, and tail like those of birds that specialized in woodpecking (Darwin 1964, 144).

Many species that seem exceptionally well equipped are uncommon, like many of the singularly adapted forms noted in the previous chapter. African hunting dogs (*Lycaon pictus*) have been found to have the highest kill rate of African hunters, but they are scarce. Cheetahs easily feed themselves, thanks to their unique speed, but lions are a hundred times more numerous (Schaller 1972b, 318; J. Gould 1982, 466).

Another problem is the persistence of supposedly inferior patterns. A new type arises and prospers on the basis of its superior organization. The older type would then, one may suppose, be able to hold on only in special environments. This is certainly true in many cases. Amphibians, for example, persist mostly as small, moisture-loving insect eaters. Flowering plants have taken over most of the earth as a result of their superior flexibility, tissues, and reproductive ability, yet ferns, horsetails, clubmosses, and their relatives, with the relatively primitive structures of ancestors of hundreds of millions of years ago, compete successfully in many places with advanced flowering plants and are even locally abundant.

Structure is not closely related to habits. Animals may change their way of life with little visible change of morphology. For example, cichlid fish of East African lakes have, in their explosive speciation, taken to many diverse diets, from predation to scraping algae from rocks, while making only minor organic alterations (Coulter et al. 1986, 171). Iguanas on the Galapagos Islands must have taken to eating seaweed strewn on the beach and then ventured into the ocean in search of more such food, but the marine iguanas look much like their land-bound cousins. The water ouzel or dipper (*Cinclus*), a small bird that finds insects on the bottom of brooks, looks like an ordinary land bird. Although it uses its wings to swim, it, like the marine iguana, has failed to acquire webbed feet (Darwin 1964, 145).

Sea otters are thoroughly aquatic in habit; they seldom haul out onto land, sleep in kelp beds, dive to respectable depths, and swim long distances. They mate, give birth, and raise their pups in the water (Riedman 1990, 52). But they look much like land animals, as seals do not. Wasps of several families find their living entirely under water and use their wings to swim, but they have all the appearances of land dwellers (Mayr 1960, 371–372). Water spiders (such as *Desis marina*) remain submerged, with an air bubble, for many days, build webs in the water, and swim out to capture prey (McQueen and McLay 1983, 383–392), yet they are very similar to their landbound relatives. Evolution is not continually fine-tuning structures.

A broader question is why land animals should be able to compete at all in the marine arena. Sea creatures supposedly have been becoming ever more perfectly fitted for an environment extremely different from the land. Yet since the waters ceased to produce important innovations about 200 million years ago, the chief additions to marine fauna have come from the descendants of animals that became adapted to the dry earth ages ago, including several families of large extinct reptiles, such as ichthyosaurs and pleisosaurs, turtles, crocodiles, snakes, and four orders of mammals.

Curiously, however, land plants have been unable to penetrate the ocean except marginally in shallow waters. It would seem that the structures they have developed would give them a large advantage over relatively simple algae. Similarly, the incredible inventiveness of insects has not enabled them to emulate the reptiles and mammals in finding or making niches in the ocean.

Selection seemingly should always go forward to more effective ways, but it sometimes turns backward. Snails have obviously done well to provide themselves with hard shells covering the soft body, but four orders of gastropods, from garden slugs to marine nudibranchs, have renounced the protective shell. Spiders over the ages developed the capacity to construct well-engineered webs, to the unending wonder of nature lovers, but not a few of them surrender much of this achievement, reducing the web to a little net or a single sticky strand, or, as in the case of the rather common bola spider, a thread with a drop of glue thrown at insects. In one case (*Pachygnatha*), the juveniles build orb webs, but the adults forsake them and return to the free-living hunting way of life of their ancestors (Foelix 1982, 146).

If one can judge, as biologists daily do, that certain traits are adaptive—that is, conducive to reproductive success—one logically should be entitled to doubt the value of other traits. We can surmise that nature often treats creatures shabbily although never so badly that they cannot continue to exist. Many are the marvelous and useful traits, but in many ways natural selection has failed to provide creatures with what would seem to be easy adaptations or has left them burdened with useless or even harmful traits.

Nonadaptation

Animals adhere, seemingly without adaptive reason, to rigid patterns; developmental pathways are constrained. In some cases, this may be attributed to the difficulty of altering traits early in the sequence of ontogeny. For example, embryonic birds and mammals have gill arches that have not been functional for their ancestors for 400 million years. The gills are supposedly necessary for the making of other structures.

Many rigidities cannot be so explained. All of over a thousand species of *Drosophila* in all manner of habitats around the world (with rare exceptions) have three bristles on each side of the head—two bent forward, one backward (Dobzhansky 1956, 337–347). Millions of species of insects, from almost microscopic parasitic wasps to lumbering 7-inch beetles, have six legs attached to the thorax. This number cannot be especially adaptive because only insects have that trait. Six legs cannot be useful for all insects, and one would suppose it fairly easy to relinquish some of them, just as cave species lose eyes and many vertebrates have suppressed limbs. On the other hand, caterpillars seem to need more than six legs. Although the embryo begins with many leg buds on its segments, the caterpillars are unable to recover lost legs of their supposed millipedelike ancestor and make do with leglike fleshy projections. Unlike legs, wings are easily abandoned; many insects, like many birds, have lost the power of flight. Arachnids have eight or more legs, although four are ample for running, and for many species a larger number must be an encumbrance. The many legs of centipedes are a poor idea for locomotion.

Zoologists can point to innumerable such limitations. For example, despite a plethora of appendages, arachnids have neither separate heads nor antennae; insects seem to find both useful. Yet very many members of the two classes have quite similar life-styles. If members of any class have diverse habits, the traits characterizing the class cannot be of adaptive value for all its members and surely will be more or less negative for many of them.

It is easy to give accumulated mutations credit for countless incredible adaptations, but this makes it the more surprising that the process has failed to endow animals with many seemingly accessible capacities. For example, no multicellular animal is known to have the ability to digest the most abundant organic substance, cellulose. The necessary enzyme, cellulase, cannot be difficult to manufacture. Bacteria, fungi, and protozoa have it, but it does not fit in the metazoan genome. Termites and herbivores rely on protozoa or bacteria in their guts; the symbiont digests the vegetable material, and the animal digests the symbiont.

It is also odd that birds and mammals, often beset by parasites such as lice, ticks, and worms, have developed practically no chemical protection, in marked contrast to the multitude of defenses evolved by plants. On the other hand, it is puzzling that plants have never invented anything like an immune system to ward off invading pathogens. It is curious that no animals, whatever their incredible and often inexplicable instincts and adaptations, have hit on the simple strategy of cultivating plants, except only fungus ants and recently humans. In view of the fact that animals frequently store or bury seeds, it would seem easier to attain something like agriculture than many an adaptation already cited.

It is reasonably supposed that a prime reason for the ability of land animals to compete in the ocean—not only cetaceans, seals, sea otters, and the like but turtles, crocodilians, snakes, and many extinct reptiles—is their breathing air; water has about 40 times less oxygen. The failure of sea creatures to develop respiration is the more striking because fish are believed to have developed lungs at an early date but mostly lost them or converted them into flotation bladders (S. Gould 1989a, 30). Instead of using the air bladders to take in oxygen, many fish have developed glands that secrete gases to balance the fish's tendency to sink.

Useful traits may be lost. Snakes have largely given up their hearing, despite its obvious utility. And why are birds toothless? Most Mesozoic birds had teeth, like their reptilian ancestors, and surely at least some modern birds would find teeth handy to grip, cut, or tear. Some fairly recent birds (*Pseudodontornis*) had horny toothlike projections on the beaks for catching fish (Fenton and Fenton 1989, 500). Chicks still have a capacity to produce enamel but not the dentin necessary to initiate enamel production (S. Gould 1983, 185). The conventional answer is that teeth were eliminated to save weight, but birds had teeth for well over 100 million years, as did most pterosaurs and all bats do. Another offshoot of reptilian stock, the turtles, have horny beaks like birds and also lack teeth. The turtles cannot be saving weight; there must be some incongruity between horny beaks and teeth.

The dinosaurs must be regarded as basically well designed— they ruled the land for three times as long as mammals have been dominant—but they never managed to develop the specialized teeth that serve mammals well. Most of them had feeble forelimbs, although their ancestors had fore and rear legs of roughly equal strength and four-legged animals can run faster than two-legged ones. The champion predator, *Tyrannosaurus*

Tyrannosaurus rex, as much as 50 feet long, was a very successful carnivore. But it was rather illogically designed. The ridiculous little forelimbs could not grapple a prey, and its long, thick tail must have been very burdensome. (Illustration from *The Fossil Book* by Carroll Lane Fenton and M. A. Fenton, © 1958 by Carroll Lane Fenton and Mildred Adams Fenton. Used by permission of Doubleday, a division of Bantam Doubleday Dell Publishing Group, Inc.)

rex, had forelimbs so tiny as to seem virtually nonfunctional. They were shorter than the fearsome head and could hardly reach the mouth; the claws were strong, but they could not reach a prey unless the jaws, with their swordlike teeth, were thrown back. They were much reduced from those of *Tyrannosaurus*'s presumed ancestors. A lion rips a zebra with its claws and would have trouble bringing down large animals without them.

On the other hand, *Tyrannosaurus* had a huge tail of questionable utility. All dinosaurs had big tails, usually thick and heavy and as long as the trunk or longer. Such tails must have been a great impediment as well as metabolic expense. The bizarre *Tanystropheus* had a massive tail twice as long and a neck about three times as long as the thorax and abdomen that housed its lungs and viscera. The 10-foot-long neck was not very flexible, single vertebrae being as much as a foot long, and the animal, with puny legs, must have had difficulty holding up its head (Carroll 1988, 266).

The tails of dinosaurs were exaggerated, but modern lizards also have tails of little or no apparent utility. Some of them take advantage of the tail's dispensability by dropping it to confuse a pursuer. Land-living salamanders and many mammals (rats, cats, lemurs, anteaters, and so forth) also have long tails—lighter than those of the dinosaurs but nonetheless more or less burdensome. The long tail is not deeply embedded in the mammal genome; guinea pigs and humans have very short ones. But it seems to be part of a common body plan. The loss of eyes by animals that live in the dark is conventionally attributed to economy, but the cost of eyes is minimal compared with that of long tails. Oddly enough, the birds have shortened the graceful tail of *Archaeopteryx* to a mere stub. Early pterosaurs had a long tail; later pterosaurs, like birds, reduced it to a stump.

The size of some dinosaurs is also incomprehensible. Many were as much bigger than an elephant as an elephant is bigger than a horse. Several species weighed 75 tons or more. *Seismosaurus* ("earthquake-lizard") was 100 to 120 feet long—possibly as much as 140—matching or outdoing the blue whale. A number of others, such as *Brachiosaurus,* were only a little less gigantic. Such an animal may have been invulnerable to predators in adulthood, but it seems poorly equipped for defense, with a very long neck, a tiny head, and no armor. The enormous body had to be unwieldy. The mass of the blue whale is manageable, being suspended in water, but the monster dinosaurs had structural problems. If such a monster lay down on its side, it would be a real problem to get up. It is an achievement for the giraffe to keep fairly constant blood pressure when it lowers its head to take a drink and then raises it to nibble 18 feet above; the problem would be much more formidable for the dinosaur, lifting its head 40 or 50 feet. Blood pressure had to be very high going from the heart to the body, although blood pressure in the lungs would have to be low to permit an exchange of gases.

In all realms there are seemingly illogical traits. The horns of ungulates, though extremely varied, are seldom well shaped for their obvious utility, fighting. Those of sheep, for example, curve backward as though to make themselves harmless, as do those of sundry wild cattle, many antelopes, and others. This deflection of the sharp point may be good for the species because dueling males do not much injure one another, but there should be strong selective pressure for horns capable of repelling an enemy or besting a rival. Why do stags grow new antlers every year? Sheep, oxen, and antelope, among others, do not. The cost of yearly renewal of antlers in terms of food consumption is believed to have brought the extinction of the Irish elk when the available fodder became less rich (Geest 1986, 54–65).

The babirusa pig (*Babirousa babyrussa*) poses a different sort of problem. It has turned two upper teeth around to make upward-pointing tusks that pierce the cheeks. A relative, the warthog, has upper tusks that grow sideways and then turn up. This is a halfway condition, but it would be uncomfortable for the pig if the tusks gradually shifted to come up through the cheek. In both the babirusa and the warthog, the upper tusks curve backward; they can only bump, not stab. In the babirusa, they sometimes even bend back into the upper jaw or penetrate the skull. The warthog's tusks are at least good for digging; those of the babirusa have no known use (Voelker 1986, 266).

The tensor tympani, a tiny muscle that tightens the eardrum to protect the human inner ear from loud noises, seems to represent an excellent adaptation for minimal need. Less understandable is the conversion of two posterior bones of the reptilian jaw, over many million years, into two small bones, the malleus and incus, of the mammalian middle ear. This triumph of evolution must have required the selection of scores of appropriate mutations at the right time. Yet it is puzzling why these bones came to be detached from the main forward part of the jaw or what, if anything, was gained from the whole process (Carroll 1988, 354). One bone to conduct sound from eardrum to inner ear is adequate, as is shown by the superb hearing of some birds. Some of the sharpness of sound must be lost by passing through two extra joints.

A minor example of evolution's doing things the hard way is the pseudothumb of the giant panda. It arose from the extension of a wrist bone, which was made movable and provided with muscles and a pad (S. Gould 1980, 22–26). But raccoons and other relatives of the panda manipulate objects quite well with normal digits. If the becoming-panda needed a stronger thumb to strip bamboo, it would seem natural gradually to improve the thumblike digit it already possessed instead of inventing a new organ. It thrusts the problem further back to observe that ordinary bears have a slightly enlarged wrist (sesamoid) bone used to improve their grip. In a bit of parallel evolution, moles have also made an extra digit of the same bone (Holder 1983, 419).

Many other adaptations seem to be pointless. Why do mother wolf-spiders (*Lycosa*) carry a hundred or so babies around on their backs for some six months, during which time the young do not eat or grow (Cromptom 1950, 90–94)? The male lion's bushy mane makes it hard for him to stalk game, a fact that may largely account for the females' making 90 percent of kills (Schaller 1972b, 360). It is not apparent that the mane is very useful in either combat or as a sexual display. Why do white-tailed deer show their conspicuous tails as they bound away from danger? A naturalist guesses that it is telling the predator that it has been detected (Bildstein 1983, 709–715), but this seems neither necessary nor useful, surely not important enough to become established by natural selection, assuming that the necessary mutations would occur by chance. The incredible aggressiveness of the African bee, which may attack a person a hundred yards from the hive and pursue for half a mile, cannot be adaptive, or if it were, the more moderate temper of other bees would be maladaptive.

Animals often do much less than they might to defend themselves. A tarantula lives by seizing and killing insects, but when a spider-hunting wasp comes around, the tarantula may passively allow it to seek the spot to insert its paralyzing sting (Smith 1984, 101). Except for the muskoxen, which stand in a circle to defend weaker members of the herd, and African buffalo, which will charge a menacing lion, ungulates hardly cooperate against enemies. Even hornless zebras could easily repel lions if they joined forces. Adult wildebeest, although powerful and equipped with potent horns, make little effort to protect their young or themselves against hyenas (Kruuk 1972, 288).

Nature is often wasteful. Avocados have a hundred flowers for one fruit matured. Marsupials may produce a dozen times more embryonic babies than they have teats to nourish. Another negative "adaptation" is the practice of many wasps and some bees and ants of eating their eggs. Founder-queen ants convert body stores into food for their first brood by laying sterile eggs, which are crushed and fed to the larvae; these trophic eggs are the equivalent of milk. But in many social hymenoptera, workers or queens eat eggs laid by nestmates or by themselves, feasting on their own substance (Brian 1981, 186). The habit is metabolically wasteful; three eggs are necessary to provide the nourishment to produce one. Some ants have developed the habit to the extent that they extrude partly formed eggs on demand (Brian 1985, 394). Another uneconomic "adaptation" is that males of social hymenoptera (with very rare exceptions) laze around as social parasites until the time comes for them to perform their sole task of insemination. Yet in many forms they are morphologically not very different from females and could certainly be useful if they had a mind to (Hölldobler and Wilson 1990, 301).

How convoluted evolution is one observes in the most abundant large animal and the greatest success (in terms of rapid increase of numbers) in the history of life. The human is ill adapted not only for the new environment we have made but for that in which the species was formed. Very long hair on the top of the head and very little elsewhere is ridiculous. Glands that keep the eyes moist are activated by sorrow or vexation. Mature males have more or less coarse facial hair, which can grow very bushy and which would make it easier for an opponent to get a grip on the head. The body is a bundle of imperfections, with sagging bellies, drooping breasts, useless protuberances above the nostrils, rotting teeth with trouble-prone third molars, aching feet, bulging buttocks, easily strained backs, and naked tender skin, subject to cuts, bites, and, for many, sunburn. We are poor runners and are only about a third as strong as chimpanzees smaller than ourselves.

The human female has most grounds for complaint. No other mammal finds it necessary to slough off the lining of the uterus periodically. She uniquely gives birth with pain and prolonged labor, with no little danger for both the child and herself. She has been endowed with functionally useless (mostly fatty) permanent breasts, although the mammary glands of other mammals enlarge only when needed. Ampler hips could surely serve equally well as a sexual symbol.

Human behavior makes even less biological sense than human antomy. Almost anything is mandated in some culture and prohibited in another. As one easily observes from reading the daily press, it seems that most of the activity of humans in industrial societies is biologically irrational. But if this animal acts illogically in terms of evolutionary theory, should we be surprised that less pretentious creatures have their unaccountable ways?

Humans demonstrate, perhaps better than almost any other creature, a truth of evolution: the way to more effective structures and instincts is paved with nonutilitarian changes that came about in unknown ways.

Infertility

In evolutionary theory, the mandate of life is to be fruitful and multiply. A large majority of plants and animals unequivocally obey. Oysters spew out millions of eggs, trees shed countless seeds, and rabbits multiply like rabbits. Everywhere one finds a host of adaptations for mating, reproducing, and caring for young.

Yet some animals and plants have far fewer young than they evidently could, or do much less for reproduction than they might, or act contrary to their procreative interests. It is unreproductive that male scorpions fight with and sometimes kill their mates. Lions may let cubs starve while the adults eat their fill. The males are the more disposed to allow cubs to share the kill, whereas the females cuff them away (Schaller 1972b, 278, 362–363). Many animals, even small ones, are in no hurry to have progeny. The common cicadas in the United States live for either 13 or 17 years underground, sucking juices from roots. But an insect of moderate size can easily grow up in a year or two, and if it reproduced after 2 years instead of 17, its reproductive capacity would be multiplied over a hundred times. Curiously, the three common cicada species in the United States all have 13- and 17-year broods, and they all emerge in the same years (Borror and DeLong 1964, 204). It is not easy to surmise either why they should be so delayed or how they could reach the same periods unless the periods antedated the separation of the species.

Numerous animals give the impression of being lazy, doing little or nothing most of the time, like most primitive humans. Tree kangaroos spend 60 percent of their time sleeping, 30 percent sitting idly, and 10 percent socializing, grooming, and eating leaves (Procter-Gray 1990, 61). "Busy as a bee" and "go to the ant, thou sluggard, and learn its ways" are inappropriate because these insects are usually loafing (E. Wilson 1971, 341). If they hunted or worked more, they could surely gather more food and produce more young.

Males are especially prone to neglect their offspring. In many species of monogamous birds, the fathers contribute dutifully, sharing the brooding and helping to nourish the young, and we take it for granted that two parents are much better than one. Yet there are many polygynous species whose males do nothing beyond insemination. Of course, it is understandable that a male might well multiply his genes by courting as many females as possible, but as soon as the mating season is done, he might well do something for his family or families, and it would seem highly advantageous for females—and for populations—to prefer helpful mates.

It is accepted doctrine that there is a payoff between having many offspring, for which the parent can do little, and having few, which can be better nourished and cared for. This is realistic. Many small animals, such as mice and rabbits, reproduce abundantly and quickly, send their young away, and go on to have a new family. Large, slow-maturing animals, such as elephants, have a relatively low reproduction rate. But many animals breed much less abundantly than they would seem capable of doing. Elephants come into estrus once in 4 years (Kevles 1986, 93). Pandas seem to have little interest in making love, at least in pampered captivity, and the females are inept mothers. The female does not become fertile until about 6 years old, has a tiny cub, and never raises more than a single one (Schaller et al. 1989, 238). Rhinos typically have one calf every four years (Moss 1975, 76). It is not surprising that the lizardlike tuatara is rare; it takes 20 years to mature and then lays one egg every second year, in marked contrast to most reptiles.

It is as though such animals are not vigorously competing to multiply their genes but are geared to sustaining numbers, producing enough offspring to replace themselves with compensation for losses. Most rodents are much hunted and are very fecund, but the well-protected porcupine female mates only one day a year and has only one or two young (Voelker 1986, 140). Grizzlies breed first at 7 or 8 years and thereafter usually at 3-year intervals (Bunnell and Tart 1981, 79). Orcas (killer whales) calve only once in 5 years. The period of lactation is ordinarily correlated with period between births, and the pilot whale is said to suckle as long as 17 years (Kevles 1986, 136).

Birds theoretically should adjust the size of their clutch to the number they are likely to be able to feed, and they seem usually to do so. But some birds are capable of supplying much more food and raising larger broods than they normally have. Gannets in some populations lay only a single egg, although they can well raise two or three young, like gannets in other populations—and as ornithologists have shown by foisting extra eggs on them (F. Gill 1990, 421; Maynard Smith 1978, 37). In humid tropics, where mortality is low and food usually abundant, birds generally lay only one or two eggs, and the male commonly neglects the family (Skutch 1987, 264).

The brood is often reduced by competition, although it seems on general grounds that mutual toleration of chicks should have survival value (Alvarez, de Reyna, and Segura 1976, 915). Bee eaters feed the first hatched. The others are pushed back and starve unless something happens to the greedy elder sibling, despite the fact that the parents usually have helpers and could probably raise several chicks with ease. In a number of quite different orders of birds, egrets, some woodpeckers, eagles, owls, boobies, and others, the oldest chick kills younger ones. In many species, especially of raptors, a second chick never has a chance (Anderson 1990, 334). This practice appears to be independent of abundance or scarcity of food. Some birds, including gulls, terns, storks, ducks, owls, and cranes, on occasion eat their own young (Welty 1982, 404–415).

The reluctance of condors to breed threatens their continued existence. The albatross does not begin to reproduce until it is at least 6, perhaps 10, years old, although it is fully grown at 2, and it lays only one egg every other year, like the big vultures. Pairs average 0.2 young per year (F. Gill 1990, 408; Lack 1966, 250, 276). Red grouse females fail to move into territories where they could mate and breed, giving rise to the speculation that they refrain from reproduction to prevent overpopulation (Wittenberg 1981, 95).

It is suggested that the albatross refrain from more abundant breeding because they suffer the stormy seas (Mayr 1988, 156), but this might be a reason to raise more young rather than less. Maynard Smith hypothesizes that the gannets that limit their families may be holding to a custom appropriate when fish were scarcer and it was more difficult to feed a larger number. But some studies indicate the birds' clutch size depends not only on ability to feed the young but on mortality, that is, need to maintain the population (Welty 1982, 365). Nothing would seem more obvious than that genes causing an animal to be devoted to reproduction should be preferentially propagated (Baldwin and Baldwin 1981, 3), but it often appears that evolution considers the well-being of the species and limits numbers to conform to conditions.

Animals and birds with the highest rates of survival are the least fertile (McFarland 1985, 86), while those most vulnerable—often the prey of long-lived species—reproduce most abundantly. Slow-reproducing birds are long-lived; albatross, for example, may live 40 years, while songbirds ordinarily do not last more than a tenth as long. Logically, there should be a very intense competition for survival in the face of adversity, leading the fittest to carry on so far as physical conditions permit. But breeders of *Drosophila* often find that the strains they have carefully selected die out, despite total pampering. When overcrowded, birds, hares, mice, and many other animals sicken, cease breeding, and may die, even though their primary needs are entirely satisfied (E. Wilson 1980, 42).

Our near relatives are also far from breeding as abundantly as would seem biologically possible. An orangutan or gorilla mother whose baby lives will not have another for 4 or 5 years, or perhaps longer (Voelker 1986, 76, 80). The female chimpanzee matures sexually about age 10 and gives birth, on average, after 5.5 years (sooner if the offspring dies). At age 21 she may have given birth three times, but not many live much past this age (Goodall 1986, 84–86). Similarly, humans in a relatively "natural" environment, that is, hunter-gatherers, have decidedly limited fertility, with late menarche (at age 16 or 17) followed by a number of years of adolescent sterility, late first birth, several years of lactation, and an interval of 4 years or more between births, followed by early menopause. This is not because of shortage of nourishment; adults on average spend only about 2.5 days per week in procuring food (Lancaster and Lancaster 1983, 43).

A very different example is even more puzzling. A study of populations of the common American pink lady's slipper (the orchid *Cypripedium caule*) revealed that most plants failed to flower for many years, and nine-tenths of flowers failed to set seed, giving a fertility rate of around 2 percent. This failure was due to lack of pollination; hand-pollinated plants produced abundant seed. Flowers are not pollinated because their beauty is deceptive; they offer no nectar and do not even lure insects with fragrance. Pollination occasionally occurs only because a naive bee stumbles in, seduced by the color. This condition should be easily remediable by natural selection, either permitting self-pollination, offering nectar, or postponing flowering from spring to summer when more insects are looking for food. A flower that furnished a lure for pollinators (the primitive condition) apparently mutated to deny it, contrary to any ordinary understanding of natural selection. This perversity is not peculiar to the lady's slipper; about 8,000 species of orchids fail to reward their pollinators. Under a tenth of their flowers produce seeds, whereas species that offer nectar generally have many seeds. The lady's slipper survives, and even prospers, thanks to very low mortality. But evolution has fashioned elaborate and complex flowers, only to frustrate their purpose (D. Gill 1989, 467–474). *Welwitschia,* mentioned previously as the antithesis of a cactus in a desert environment, also exemplifies low fertility with high survivorship: only about 0.2 percent of its seed are germinable, but the plants may attain 1,000 years or more (Bornman 1978, 59, 65).

Restricted fertility presents another problem. If organisms produce a large number of offspring, the intensity of selection is high. This makes it possible to eliminate unfavorable mutations, and a larger number of offspring increases the possibilities for favorable mutations to become established, that is, for the animal to improve itself or adapt to environmental changes. For example, small birds are not much bothered by DDT, but infecundity makes it difficult for eagles and other big birds to develop resistance to it. Under the best of conditions, useless mutations are sure to occur, and they are frequent—in the human, probably one to several per individual in each generation (Neel 1989, 301). If an animal (or plant) has few offspring, deleterious mutations can with difficulty be eliminated, while there is little chance for positive mutations to occur and become established. If there were only enough young to replace the parents, make up for natural losses, and maintain

Different species of orchids are much alike in stems and leaves, but their flowers are works of fantasy. (Illustration by Ernst Haeckel, *Art Forms in Nature*, © 1974 by Dover Publications, Inc. Originally published by the Verlag des Bibliographischen Instituts, Leipzig and Vienna, 1904.)

a stable population, the species would surely deteriorate as rapidly as mutations occurred and would presumably die out.

Since most deaths are more or less accidental and mortality is highest among the young, the limited fecundity of the condor, chimpanzee, or precivilized human leaves very little margin for natural selection. Although large birds are more vulnerable to environmental changes, it is not apparent that they are less well adapted than abundantly reproducing small birds under strong selective pressure.

In sum, it is commonly unclear what factors restrict reproduction. The idea that organisms are geared to maximize their reproductive success is logical, but it does not answer the facts.

Premature Death

It is not surprising that many small, rapidly growing animals die, as annual plants do, after a single procreative effort. But small size does not necessarily mean brief life; honeybee, ant, and termite queens may live about as long as dogs and cats— perhaps more than 20 years (Hölldobler and Wilson 1990, 267). However, colonies of many kinds of wasps and ants come wastefully to an end when the queen dies (Evans and Eberhard 1970, 156–157).

Death after reproduction is more impressive in larger animals; a big body is a big investment. The demise of the salmon after spawning is familiar. It is not that the body's substance is exhausted by the effort; males as well as females die even if fed and protected (Kevles 1986, 65). Their death seems to be related to imbalance of pituitary hormones, but it cannot obey any deep necessity because closely related steelhead trout make repeated round trips to the ocean. Even some salmon are exempt; king salmon achieve royal size because they survive spawning and return to the ocean to continue growing. Brook trout (*Trutta fario*) spawn repeatedly in brooks; the same species living in lakes spawns only once or at most twice (Mayr 1963, 355). Most lampreys have a life cycle like that of the salmon, from brook to ocean and back, with death upon spawning. Some do not eat after reaching maturity. Eels, which migrate oppositely from salmon, from the ocean to freshwater, also die after laying eggs in the Sargasso Sea. Both male and female octopi cease to eat on reaching maturity, and they die after reproduction. A sidelight is that the octopus population (at least in Hawaiian waters) is not limited by food supply (Wells 1978, 109–110).

Many mammals are subject to a similar fatality. Males of the marsupial mouse (*Antechinus*) enjoy only a single season of love (Cockburn and Lee 1988, 43). Female shrews have one or at most two litters. The life span of some is measured in weeks. Many voles, mice, gophers, moles, and other small mammals also live only a year or two at most and have only one reproductive season. Not only is the parental body thrown away; there is little overlapping of generations for learning or socialization that should greatly help the animal to cope with a hostile environment (Bouliere 1975, 6–7).

Short life and high fertility are associated with severe predation. It is as though animals much preyed upon, such as mice or voles, reproduce maximally as rapidly as possible. Since they are likely to be snapped up soon in any event, there is less utility, it is argued, in the genes' endowing them with longevity. However, it should always be advantageous for the mature animal to survive to leave as many progeny as possible, and it is not clear why continued vigor is incompatible with fecundity. Shrews and moles, small animals not much subject to predation, are also short-lived.

Short life is also associated with high metabolic rate, but tiny hummingbirds, with metabolism among the highest in the animal kingdom, can last 5 years (Welty 1982, 425). The opossum, a medium-sized animal with low metabolism, generally has only a single reproductive season; perhaps 10 percent survive into a second, with signs of senility (Austad 1988, 103).

Where cold or drought prevents growth during much of the year, a host of plants grow rapidly, produce seed, and die. It is less understandable that some long-lived plants reproduce only once. A Jamaican tree has the flowers that give the name "mountain pride" only once (Huxley 1974, 44). A deep root system in a dry climate is a vested position not easily acquired, but a century plant (*Agave*), having grown for many years, ordinarily does not live beyond a single flowering.

The die-off of bamboo, following simultaneous flowering throughout a region, is more puzzling. This phenomenon afflicts over a hundred species, including all economically important Asian species. It was much publicized when the great pandas of southern China faced starvation after the bamboo on which they subsisted seeded and died in the 1970s.

Various species flower at multiyear intervals, typically about 15 to 60 years. Several wait a century or more; *Phyllostachus bambusoides* has a cycle of 120 years (McClure 1966, 84; Janzen 1976, 347). The plant then dies of exhaustion, and the new crop of seedlings can flourish in the open (Janzen 1976, 347–391). Some bamboos, however, flower more or less continually (often without producing seed), spreading by shoots or rhizomes. Gardeners hardly need fear their bamboo clumps will bloom and die because the most commonly cultivated species have never been known to do so.

The bamboo plants do not invariably follow the regimen exactly, but millions or billions of individual plants flower in close unison after very long periods. A cutting rooted separately goes on counting as though nothing had occurred, and diminutive plants flower on schedule along with tall ones. It appears that timing depends on length of day because bamboos near the equator do not have the habit of gregarious flowering. It seems difficult to increase the period by small increments; apparently periods have been doubled, or the counting mechanism has been slowed by half (Janzen 1976, 372). Lengthening the period by less than double would make one or a few plants flower precariously alone, but how or why the clock could start ticking at half speed is enigmatic.

The presumed reason for delayed simultaneous flowering is to satiate seed eaters, mostly birds and rodents. This desideratum is partly frustrated, however, by the fact that the harvest continues long enough for small rodents to multiply and for birds to flock in large numbers to the area of abundance. It might make sense for the bamboo to pile up stores for large-scale production of seed, but for this purpose far fewer than 60 years should suffice. To make a very large number of seeds all together has disadvantages: seedlings may occupy all available spots and choke one another; sometimes the seed is wastefully piled up inches deep.

Why must the bamboo die upon flowering? It does not expire of simple exhaustion; if so, it would shrivel up immediately. It may be a year or two surrendering life; if it manages to survive longer, it may return to health. Dying means the loss of a large amount of structural material and root systems, with the ability to spread vegetatively. The little plants that spring up from seed have to make a new start while competing vegetation prospers (Numata 1979, 253).

There seems to be something pathological in the habit. In the year before flowering, no shoots or only stunted ones may be produced (Numata 1979, 244). Moreover, while seed production may or may not be abundant, it is likely to be unviable, as though the plant lost fecundity during the long vegetative period. There may be a need to go through a process of sexual renewal; this being costly, there is a tendency to extend the period.

Whatever the utility of the habit, long-period bamboos are extremely successful, the dominant vegetation of large areas of Southeast and East Asia. Some other subtropical plants behave similarly. A South Asian shrub, *Strobilanthes,* and a palm, *Corypha umbracilifera,* have periods of 3 to 16 years and 37 to 44 years, respectively, with subsequent die-off (Janzen 1976, 379–381).

Evidently much of the diversity of living creatures is not directly adaptive. A partial explanation is that traits may be indirectly adaptive: genes have multiple effects (pleiotropy), and a valuable trait probably has secondary neutral or even negative consequences. The gene that makes Siamese cats white makes them cross-eyed. Explanation by supposed linkage of traits has the logical defect that it can be taken to account for anything. Instead of asking how a trait contributes to reproductive success, we are reduced to asking the more abstruse question of why it is related to more important traits or why reproduction often entails seemingly unnecessary death.

In sum, the fit between adaptation and environment is loose. Plants and animals have dynamics, directions, limitations, and potentialities given by their entire nature formed by the accidents as well as the needs of many million years in ways much beyond our understanding and for which there can be no simple explanation.

But even when we have what seems to be a simple and logical explanation ("cheetahs developed by natural selection for hunting by rapid chase"), it is really only a more or less plausible hypothesis, which might be confirmed or falsified if we had a time machine to observe how the adaptation came about.

6

The Question of Sex

The Role of Sex

The only clear utility of sex is to make life and biology interesting. It seems unnecessary, and accepting or requiring a set of genes from another organism contradicts the fundamental principle of self-multiplication. Sexual reproduction has great costs and complications. Many plants and animals dispense with it, although birds and mammals are quite unable to do without. But sex seems to be essential for evolutionary progress.

Some bacteria occasionally have a sexlike process, or conjugation: a donor passes part or all of its DNA to a recipient, which splices it into its own, discarding duplicated material. The recipient may in turn become a donor. This practice may have arisen to repair DNA when there was no protective ozone layer (Margulis and Sagan 1986, 35–36, 49). If this is so, it remains unclear why the habit persists, contrary to the idea of the "selfish gene."

Sex in the ordinary sense implies meiosis, the process by which the double (diploid) set of chromosomes normal to eucaryotes is split, making haploid sex cells, which fuse to make a new diploid individual. This complex process apparently originated in early protists, or one-celled eucaryotes. It may or may not be related to bacterial conjugation.

Protists show a great variety of sex or sexlike behavior. All are capable of asexual reproduction, but in most species individuals fuse from time to time, sometimes in complex ways. Many have a process (autogamy) for renewing the nucleus without fusion or division of the cell. For protists, sex is independent of reproduction; the important fact is restructuring the nucleus, not increase of numbers. There may be plural nuclei. Some species have multiple mating types, each capable of conjugating with a different type (Fenchel 1987, 7; Sleigh 1989, 84–85). When ciliate *Stentor*s yield to the temptation to engage in sex, both partners die (Margulis and Sagan 1986, 220).

In multicellular animals, individuals of one sex provide small cells, spermatazoa, to fertilize a much larger egg cell, or ovum; this is the definition of male and female. Such a process is believed to have been necessary for the evolution of cell differentiation and multicellularity. It is found in all phyla of animals and is the more prevalent, apparently more necessary, in the more highly evolved classes. Many invertebrates reproduce asexually part or most of the time. Rotifers, tiny ciliated water animals, are mostly sexless, as are many worms and other relatively simple creatures. Some flatworms (planaria) spread only by fission or budding (Pearse et al. 1987, 225). Jellyfish and other cnidarians alternately reproduce by asexual budding and by sexual process.

A number of insects, such as aphids, go through a series of asexual generations before producing males for a sexual phase. Typically a winged female flies to a new plant in the fall and bears diminutive males and females, which are born half-grown. These mate and lay eggs, which hatch in the spring to produce larger wingless individuals, which reproduce asexually through the summer and in the fall give rise to winged females to start new colonies. Some aphid species, like some beetles and a few hymenopterans, are fully parthenogenetic, that is, have no males (Evans 1984, 111–112). This also the case with a few fish. Some 30 species of lizards, including 15 of 45 species of whiptails (*Cnemidophorus*), are unisexual. One small burrowing snake (*Rhamphotyphlops bramius*) is known to be parthenogenetic (Mattison 1987, 126).

All asexual higher animals, however, are recently derived from sexual ancestors. The uniparental lizards, for example, are closely related to lizards with conventional habits, having been separately evolved, perhaps by hybridization (C. Cole 1978, 57). Some of them need a mating ritual to lay eggs; at the beginning of their cycle, females behave like males and go through the motions of copulation (Crew 1987, 116–121). A parthenogenetic fish, the Amazon molly (*Mollienesia formosa*), requires the stimulation of sperm in order to lay; she lures males of a related species into "mating" (Forsyth 1986, 147).

Many animals, such as earthworms and snails, are hermaphroditic, individuals being both male and female. The hermaphrodite may or may not be self-fertile. The sexuality of one worm species (*Diplozoon*) is intimate; the male and female merge into a single organism (Pearse et al. 1987, 235).

The determination of sex is usually genetic, one sex having a pair of dissimilar chromosomes. Among invertebrates, the male usually has the different chromosome. Vertebrates vary. In mammals, the male has one X and one Y chromosome, the female a pair of Xs. In birds, the reverse is the case. Practice is variable even within restricted taxonomic groups. In some flies, for example, it is the male, in others the female, that has the odd chromosome (Stearns 1987, 21). It is theorized that the change could come about by adding a new factor to override the previously determinant chromosome (Bull 1983, 87), but it is not easy to understand why this should occur or what could be gained by the shift.

In several orders of insects and some rotifers, the male is haploid, with a single set of chromosomes, whereas the female is diploid. Haplodiploidy may have the advantage of eliminating deleterious genes (not being shielded by heterozygosity) (Mayr 1963, 409), but this utility would seem to make it impossible for haplodiploidy to evolve: mutation to male haploidy would cause the expression of harmful genes. Yet it has arisen at least a dozen times. However, it is unknown in vertebrates (S. Gould 1983, 59). Something in the vertebrate makeup must exclude this type of sexual determination.

An equally unaccountable mode of sex determination is environmental. Some lizards, most turtles, and all crocodilians take on one sex or the other according to the temperature at which they are incubated. Although many must be near the borderline, intersexes are rare (Deeming and Ferguson 1989, 33). Lower temperatures make crocodile and some turtle hatchlings male; higher temperatures, female; in lizards, alligators, and other turtles, the reverse is true.

How turtles came to leave the sex of their offspring up to the weather is a riddle. To make the transition from genetic to environmental determination of sex, it would seem that at certain temperatures some females had to become males—an unlikely event of no apparent benefit to the individual or the species (Bull 1987, 109). A leading authority, J. Bull, believes that temperature sex determination may be ancestral in reptiles, chromosomal determination having arisen more recently and separately in the families where it occurs (Bull 1983, 120).

In some fish also and in numerous invertebrates, temperature during early development or other environmental conditions determine sex, alone or in conjunction with chromosomal determination. An environmental factor may override the genetic. In many species of fish and crustaceans, individuals change sex one or more times during their life. All members of a school of blue-headed wrasse are female except for the dominant male. If he disappears, the dominant female within hours turns male. If he should return, she reverts to female status (Kevles 1986, 70–71). Female frogs in a largely female population similarly can become male (Cowen 1990, 134).

In such cases, environmental sex determination seems adaptive. Where there is a surplus of females, an individual gains by becoming male. But leaving the sex of the offspring to the weather seems unhelpful. An unusually cold or hot season would cause an imbalance, and a precise cutoff is essential to avoid intersexes.

The role of sex, or pollination, in flowering plants is quite varied. A majority require fertilization, and many, such as orchids, go to extraordinary lengths and make incredible flowers to ensure cross-fertilization. In most plants, male and female organs are on the same plant, commonly in the same flower, but many species have male and female individuals. Some plants have one kind of flower for cross-pollination and another for self-pollination. Others are sometimes or always self-fertilizing. Many reproduce vegetatively, by runners or the equivalent, although they may also have seeds. These include bamboo and other grasses, probably all water plants, and some trees, such as the aspen (Richards 1986, 370–374). Garlic produces little bulbs on a stem in lieu of seeds.

Many higher plants are sexless. Dandelions, hawkweed, blackberries, and sundry other genera make seed without any sexual process. Some such plants, like the dandelion, have retained showy flowers, even offering nectar and pollen. In some, pollen is needed to start seed production, although it has no genetic role (Holm 1979, 140–141).

It is a curious contradiction that plants are more likely to be dioecious (with separate male and female individuals) in the tropical forest, where there is an enormous variety of species and the distance from one to another of the same species is likely to be great, than in temperate regions, where species are fewer and mates are probably closer (Forsyth and Miyata 1984, 66).

Some flowers specialize extraordinarily. An orchid, for example, pretends to be a female wasp in order to induce males to attempt copulation, although they tend to learn that they are being fooled. Such specialization would seem to risk dangerous dependence on one or a few carriers of pollen (Ghiselin 1974, 44), and it would be much simpler to offer a little nectar. However, many orchids do not seem to care much whether they are pollinated.

Different kinds of flowers on a single kind of plant (heterostyly) are also puzzling. Many species have two or even three forms, with blossoms differing chiefly in length of stamens and pistils, as in the primrose, *Primula*. They are so fixed as to place pollen on or receive it from different parts of an insect body. Darwin wrote one of his last books, *Different Forms of Flowers on Plants of the Same Species*, about this phenomenon. He found that fertilization of a flower by pollen from another type produced normal seed, but fertilization by the same type resulted in partial or complete sterility. There are thus not only two sexes (joined in the same plant) but two (or three) mating classes; offspring belong to each class in roughly equal numbers. Darwin treated heterostyly in terms of outbreeding. Natural selection did not enter the discussion except by inference, but he appreciated the difficulty of postulating natural selection for sterility.

It would seem difficult for heterostyly to arise. Self-sterility and obligatory outbreeding would be advantageous only if the species were burdened by many recessive negative genes, but self-fertilizing plants eliminate negative genes as they double up. The arrangement of different forms of flowers with corresponding sterility barriers is genetically complicated, yet distyly appears in 25 families, independently in about 18 of them (Robert Ornduff, pers. com.). Tristyly, which may have quite complex relations of fertility and sterility among the three kinds of flowers, seems to be even more difficult to reach. However, it has come about in 3 widely separated families. The arrangement would seem to be costly because half or more of the neighbors of the heterostylous plant are inacceptable mates. Whatever may have been the reasons for the appearance of the trait, it shows frequent tendencies to break down with reduction of self-incompatibility (Ganders 1979, 628–631; Charlesworth 1979, 486–498).

Sexual Practices

Myriad complications are woven around the essentially simple operation of making sperm available for fertilization. The damselfly, instead of copulating like the great majority of insects, transfers sperm to a special organ, from which it is given to the female. This requires not only a receptacle on the fore part of his abdomen but special claspers to grip the female's neck, an apparatus that varies markedly from species to species but is comparable to nothing else in nature. It represents no evident improvement over the more direct method (Cloudsley-Thompson 1988, 71). A four-eyed fish (*Anableps*) has a lateral sex organ— sometimes on the right, sometimes on the left— so that a right-sided male can conjugate only with a left-sided female (Kevles 1986, 79).

Sexual behavior is diverse for no apparent reason. In one salamander (*Ambystoma*), the males function in the ordinary amphibian way; in a closely related species, a cluster of males deposit spermatophores at a distance and compete in luring the female to them (McFarland et al. 1985, 226). Most birds are monogamous; a widow or widower swan is said to refuse to take a new mate. Others are polygynous, polyandrous, or more or less promiscuous, as are most orders of animals. Commonly, females seek multiple matings, as would seem desirable both to ensure fertilization and to improve chances for the offspring by providing greater variation. But many fish and arthropods make or accept a vaginal plug, like a chastity belt, perhaps developing structures to hinder a second mating (Kevles 1986, 106, 109). So do a few mammals, such as insectivores and chipmunks (Voelker 1986, 116).

Among bluegill sunfish, there are large territorial males, which try to keep females for themselves, and also small males with female markings. The latter try to sneak in and fertilize the eggs when the boss is napping. Sundry other fish, such as topminnows, also have two kinds of males. It seems to be logical for animals with external fertilization and dominance among males to snatch an opportunity for paternity from the top male. Salamanders have a similar strategy, as do some birds, mountain sheep, and hyenas (Alcock 1989, 406, 411–412).

Many land arthropods produce sperm packets that the male, often after much maneuvering and dancing, makes appropriately available for the female. Variants of this procedure are followed by arachnids (spiders, scorpions, and others), land crustaceans, and a few insects. A male spider exudes sperm onto a little web he makes for the purpose and then draws it into special organs on his pedipalps, which he inserts—if the female allows—into her genital orifice. A female millipede (*Scutigerella*) comes upon one of the spermatophores the male has left around, takes it into her mouth, and stores the sperm in a cheek pouch. On producing an egg, she sticks it onto a suitable spot and smears it with sperm from her mouth (Pearse et al. 1987, 571).

Insects have nuptial dances (such as fruit flies), call their mates (as katydids), or signal (as fireflies). Among the more eccentric are scorpion flies, the males of which have three distinct strategies: using a food resource, such as a dead insect, to please the female; placing saliva on a leaf for the female to lap up; and forced copulation, which ethologists refer to as rape (Barash 1982, 268). In addition, a male that spots another carrying an offering will go up to him and pretend to be a female in order to get the offering (Thornhill 1979, 412–415).

If the males of some species have shown much ingenuity in getting their sperm to the female, a few have developed means of removing sperm deposited by a previous suitor. Male damselflies, sharks, and some birds do this, with appropriate apparatus. Damselflies have a scrub brush on the penis; the shark has a tube for giving his mate a douche of seawater (Alcock 1989, 418–419).

The bedbug's mating is perverse. Instead of inserting sperm into a duct leading to the female reproductive organs, the male punctures his mate's abdominal wall. The sperm goes into the body cavity, is carried to the heart, and is spread throughout the insect. In some species he stabs her more or less randomly, and she shows a scar for each sex act. In other species, the female has areas of the abdomen designed to be pierced, with columns of tissue to guide the sperm to the ovaries—a new set of genital organs, imperfectly reverting to the ancestral procedure (Eberhard 1985, 95). Males of some species that prey on bats (*Afrocimex*) develop patches for puncture like the females, thus inviting homosexual intercourse. Males may even have more developed structures for receiving punctures than the females. If a male of one genus (*Xylocoris*) mounts a female, he can expect another male to mount and copulate with him (Lloyd 1979, 17–34).

A tendency of bugs using the normal route to miss their aim could hardly be permitted by natural selection, and fertilization by injection requires modification of the apparatuses of both male and female. It has been hypothesized that nutritional transfer led the way, as it contributed to the nourishment of the offspring. But this could be accomplished by delivering seminal fluid by the normal route. The female in any case needs a meal of blood in order to lay; the gain from semen would be trivial (Hinton 1964, 101–103). Observation indicates that the female receives no benefit and that she may even die from the wound (Eberhard 1985, 98–99). Yet this mode of insemination probably developed several times independently because the location of the receptive sites varies greatly.

The caprice of sex is also shown by the enormous diversity of interactive genitalia. According to William Eberhard, "In many species, the male genitalia are structurally the most complex organism in his entire body; in some they are also incredibly large, as in nematodes and flies, whose intromittent organs are longer than the rest of the body" (Eberhard 1985, 12–13). A male spider, *Tidarren fordum,* has such grossly enlarged pedipalps (used for transfer of sperm) that he breaks one off upon reaching maturity (Eberhard 1985, 74). For no apparent reason, the spermatozoa of insects are quite varied, even within a genus; some are gigantic by cellular standards, the better part of a centimeter long (Sander 1983, 154–155).

Sexual organs, especially of the male, sometimes of the female, are the most distinctive parts of the anatomy in many invertebrate and a few vertebrate genera. Taxonomists look to them to separate outwardly similar species. This is contrary to the logical expectation of natural selection: species should be separated by adaptive traits. But many species that are otherwise indistinguishable are quite distinct in organs performing the same simple function: transfer of sperm. A good example is that of the species-swarm, approaching a thousand, of Hawaiian Drosophilidae. Despite a wide variety of ways of life, a large majority of interspecific differences relate to courtship and mating; males differ, sometimes strikingly, but females are much more alike (Carson et al. 1970, 481, 484).

The commonest explanation is that sexual peculiarities serve as barriers to unsuitable crossmating, and there are sundry other guesses, all of which Eberhard effectively rebuts. His hypothesis is that it is a question of female choice, as she may or may not carry through her part in making insemination effective. This opinion is supported by the fact that diversity appears in the ways the male stimulates the female, including not only genitalia but spermatophores. In a number of families where courtship is complex and species specific, genitalia are simple (Eberhard 1985, 22).

It is not clear, however, what is gained by more differentiation than necessary to ensure that the partner is of the correct species. Moreover, if the signal—some detail of the way the male stimulates the female—is vital for reproduction, it should be quite stable. To alter it by mutation would be to frustrate reproduction.

The question remains unanswered whether species develop barriers to prevent deleterious crossbreeding or whether the barriers cause the species to become separated. The latter would seem to be implied by the fact that females of different species may be difficult or impossible to distinguish morphologically and may have identical or very similar habitats; for example, a half-dozen or so species of Hawaiian *Drosophila* may live on a single flower. But it could hardly be advantageous for the female to insist on a particular curvature of the male's hind legs or for the male to have such a curvature unless the female insisted on it.

It seems significant that there is in higher animals a good deal of nonreproductive sex, both hetero (in times of infertility) and homo, both male and female. In this connection, the imitation male external genitalia of the female spotted hyena are interesting. So much alike are male and female that it was once speculated that the animal was hermaphroditic; unless the udders are enlarged, distinguishing the sexes by external appearances is difficult. The sex organs are prominent in greeting ceremonies, including erections on the part of both sexes. Whether the female's masculine equipment is related to this use is unclear. However, she has a high level of male hormone, testosterone, which enables her to grow larger and stronger than the male. She guards territory, leads in the hunt, takes priority in eating and resting places, and dominates the clan (Kruuk 1972, 228–234). How she retains female reproductive capacities while becoming otherwise male is unknown, but the spotted hyena is one of the most abundant of carnivores.

In view of animals' whimsical ways, it is not surprising that humans are inventive in sexual matters. Our nearest wild relative, the bonobo (pygmy chimpanzee), is probably the erotically most versatile of nonhuman animals (de Waal 1989, 180–181).

In sum, evolution is exceptionally creative in matters related to sex. The interaction of male and female implies more reactive feedback of behavior and stimuli than any other aspect of life, hence most readily leads to unforeseeable relationships, including sexual selection. For this reason, sex complicates and confuses the evolutionary picture. Would Darwin have proposed his theory of *The Origin of Species by Natural Selection* if he had known that in many groups, especially but not only invertebrates, interspecific differences relate mostly to ways and organs of copulation with little or no relevance for adaptation to environment? Malthus suggested no such thing.

Why Sex?

The reason usually given for the prevalence of sex is that it permits recombination of genes from slightly different individuals, thereby making more patterns from which the fittest can be selected for improvement of the species and adaptation to environmental change (Stearns 1987, 18). There is no real evidence for this widely repeated explanation, however, which seems to be held because it apparently makes sense and there is no other answer. Species are poor trackers of environmental change. They may shift, moving toward the equator in an ice age, for example, but they are usually little altered until, after a million years or so, they are replaced. Most species seem to have had little use for variation for millions or tens of millions of years.

Moreover, sex is not needed to produce variation, and it is a costly method. Errors enough occur in the replication of chains of many million bases and their reassembly in new nuclei, and if more variation were desirable, the organism would need only to relax the error-correction mechanisms. It is not clear how much genetic variation is desirable; vigorous and wide-ranging species thrive in markedly varying situations. A higher mutation rate is desirable only under conditions of stress, which ordinarily increase genetic variation (McDonald et al. 1987, 258).

In any case, there is no evidence that sexual organisms are more adaptable than asexual (Margulis and Sagan 1986, 4). The sexless dandelion flourishes on lawns across the United States and in its native Eurasia in many varieties. A sexless fern-ally, *Salvinia molesta,* chokes lakes and waterways throughout the tropics (Barrett 1989, 97). Although a large majority of rotifers lack a sexual process, they have produced hundreds of diverse species (or distinguishable clones) in a great variety of habitats (Stanley 1979, 225). While freshwater rotifers, subject to vagaries of cold and drought, are predominantly parthenogenetic or have asexual phases, marine rotifers, in a more stable environment, are sexual. About half of American species of earthworms are parthenogenetic or nearly so, supposedly to meet varied environments (Pearse et al. 1987, 306, 408). It appears that sexlessness in plants occurs most typically in unstable environments (Huxley 1974, 147).

Mixing chromosomes through sex creates new combinations but breaks up existing ones. Any chance improvement in the genome of an asexual animal is passed on intact and may be favored by natural selection and lead to an improved form. But if a recombination improves a sexual animal, its mate will probably not possess the same combination. Even if it should, only a fraction of the offspring will have the desirable combination.

Even if sex is somehow good for the future of the species, it is burdensome for individuals (Lewis 1987, 34–37). Costs include the possibility of not finding a mate in a sparse population, and seeking a mate exposes animals, especially males, to predators. Not only female frogs but frog-eating bats hear male frogs croaking. Female sticklebacks take note of red-bellied suitors; so do trout (Kevles 1986, 12). In addition, many male mammals engage in exhausting combat. Not a few—such as stags and seals—are wounded and sometimes killed. It is equally uneconomic that whatever is expended on sexual display, such as cumbersome feathers or continuous vocalizing, is wasted.

Sex slows reproduction. An animal coming into an abundance of food does well to multiply as rapidly as possible to take advantage of it before competitors do; for maximum fecundity, all individuals should produce young. When food runs low, then sexual reproduction is in order. Very many protozoa are asexual as long as they can grow and divide readily; when food becomes scarce or the environment adverse—their puddle drying up, for example—they have a sexual phase and form cysts to wait until times are better. Numerous animals, such as some rotifers, daphnia, and sundry insects, also follow this rule.

The biggest cost of sex is the limiting of reproduction to the female, which nearly always bears most or practically all of the burden. If a sexual female has two young, she has in effect reproduced herself once, projecting half her genes to each offspring. An asexual female with two young has doubled her genes. The same is true of plants that separate the sexes: the female dioecious plant has to produce twice as many seeds to keep up with hermaphroditic species.

In the economy of nature, males are usually useless; birds are unique in the extent to which the male (in 90 percent of species) is helpful (Hapgood 1979, 16). The capacity of the male to contribute materially to the next generation is more or less wasted. In the extreme, the drone bee or female ant is a parasite entirely dependent on the labors of his sisters. The situation changes if the male contributes to raising the young, as many bird fathers and some mammal fathers do. But the asexual mother, if she can manage by herself, should have a huge advantage. Yet parthenogenetic lizards do not displace kindred sexual forms in the same region, despite their double reproductive capacity (Barash 1982, 309).

Mutation to parthenogenesis does not seem difficult, as it occurs in very many different groups. A number of kinds of ants, for example, are occasionally parthenogenetic. One, the common Japanese ant *Pristomyrmex pungens,* is fully so, rarely producing males, and having no queens; ordinary workers do the egg laying (Hölldobler and Wilson 1990, 188). Yet only a few dozen vertebrate species are parthenogenetic. They include no mammals or birds, although unfertilized turkey eggs sometimes hatch (F. Gill 1990, 328). Possibly the failure of animals to avail themselves of the possibility of greater multiplication of genes through parthenogenesis is somehow related to a failure to maximize reproduction generally.

Hermaphroditic animals should have a big advantage because parents contribute equally. Hermaphroditic animals (and plants) would seem to have the full benefits of sex, whatever they may be, at less cost. One might expect hermaphroditism to be universal, but it is unknown among arthropods and vertebrates except a few fish.

There are indications that the meiosis and fertilization may be necessary to correct errors in the genome (Stearns 1987, 25). The need is not for different genes but for genes from different sources. Self-fertilization is very common in plants, and many lower animals, notably mites, have "incestuous" matings among siblings or between mother and son for generation after generation. In mammals it does not suffice to insert a set of genes in an egg cell. Development of an embryo requires chromosomes from both sexes, and it matters which parent a particular chromosome comes from (Nicholls et al. 1989, 281), as the mule and hinny show.

The mule, offspring of a jack donkey and a mare, differs from the hinny, offspring of a stallion and a jenny donkey. The male mule and male hinny can be expected to differ because they have a Y chromosome from horse or donkey, respectively; the females also differ, however, although their chromosomes should be the same except for their origin in one species or the other (Jones 1971, 106). This is a general rule: reciprocal crosses differ depending on whether a trait comes from the father or the mother (Sapienza 1990, 52–60).

One hypothesis is that sexual union is necessary to correct alterations in the methyl groups that, attached to DNA, signal where to start or stop the transcription of genes. These methyl groups are lost from time to time, but they can be restored by virtue of differences between male and female gametes (Holliday 1988, 46; Holliday 1989, 68). That sex is needed for some kind of normalization is supported by the fact that many animals that can reproduce asexually revert to a sexual process after a number of generations. For example, paramecia degenerate and die out if they do not conjugate from time to time (Pearse et al. 1987, 54).

Although a need for correction may well be part of the answer, it is argued that the existence of paired chromosomes (diploidy) should provide adequate means of repair, which is continually being carried on by various mechanisms (Shields 1988, 256). Sexless forms show that sex can be dispensed with, at least for very long periods. Unless sexuality carries some general short-run advantage, selection should overwhelmingly favor mutants capable of parthenogenesis in any sexual population. Whatever the utility of sex for the species, individual advantage would soon break it down; sex should never become established with inequality of burden sharing in reproduction.

The adaptive advantage usually attributed to sex—that it generates greater diversity of offspring—could hardly be selected for. Variability is generally deleterious in a well-adapted species. If the sexual process throws up random new combinations, the likelihood that they will be positive is slight; any deviation from average is probably negative (Mayr 1963, 280).

Yet sex seems to be, if not always necessary, generally desirable. Not only does it seem positive to have a process of genetic conjugation; it seems to be preferable that genes come from slightly different individuals. This may be related to the fact that plants and animals usually favor outbreeding, although it is not known why inbreeding should necessarily be harmful. The ordinarily cited reason is that inbreeding brings deleterious recessive genes to the fore, but it also gets rid of them. Strongly inbred lines, having eliminated harmful genes, should be exceptionally vigorous; they are not. Moreover, plants and animals that are haploid or have a haploid phase automatically eliminate injurious recessive genes, but they generally have a sexual process.

Cheetahs, rather infertile in the wild, are hard to breed in captivity, have a high proportion of defective sperm, and suffer high mortality of cubs. This is attributed to a high degree of genetic uniformity, perhaps arising from a reduction of the population for some reason to a very small number a few thousand years ago (O'Brien et al. 1986, 84–92).

Why uniformity should be so bad is not explained. The selection-variation theory would apply much more neatly to a sexless world, and evolution should proceed much more rapidly. By Darwin's theory, the improvement that appears in any individual should be passed on as a unit to its descendants; the formation of sex cells by meiosis, leaving behind half the chromosomes, presents a fundamental problem. The difficulty is increased by the fact that most improvements are recombinations, which are automatically broken up by the reassemblage of chromosomes (Maynard Smith 1988, 225). (Anyone who doubts this might try producing avocados from seed without grafting.) But asexual individuals vary slightly, as Darwin postulated. As the more successful make more copies of themselves, positive mutations would easily become established, and the species would slowly change. More exactly, there would be no species, only a large number of varying clones; it is the mixing of heredity that gives unity to populations and helps to define the species.

But for life as we know it, sex is evidently necessary. There are parthenogenetic fish and lizards, not to speak of unnumerable invertebrates, but no family is exclusively asexual. Asexual forms have recent sexual ancestors. This must mean that sex is prerequisite for major changes—for transitions between forms and realignments of the control system. Sex is not necessary to permit genetic change, which comes all too easily to the complex dynamic systems called organisms, but it seems somehow to make it more possible for organisms to change usefully.

To understand why sex is necessary for evolution, one must look for non-Darwinian factors. In one opinion, "Sex represents the most important challenge to the modern theory of evolution" (Halvorson 1985, 5). This is slightly exaggerated—as a mystery, sex must take its place alongside many others—but sex may well have fundamental significance, and understanding its role seems essential for deeper comprehension of evolution.

7

Sociality

Social Animals

There has been much discussion of what biologists call altruism—behavior that benefits not the individual and its lineage but the group. It is not easy to explain how natural selection can bring about traits that do not directly assist the survival and reproduction of individuals bearing them.

Sociality rests ultimately on the remarkable faculty that all animals possess of recognizing their kin (Sibatini 1989, 25). Everywhere there are aggregations: flocks of birds crowding in limited nesting space, schools of fish that swim in formation with military precision, clouds of locusts, and congregations of migrating butterflies or ladybugs. Cichlid fish (*Tropheus*) form gangs whose members know one another "personally" and fight outsiders (Crook 1972, 318). Hundreds of rattlesnakes winter together in a den to which they return in the fall. Coati bands forage together for mutual defense.

Although social existence has negative aspects, such as the spread of disease and parasites and the need to share limited resources (Vaughan 1986, 39), higher animals usually seem to be in company. Herbivores congregate, probably to reduce the vulnerability of individuals; it is harder to surprise the member of a herd. Social mammals sometimes feed the sick or injured. A healthy African bull buffalo acts as guardian for an impaired bull (Miloszewski 1983, 199–200).

Carnivores, like the tiger, are commonly solitary, coming together only for mating; but some, like members of the dog family, live and hunt in small groups. It is especially profitable to do so when the prey is large: a single wolf cannot take a healthy adult moose, and a lone hyena cannot tackle a zebra. Lions hunt and rest in small groups (called prides in deference to the king of beasts), grudgingly share the kill, have no fixed hierarchy, and occasionally fight (Schaller 1972b, 135–136). Several lions together stalk and ambush prey. Two or three can bring down a bull buffalo or perhaps even a giraffe; they can use a carcass that would be too big a feast for a single animal and defend it from the hyenas, jackals, and vultures. The solitary leopard cannot take an animal much bigger than itself, and it has to drag the carcass up a tree to have a fair chance of keeping it. Another important benefit of group living for higher animals is the opportunity for learning. This advantage was crucial, of course, for the emergence of the human species.

Monkeys, though they are very social, do not cooperate a great deal. Their customs are irregularly varied. Brown capuchin monkeys are aggressive toward their fellows, and the leader keeps females in estrus for himself so far as he can; white-faced capuchins share fruit and mates (Janson 1986, 45–52). Olive baboons and hamadryas baboons are similar and sometimes interbreed, yet males of the former are totally proprietary toward females, and those of the latter bother with them only in estrus (E. Wilson 1980, 9). Baboons sometimes hunt cooperatively under a leader, which apparently commands the troop (B. Campbell 1985, 380). They have cliques and friendships within the troop. Males form partnerships for courtship activities, and a male may become associated with a female and her young, whether they are his offspring or not (Moss 1975, 206–210).

Much mammalian sociality is a broadening of parental instincts. Communal care of the young is found in diverse orders, including several rodents, one seal, some whales, two bats, a dozen or more primates, various carnivores, wildebeest, and elephants (Gittleman 1985, 189–191). In the elephant herd, cows not giving birth will lactate for the benefit of others' calves. If a mother is lost, other females will take care of the calf (Eisenberg 1972, 17). African wild dogs (*Lycaon pictus*), like wolves, take food back to the young and to adults that stayed behind as guardians or because of infirmity. When the pups are large enough to go out with the hunters, they are given priority at the kill (Schaller 1972b, 83). Brown hyena pups, when somewhat over two months old, are taken to a communal den where all the females suckle them indiscriminately and bring food to them (Vaughan 1986, 404).

Perhaps the most altruistic of mammal societies are those of the vampire bat (*Desmodus rotundus*). Vampires have to feed frequently; although blood is nutritious, they can survive only one night without a meal. The response of evolution has been not to provide reasonable storage facilities—many small birds can fly for much longer times without eating—but to engender social support. Those that return gorged to the roost regurgitate for the benefit not only of the young but of hungry roostmates, relatives or not (Wilkinson 1990, 76–82).

Dwarf mongooses feed on insects, small animals, and the like, yet their social life is much more advanced than that of the wolves. When a mongoose clan of a dozen or so goes foraging, one individual, somehow designated or volunteering to protect the others with heads to the ground, climbs a tree to watch for hawks or other enemies. Only the dominant pair breeds, but all cooperate in caring for the babies; nonbreeding females lactate for their benefit. Subordinates may copulate but do not conceive (Vaughan 1986, 403). Immigrants to the pack are equally helpful, although only with much luck will they ever reproduce (Rood 1988, 41–46).

The mammal with most complete social subordination is the hairless African mole-rat (*Heterocephalus*). Colonies of as many as 50 have a single breeding queen, which not only suppresses the sexuality of other females by a pheromone in her urine but bullies and sometimes kills them. Even if the queen is removed, older members of the colony remain infertile. The subjects are divided into castes: small and hard working, larger and lazier, and largest and nonworking (Jarvis 1981, 571–573; Kevles 1986, 217–219). The workers coordinate digging, operating somewhat like a conveyor belt for casting out dirt. This may be an efficient system, but the root-eating rodents have little to gain by joining forces under a dictatorship. Solitary gophers are much more numerous.

Among birds, as among mammals, there are all degrees of sociality, from simple congregations to closely ordered groups. Migrating geese (except the lead bird) save energy by flying in the turbulent wake of the one ahead. One crow stands sentinel while the rest of the flock feed and may call a predator's attention to itself by its warning cry. Up to 15 Arabian babblers live communally; only one female lays, but all feed and protect the nestlings. The dominant babbler gives food to subordinates

(Zahavi 1974, 84–87), as though to show superiority by generosity.

Young adults of more than 140 species of birds, in all manner of habitats, help raise chicks not their own, usually but not always siblings (Skutch 1987, 253). A pair of Florida scrub jays is quite capable of rearing a clutch, but they may have three or four helpers. The young, although able to breed when a year old, rarely do so for another year. Instead they stay with parents or other kin and are more or less helpful with new clutches (McFarland 1985, 138).

The white-faced African bee eater (*Merops*) lives in extended families of one to five mated pairs, plus as many as a dozen unpaired birds. All cooperate in digging the nest, hunting and resting together, incubating the eggs of the senior birds, and feeding the chicks. In most cases, young adults help raise full or half-siblings; sometimes they serve a breeding couple to which they are not related (F. Gill 1990, 395; Emlen 1990, 489–526).

Such sociality may have utility for defense, but it has no simple cause. The helpers are not necessary; it has been found that they may add little to success in raising young and sometimes even destroy eggs in their own family (Skutch 1987, 265). In compensation for postponing reproduction, helpers may gain experience contributing to subsequent success, and a shortage of nesting sites may force the young birds to stay home. But there is no obvious relation between bird sociality and way of life or means of subsistence.

Social Insects

Sundry arthropods show beginnings of social life. Some beetles, for example, care for their young. A few kinds of spiders, such as *Stegodyphus*, make webs shared by hundreds of individuals that cooperate to capture relatively large prey (Griswold and Meikle 1990, 6–11). But a highly developed social existence, with specialized castes, has come about only in insects.

The superlative examples of animal cooperation are the social hymenoptera and termites. Many species, especially of ants and termites, form gigantic families of hundreds of thousands or millions, all totally subordinated to the community. Individuals labor for the hive or colony, usually for a single mother-queen, and will sacrifice themselves for her. Honeybees, attacking an intruder, leave their barbed sting and the poison sac in its flesh. Soldier termites rush out to fight attackers while workers behind them repair the breach in the wall of the nest, abandoning the defenders to die like Leonidas' band at Thermopylae.

In the absence of transitional forms, stages on the way to the fully evolved termite society can only be conjectured. There is, however, a complete sequence from solitary wasps and bees to the fully socialized hive or colony. Primitive wasps lay eggs in insects or other prey. The next step is to paralyze a prey, typically a spider or a caterpillar, place it in a nest, and leave an egg on it. Next, the mother wasp returns to the nest with additional provisions and cares for the larva or makes chambers for several larvae. She may continue looking after the brood when the first daughters have matured; these, in turn, may join in feeding and caring for the young in numerous cells (Evans and Eberhard 1970, 114–124). At the highest stage, the helpers largely or entirely cease to reproduce and become specialized as workers in an ordered colony. Bees show the same stages of socialization, from solitary nesting through group nesting to full sociality (West-Eberhard 1988, 172–174).

Paper wasps (*Polistes*) exemplify a middle stage of this process. They are fully social in that infertile individuals do the work, but there is no visible physical difference between the queen, which does little but lay eggs, and the workers. A colony is founded in the spring by either a single female or several, perhaps sisters. If there are more than one, they fight, mostly by threat or display, to establish dominance (Evans and Eberhard 1970, 131–132). The victor becomes the queen and reproductive for the colony; losers become workers and more or less repress egg laying. If they occasionally lay, the queen will eat their eggs. Newly raised females become workers until near the end of the season. Then some of them, selected by means unknown, desist from work, mate, and eventually hibernate.

Bumblebee colonies are also partially disciplined. The queen, which is little different from the workers, has to be alert to keep them from egg laying (E. Wilson 1971, 299), and she may injure or even kill workers before they become reconciled to servitude. Workers, although they do not mate, sometimes contend with the queen for the privilege of producing unfertilized eggs to make males (Rank 1987, 325).

In more evolved hymenopteran societies, the workers differ physically from the queen, and they are chemically controlled. For example, the honeybee queen governs the workers by means of pheromones. Workers cannot mate, and their ovaries are much reduced. They can produce unfertilized eggs, which become drones, but they do not often do so as long as they are inhibited by the queen, and their eggs are eaten. The slaves, however, sometimes decide that a queen has outlived her usefulness. They then raise new queens and prevent the old queen from killing them; they often do away with superfluous princesses (Michener 1974, 114; Brian 1985, 395). Army ant workers are sterile and much smaller than the single queen, yet there is a bit of democracy: the workers decide when it is time for the old monarch to go and which of her daughters is to succeed her (Hölldobler and Wilson 1990, 584).

Among social bees, wasps, and especially ants there are all imaginable varieties of subordination, competition, and sharing of reproduction, methods of queen control and determination of castes—one queen or several; frequent, occasional, or no worker reproduction; and so forth—with little apparent relation to food source, climate, or any other external influence or adaptation (Brian 1981, 121–222). Within a single genus, there may be varying ways of founding colonies, single or multiple queens, territoriality or nonterritoriality, all in approximately the same situations (Alloway 1988, 172–174). Some ants have hierarchies among workers, the dominant among which lay unfertilized (male) eggs (Cole 1981, 83–84). In species such as the Argentine ant (*Iridomyrmex humilis*), it may be unclear where one colony ends and another begins, and queens are accepted from outside.

In the more primitive species of ants, workers and queens are much alike, differing most obviously in that the queens are equipped with wings for the nuptial-migration flight. But more advanced species have carried differentiation very far. Queens are sometimes hugely larger than the smallest workers, and workers vary in size by a factor of ten or a hundred or more and are specialized into castes, sometimes so different as to seem to belong to quite different species.

Insofar as workers have lost reproductive capacity, the colony becomes the unit of selection. The fitness of the queen (and that of her mate or mates) depends on her ability to produce effective workers to make possible a large output of reproductives to form new colonies. Individual workers are thus comparable to the somatic cells of a multicellular animal, and the colony may be regarded as a superorganism. The acme of such sociality is reached by army ants, leaf-cutter ants, and termites, whose colonies are as populous as a large human metropolis. All members are probably daughters of a single huge egg-laying machine of a queen. They are so numerous as to be important factors in tropical ecology.

Although termites are more destructive than ants and more numerous, especially in the tropics, they seldom emerge from their nests in daylight and consequently are harder to study. Their nest mounds, as high as 10 feet, dot tropical landscapes, especially in Africa, and they nourish many animals that break into their fortress. The biology of termites is rather different from that of ants. Their diet is oddly limited to wood and related vegetable material, digested by protozoa and bacteria in their guts. Like most ants, termites ordinarily have only a single queen, but after mating the male lives on as royal consort. Whereas in hymenopteran societies workers are all female, termite workers are either male or female.

Termite and ant societies are convergent in important ways, however, despite belonging to different orders and having different modes of existence. Both have many sterile workers and a larger, usually single, queen. Both use pheromones to keep order. Both have different castes of workers, or workers and soldiers; termite soldiers are more specialized in that they cannot feed themselves. Both usually found colonies by sending out winged sexual forms on a nuptial flight, after which the female, or in termites the couple, establishes a new colony. Social life seems to have similar implications for very different animals.

Theory of Altruism

That individuals sacrifice their interests to those of others or of a group, large or small, especially if they renounce reproduction, is not easy to reconcile with principles of selection and survival. Even if the group gains more than the individual loses, it would seem difficult to establish the group trait or to maintain it if at any time a deviant individual could gain reproductively by looking to its own interests. Australian bellbirds eat the secretions of leaf insects on eucalyptus trees but refrain from consuming the insects (Loyn 1987, 54–60). Any bird could raise a bigger brood by consuming the collective livestock.

If the overriding aim of the animal were reproductive success, all adult members of the wolf pack or beehive should procreate to the maximum of their capability. They do not. Sooner or later, the alpha individuals in a pack of perhaps a dozen wolves will be displaced or disappear, but most wolves, male and female, end their lives without having propagated their genes. It is not even certain that this renunciation of the primal urge by most wolves is very advantageous for the group (Kruuk 1972, 30). African hunting dogs, with a social order much like wolves, are much less numerous than hyenas, which have their own families.

To account for behavior whereby the interests of individual animals are sacrificed to those of the collective, Darwin advanced the idea that the group, tribe, or community would prosper and increase if its members were prepared for such self-subordination (Darwin 1971, 93). This led to a theory of group selection: groups compete among themselves, and those with more effective patterns of cooperation prevail, thereby increasing the genes dictating altruistic behavior (Starr 1981, 39–42). However, little evidence has been adduced in favor of group selection among freely breeding animals, and it is unclear how individuals are restrained from reverting to selfish behavior. The altruistic trait could be expected to prevail only if the group advantage were decisive, the individual disadvantage were small, and groups were readily multiplied or brought to extinction.

To overcome such difficulties and particularly to account for the strikingly socialized hymenoptera, William Hamilton developed (after J. B. S. Haldane) the concepts of "relatedness," or gene sharing, and "inclusive fitness" (Hamilton 1964, 1–52; Hamilton 1972, 193–232). He noted that offspring get half their genes in a random assortment of chromosomes from each parent. This means (as it was stated) that a parent and child share half their genes; siblings similarly have on average half their genes in common. By extension, aunts and nieces share a quarter, first cousins one-eighth, and so forth. Thus an individual's saving two siblings or eight cousins would be genetically equivalent to saving itself. It might do more to promote its genes indirectly, by helping a number of relatives, than it could do directly by seeking its own reproductive success. Haldane, in his devotion to this idea, volunteered that he would sacrifice himself for two brothers or eight cousins (Haldane 1955, 34). (Happily he was not called upon to do so and was spared facing the question whether the brothers or cousins shared his better or less admirable genes.)

By this concept of relatedness, apparent altruism could be interpreted as indirect self-interest and thus reconciled with natural selection. Social or cooperative behavior would be widely advantageous, and the self-sacrifice of many animals, especially social insects, would become understandable. "Animals can be expected to help other individuals, even at cost to themselves, providing those individuals are closely related to them" (Evans and Eberhard 1970, 200). This idea was incorporated into the notion of survival of the fittest by calling helping one's own lineage "direct fitness," while helping other relatives was termed "indirect fitness." Together they constitute "inclusive fitness," which means overall reproductive success (Barnard 1983, 120–121).

Hamilton's work was hailed as a major advance in Darwinism in the 1970s. As an austere application of gene theory, the approach has been much used, especially by mathematically inclined biologists and sociobiologists. But kin selection and inclusive fitness are not without problems. What is shared, in the relatedness concept, is not genes but the probability of receiving genes from immediate ancestors; the great majority of the genes of any two individuals of a species are the same. It is not explained why sharing genes by immediate inheritance is more important that sharing genes by distant inheritance, or genes that are structurally alike regardless of ancestry. The more homogeneous the population, the less significant is the familial relatedness of individuals. Going back a few generations, one would find that any two individuals in a population are related to an indeterminate extent. For example, tracing descent for 20 generations, one would find almost all native British related. Far enough back, we are all related. All humans are said to have had a single female ancestor in Africa over 100,000 years ago.

A biologist pardons a scorpion (*Paururoctonus*) of the American Southwest for subsisting to a large extent on small members of its own species on grounds that the victims are mostly unrelated; the scorpion is hence supposedly not injuring its own inclusive fitness (Alcock 1990, 153). But the scorpions of a tract of desert must have been sharing genes for many thousand years. Except for mutations, all must have almost exactly the same set.

The importance of relatedness would thus seem to be limited to the probability of sharing mutations occurring in the family or unusual genes not frequent in the general population. But this would be rather marginal. Gene relatedness also overlooks the fact that much evolutionary change comes not from new genes but from recombination. Siblings are likely not to share combinations, and biologists write of sharing genes, not combinations.

No one is likely, however, to say, "I will sacrifice myself for two persons who happen to share mutations with me." The concept of the gene's promoting the reproduction of not only itself but of its homologues in other organisms is very fuzzy. Does it add to inclusive fitness if a gene favors another that is in itself identical but is differently linked and has a somewhat different effect, or a gene that is slightly different in its sequence but the same in physiological effect?

The idea of inclusive fitness is at best an approximation that blurs the theory of evolution by variation-selection. It means taking the gene, not the individual, as the unit of selection. Darwinian competition deals with individuals, not their interest in others, and takes no account of their relatedness except so far as the family becomes something of a unit of selection. Individuals and lineages have their own destinies.

The Darwinian "struggle for survival" may be most intense between those most alike because they compete most sharply for the same resources (Godfrey 1985, 56). According to Darwin, "But the struggle almost invariably will be most severe between the individuals of the same species" (Darwin 1964, 75). It may not be "almost invariable," but competition within species is usually more intense than between two species using the same resources (as seen in ants by Hölldobler and Wilson 1990, 424). An ecologist defines species as populations that compete mostly among themselves (Colinvaux 1986, 52). By the same thinking, competition should be most severe within small populations, within closely related groups, or among siblings. This applies especially to humans, who compete with other humans and most often with close associates.

The theoretical relatedness between siblings is the same as between parents and offspring, but their relations are emphatically different. The behavior of siblings toward one another varies from murder and cannibalism to amiable cooperation. Fratricidal baby birds do not seem to realize that they are reducing their inclusive fitness. But such conduct is understandable under natural selection: if one chick in a clutch has a murderous gene, it should be strongly favored. The vice is rather common among birds, especially raptors. It seems to be widely latent. Chicks of various nonparasitic species introduced into magpie broods crowded out or killed their foster siblings (Alvarez, de Reyna, and Segura 1976, 907–916). Mammals lack this vice, although it might well be advantageous for the suckling to do away with rivals for the limited milk supply.

Relatedness and inclusive fitness have seemed most cogent in regard to the ants, bees, and wasps, for which the approach was primarily designed. This is because of the way their heredity is organized. In hymenoptera, as in a few other orders of insects, males are produced by unfertilized eggs. Thus they are haploid, whereas females are diploid. This means that all daughters receive the same chromosomes from the father, along with a randomly assorted half of the mother's chromosomes. By the calculus of relatedness, they share not 50 percent of their inheritance, like siblings of a monogamous diploid union, but 75 percent. Consequently they should have more to gain, in terms of reproductive success, by helping their sisters than by having offspring of their own. Hence the workers feed their younger sisters, defend the colony, and so forth.

This calculation presupposes, however, that the queen mates only once. Hymenopteran females are not necessarily so chaste. Honeybee queens may mate 20 times. Females of various species of ants mate numerous times, as they perhaps must in order to store enough semen to produce millions of workers (Hölldobler and Wilson 1990, 154). If the workers have ten fathers, the female offspring share not 75 percent but 30 percent of their inheritance and so are less related to one another than siblings produced by a diploid couple. The situation is worse in colonies with plural, sometimes many, queens. The actual relatedness between worker ants is thus usually much less than the 75 percent called for by the theory, ranging down to 10 percent or less (Hölldobler and Wilson 1990, 187). The kin selection-relatedness approach thus becomes an argument against sociality in hymenopterans. Nonetheless, writers on ants freely speculate on relatedness of workers, queens, and males as causes of such matters as the relative expenditure on production of males and females. And the Hamiltonian theory is accepted uncritically, especially by nonzoologists (as by Kitcher 1985, 85; Bonner 1980, 97; B. Campbell, 1985, 21).

The variety of social relations among ant workers and queens is great: there may be one or plural queens; domination of the queen over workers may be near total or incomplete; workers may lay occasionally or frequently, producing male or, less often, female eggs; in some genera there are diploid males; in some species, males stay home to copulate with sisters, although in most they fly away to find unrelated mates; a few genera of relatively primitive ants have no queen caste (Hölldobler and Wilson 1990, 185, 188). Genetic relatedness can be only a very partial determinant.

There is no need to search for a genetic explanation for such workers' "altruism." Since they are chemically controlled, it is no more necessary to credit the workers with a desire to serve than to seek a genetic basis for their dedication to parasites with attractive secretions. They are usually incapable, physiologically and anatomically, of mating and, commonly, of egg laying. One might say that the queen enslaves her daughters in what biologists call parental manipulation. It lies in the endowment of the queen to be able to produce sterile daughters to work for her.

Insofar as reproduction is the prerogative of the queen, the colony has become the unit of reproduction and selection, as Darwin observed long ago (C. Darwin 1964, 237). The genetic relatedness of workers to workers, queens, or males becomes irrelevant; if they do not reproduce, natural selection does not affect them, only the queen or the colony as a whole. The same genes have different consequences in workers and queens, and in the workers they come to a dead end (aside from males produced by exceptional worker egg laying). What genes the workers may have is immaterial; it would not matter if the queen passed to her slaves only a reduced set of those genes needed to equip them to serve her. Natural selection does not apply to them.

This leaves open the question of how social behavior has become established. In large part, it must be an outgrowth of the family. Social spiders are unusual in that the mothers feed their young, first by regurgitation and then by bringing them pieces of prey (Griswold and Meikle 1990, 8). Hymenopterans are outstanding among insects for the intricacy of their instincts, the construction of nests even by solitary species, the extent of care for their young, and the frequent overlapping of lives of mothers and daughters—traits that form the basis for social life.

All degrees of sociality are represented by living species of bees and wasps, from the completely individualistic wasp or bee to those that may share burrows or gather in small colonies, to those that form colonies without a caste difference, and to those that can live only in colonies with specialized castes. There are two paths to socialization, which may overlap. One is the cooperation of several individuals in nesting, with some becoming specialized reproductives and others contenting themselves with working for the society. Why they would do this is not entirely clear because in semisocial species individuals seem to have more offspring per female than members of colonies; defense against predators may be part of the answer (Michener 1974, 45–47). Alternatively, generations may overlap. One can imagine that a hunting wasp or pollen-gathering bee, instead of capping a nest containing an egg and its food store, keeps building more cells together and feeding the young, some of which may stay around for a time after hatching. It may be advantageous for the family if the genetic program is such that those hatching at the beginning of the season dedicate themselves to the care of brood on hand rather than setting out to make a new nest, much like bird helpers. Their subordination ultimately should make possible the production of a larger number of reproductives.

Eventually the workers would become a distinct sterile caste, and the colony would be eusocial. This evolutionary trend seems natural, but it is not strong. Far more kinds of bees have stayed at various lower levels of sociality, and many have gone up and down the ladder. The matter, moreover, is confused. In many species, individuals not closely related may join to make a semisocial colony, some of them leaving egg laying to others. It is also contrary to genetic theories of sociality that eusocial workers (including honeybees) often drift from one colony to another, promoting the "fitness" not of their relatives but of their rivals (Michener 1974, 47, 233–253).

A key element in the evolution of sociality is the restriction of reproduction. This is evidently not difficult; many species procreate much less than they easily might. The ruling principle in animal societies seems to be much less altruism than domination. Reproduction is linked to social status; both males and females compete for the privilege of raising young, and dominant individuals inhibit inferiors. A dominant male mouse excretes a pheromone to prevent sexual maturation of young males and to terminate the pregnancy of females fertilized by other males, and dominant females repress breeding by their subordinates (Voelker 1986, 132).

Among baboons, many kinds of monkeys, several kinds of mice, marmots, gerbils, mongoose, elephants, social woodpeckers, and other animals, subordinate females fail to reach sexual maturity or to come into estrus (Kevles 1986, 216). High-ranking females of some species, such as wolves, monopolize the attentions of dominant males when not in estrus, keep lower-ranking females from mating (Fox 1978, 29), or kill and very likely eat their pups. Among such species as acorn woodpeckers, anis, and ostriches, superior female birds remove subordinate eggs from the communal nest. Primitive social wasp queens eat eggs laid by workers.

The importance of domination in the evolution of social behavior has no doubt been underestimated. There has also been more emphasis on male rivalry and hierarchy than on the usually less conspicuous female rivalry. But female-female competition may well be stronger than male-male because of the greater responsibility of the female for the young (Koenig, Mumme, and Pitelka 1983, 250).

Domination is reinforced as the female more successful in checking the breeding of others leaves more descendants. It is only necessary that the dominant female transmit a behavior pattern such that her offspring, whether subordinate or dominant, are prepared to fit into a social order. In this manner, differences of temperament favor pack formation; pups of wolves are much more varied in temperament, from timid to aggressive, than unsocial foxes (Fox 1988, 32). The interplay of dominance and submissiveness makes possible the wolves' social order (Clutton-Brock 1989, 41).

"Inclusive fitness" is an effort to find an exact formulation for the reality that a certain amount of altruism is advantageous for individuals and the family. By itself, natural selection of individuals with the greatest reproductive success would never have permitted social animals to evolve. Similarly human society would have been in the past and now would be impossible if its members were solely motivated by reproductive "fitness."

Social Specialization

Cooperation lies as much in the nature of life and evolution as does competition, and there seem to be tendencies toward sociality beyond environmental conditions and adaptation. The self-subordination of the mongoose, not to speak of the mole-rat, has no obvious relation to its ecology. Notably successful social creatures have quite different bases of existence: honeybees gather nectar, fungus ants grow gardens, army ants are mass predators, termites chew wood, and humans do many things. Sociality has an evolutionary dynamic independent of environmental adaptation.

In forming a community, animals become in effect more intelligent and more capable. They do this primarily by specialization—by taking on different functions, which are somehow allocated to different individuals. This is the case most clearly of multicellular animals, which are colonies of cells with hereditary integration.

Very simple multicellular animals can do much more than protists both to obtain nourishment and to protect themselves, and the problems of achieving multicellularity were parallel to those that had to be overcome in forming highly structured animal societies. Primarily, for cells to become specialized into tissues and organs, it was necessary that they give up reproduction for themselves. This renunciation is incomplete in primitive animals, as a small part of a jellyfish or flatworm can grow into a new individual. It is total in higher animals, the germ line being segregated by the Weissman barrier.

On a higher level, the group, pack, or colony can do much that the individual cannot; it becomes, to the extent that members are submerged in the entirety, a superorganism. More effective specialization implies more complete renunciation of reproduction by the "cells" of the colony. Thus in the more advanced ants, the workers are not only much differentiated but totally sterile.

It seems, then, that social insects might continue toward a greater degree of specialization of individuals, perhaps achieving differentiation corresponding to the hundred or more kinds of cells in an animal body. This would certainly be advantageous; the beehive or ant colony with many different kinds of workers specially equipped for many different occupations should be a formidable entity. But it seems impossible; the advanced hymenoptera are not believed to have changed importantly for tens of millions of years. Perhaps the genetics required to attain a much higher level of differentiation are too complicated. To blueprint the formation of many cell types and instruct them to build the organs composing the higher animal is a phenomenal achievement, hundreds of millions of years in the making. It is a further accomplishment for the genome to convey instructions for making different bodies for males and females, or for young and adults (as caterpillars and butterflies), not to speak of different "organs" of colonial animals such as siphonophores.

Wasps and bees have managed to develop only rather slight differences between queens and workers, and workers are capable of laying eggs though probably not of mating. Ants are more advanced in this regard; in most species queens and workers look like quite different species, and workers are incapable of reproduction. In many, workers differ widely in size—by as much as hundreds of times—and morphology, for example with jaws suited for fighting, grinding seeds, or simple household tasks. Species with such specialization, including army ants and fungus growers, are among the most successful, but their differentiation is slight by comparison with what it might usefully be. Social insects have reached a plateau beyond which they apparently cannot advance. Programming many different sorts of independent bodies, each with specialized tissues and organs, seems to have been too much for evolution to master.

Birds and mammals have not evolved any genetic specialization of castes and hence have remained at a lower stage of socialization. Only humans have been able to achieve more specialization by advancing from genetic evolution to cultural evolution. Indeed, they have managed to surpass indefinitely the degree of diversification of cells in the body, thanks to the immense flexibility of the brain and the storage of knowledge. High intelligence and the civilization that it makes possible, multiplying its capacities, have made a new kind of society with infinite potentialities.

In human society, the control of reproduction is no longer crucial because individuals are as much culturally as genetically motivated. They do not strive merely for fitness in the evolutionary sense; genuine altruism is possible, and a dose of it is indispensable for the good order of the society. There can be no such subordination as keeps order among ants or termites. Dictatorships have tried to impose collective purpose—or the purposes of the dictatorship—to make all wills conform to goals imposed from above, but humans always cheat tyranny, and stiffly ordered societies fail to fulfill the human potential.

We do not and perhaps cannot really understand this new kind of society in which individualism and community are precariously joined. The quality of social structures and the ability of the conglomerate to direct and govern itself will determine the future of the species and its civilization. This becomes more critical and more difficult as the powers of technology increase and cultural structures become ever more dominant over the biological basis of existence.

8

Dynamics of Evolution

The Genetic Medley

By standard genetic theory, as developed from the 1940s into the 1960s, genes are like beads strung along the chromosome, each about 1,000 to 10,000 bases long. Uncoiling DNA allows parts to be copied, making molecules of messenger RNA. These proceed to ribosomes in the cytoplasm, and transfer RNA brings up amino acids in the order dictated by the messenger RNA to make the encoded protein. The only real complication would be how genes are turned on and off to make a self-regulating entirety.

But the affair is far more complex. The strings of DNA are only the central part of a wonderfully buzzing world of the cell, with a ceaseless dance of movements and thousands of substances and structures. The average cell carries on about 500 metabolic processes involving about 10,000 proteins (Endler and McLellan 1989, 401). There are compartments and vessels, organelles for various purposes, and means of conveying substances from where they are made to where they are to be used, with filaments to guide them. The nucleus holds a tangle of threads that form the chromosomes. A smaller body, the nucleolus, with nucleic acid of its own, helps to make the ribosomes that make proteins according to instructions from the DNA. Self-reproducing mitochondria carry on metabolic processes. Other bodies finish proteins only partially manufactured by the ribosomes and digest food sucked into the cell. An outer membrane controls the passage of materials in both directions.

The work of the genes themselves is confusingly involved. The active parts of genes, exons, are split by perhaps dozens of inactive sequences, introns, that have to be excised when the gene goes into action. Rarely is a eucaryotic gene a single segment of DNA (Oettinger et al. 1990, 1522); the typical gene consists of various perhaps widely scattered pieces. On the other hand, two or more genes may code the same protein. DNA is subject to transposition, duplication, or deletion; nucleotides are constantly changing in its unexpressed stretches (Rothwell 1983, 356; Pollard 1988, 64–67). Most, perhaps all, genes have multiple effects; most proteins are made not by single genes but by multigene families. More or less homologous genes, linked and perhaps overlapping, in as many as a thousand copies, code for parts of a protein or for related proteins, and the family may be turned on or off by a single gene. Much DNA may do nothing more than survive in the intracellular competition (Orgel and Crick 1980, 604–607), but it adds to the potential for change.

The cell controls the genes that control the cell. A large number of protein repressors and activators not only switch genes on and off but relate genes to one another (Ptashne 1989, 40–47). At least 20 gene-processing enzymes mediate a dozen or so types of alterations (J. Campbell 1985, 135). Alongside "structural" genes that direct the production of proteins, there are integrator genes and sensor genes, with overlapping scope and functions, working with different RNAs, influencing all stages of the process of manufacture and use of proteins (Kauffman 1985, 173). Most important, regulator genes by means of enzymes control one or many near or distant structural or regulatory genes, such as those that switch patterns in butterfly wings.

Proteins binding to DNA and mediating the start and stop of transcription are blocked or facilitated by small side groups directed by specific enzymes (Holliday 1989, 60–64). Restriction sites, promoters, operators, operons, and enhancers play their part. Not only does DNA make RNA, but RNA, aided by an enzyme suitably called reverse transcriptase, makes DNA (Lewin 1983, 1052–1054).

Most genes are inactive; few are always active. The human has about 3 billion base pairs, 600 times as many as a typical bacterium, although bacteria produce most known enzymes (Gilbert 1985, 349). Much of the difference stems from the fact that bacteria have little inactive, or cryptic, DNA. As little as 1 percent of human DNA actually directs the manufacture of proteins, and some sequences are repeated hundreds, even millions of times. In eucaryotes, the size of the genome does not have much to do with the complexity of the organism. Some bony fish have 350 times as much nucleic acid as others; salamanders may have 10 times as much as mammals (John and Miklos 1988, 319; Gilbert 1985, 350). Certain protozoa have hundreds of times as many base pairs as humans, or as many as a trillion (Wöhrmann 1990, 15).

Mutation of protein-producing genes is incompletely understood. The error rate in replication of DNA is so high that the simplest organism could not reproduce reliably without corrective mechanisms. One set of enzymes monitors that the right amino acid is put in place; another set checks that the newly forming DNA corresponds to its template and cuts out and replaces defective sections; a third confirms the finished product. Repair is so effective that the error rate may be no more than one in 10 billion bases at each replication, although it is 100 million times as high without correction (Radman and Wagner 1988, 40–46). Each human gene has a likelihood of mutation of about 1 in 100,000 per generation (implying about one mutation in each new individual), and roughly this rate seems to prevail among animals. It is variable, however; a mutator gene in *Drosophila* raises the rate tenfold (Grant 1985, 48).

It is unclear why a few errors slip by. Many or most mutations are mediated by enzymes and are consequently limited to particular conformations, as they duplicate, switch, express, repress, excise, or otherwise alter genes (J. Campbell 1987, 287). Genes differ by hundreds of times in mutation rates. Mutation of some genes increases the frequency of mutation of others, and the rate is generally raised by malnutrition or other stress (Rothwell 1983, 396–399), a frequent condition of living things. Changes of structural genes are almost always indifferent to harmful; as few as one in a million may be advantageous (Loomis 1988, 103). A further possible complication is that a gene may be injected by a virus, with unforeseeable consequences. Such an origin is surmised for the homeobox-carrying genes discussed in chapter 9 (De Robertis, Oliver, and Wright 1990, 52).

Most genetic variation comes not from altering a particular base or two in the structural genes but from shifting elements and interrelations. The active parts of genes—exons—may be reshuffled. Genes may be repressed or inactive genes expressed. DNA is transposed, duplicated, or altered by neighboring DNA. Sequences are capable of promoting their own multiplication and dispersion, in what is called "molecular drive." There are several mechanisms of genetic turnover, including transposition, gene conversion, and unequal crossing over, which can gradually spread mutant sequences through a population. Molecular drive is due to such internal mechanisms, as opposed to the external factors of natural selection and neutral drift. All three processes may interact in the evolution of a given trait (Dover 1986, 160).

It has been suggested that the inactive genes are there because, in line with the "selfish gene" theory, it is the business of genes to make copies of themselves. The self-propagation of genes and consequent redundancy may be such as to hamper the functioning of DNA, but gene amplification, as it is called, can increase the output of a needed substance. Moreover, if a gene is duplicated, one copy can be turned to a new use while the other continues to perform the old function (Hedrick 1983, 501). Unused genes serve as a reserve for possible innovation; perhaps slightly changed in time, they are vastly more likely to prove functional than an arbitrary string of nucleotides.

Thousands of enzymes and structures are intermingled in the cell, none quite independent of others. The entirety, organized for a high degree of stability with possibilities of change, represents determinism coupled with unpredictability. The genome mixes fixity with flux, self-organization with chaotic turmoil, and different levels of stability. The interconnected elements are figuratively, perhaps actually, in motion, renewing themselves and dynamically interacting with feedback loops.

As S. J. Gould notes, "The genome is fluid and mobile, constantly changing in quality and quantity, and replete with hierarchical systems of regulation and control (S. Gould 1987b, 159). Its ways can be paradoxical, even in "simple" organisms. A strain of *Escherichia coli* was forced to become adapted to 42 degrees C; it proved more fit at 37 degrees C than the strain that had always been kept at that temperature. A yeast, *Saccharomyces cerevisiae,* was brought (under constant conditions) to a superior adaptation, A; then to another, B, more fit than A; and to C, more fit than B. But C turned out to be less fit than the starting strain. In algebraic terms, fitness is not transitive (Dykhuizen 1990).

The basis of life is a tangled complexity.

The Problem of Information

A living being represents order-creating order, capable of multiplying, and perhaps improving, itself. In terms of the ordinary behavior of matter, this is incomprehensible.

Higher animals should be impossible because approximately 100,000 genes are far too few to describe fully a body like the human. It would be ridiculous to suppose (if it were not true) that an assortment of proteins could engineer the intricacies of kidney, liver, and so forth. We have no idea what huge number of words would be needed to tell a biochemist just how to make 150 to 200 different kinds of cells and put all of them in their proper places in a body of some 75 trillion cells. Even something outwardly simple like the human skin holds at least 15 different organs, from various glands to a half-dozen kinds of sensors, and many genes are needed just to make a hair follicle. It is particularly puzzling how genes can give instructions for making the brain, with 100 billion or so intricately ordered neurons of many kinds.

The consummate accuracy and incredible detail with which heredity operates is demonstrated by the near-total similarity of persons with the same genes, identical twins. Even when raised apart, without knowledge of one another, they have nearly the same pulse, blood pressure, and brain waves; they have been found to prefer the same arts and school subjects and to have similar tastes (Holden 1980, 1323–1328).

This accuracy is possible only because the genome is not a blueprint describing everything about the body but a set of instructions, and instructions can be much briefer than a description. If one folds a square paper along its diagonal, refolds twice, cuts off the corners, and unfolds, what appears is a shape easily made but describable only by many words. Its complexity is due to the fact that a single cut has multiple effects, much as a single gene or protein has multiple effects in reacting with many other proteins. A better example is the Mandelbrot set mentioned in connection with fractals and chaos. The computer is given a simple program: to test whether an iterated function goes to zero or infinity at all points in a space. To do this requires an indefinite number of calculations, and the result is a fractal filigree of complexity limited only by the fineness of detail of which the screen is capable. This example is the more suggestive because of the many fractal structures in an advanced animal. It is not necessary for the genome to encode all the branches, only the instructions for branchings into fractal-like anatomical structures, bronchial, vascular, neural, and renal.

To follow a recipe, it is necessary to have a general understanding of materials and procedures—in the example, a square sheet, folding, and cutting. For the gene, this "understanding" must consist of channels of development that are possible on the basis of structures already present, using or slightly modifying substances available. It is as though the genome consisted of certain basic parts and plans of action, laid down and difficult or impossible to change, others less fixed, still others partly assembled, and a number of small pieces that may be used or not but can be fitted together only in definite ways.

Consequently, the organism can be much more detailed than the actual specifications of the genes. Twins may be so closely alike that the mother has trouble distinguishing them, but twins with identical heredity are not absolutely identical. The genes decree the making of ridges on the skin of fingers, not the precise location and direction of arches and whorls. The brain has to be largely self-organized; genes can give only directions for putting together information-processing modules. In the growth of the embryo, parts stimulate others in a cascade of differentiation that must be slightly different in every case.

A limited number of genes or proteins can be assorted in a very large number of different combinations, greatly increasing the potential for information. Most of these would not be viable, but the number of useful combinations can be immensely larger than the number of genes. The sequence of amino acids dictated by the gene or genes forms a protein that curls up into a complicated structure, capable of adhering to or affecting others. Each protein is like a special building block, with convolutions and regions of different reactivity, such that it can combine with certain others in particular ways. Complex aggregations would be produced simply by regulatory genes' disposing that certain blocks should be made to link with certain others. Fairly simple structures fitted together make very intricate ones, forming integrated functional entities with qualities indefinitely beyond those of the components, just as the qualities of compounds are indefinitely beyond those of the component elements.

The body also needs less detailed instructions than would seem to be implied by its complexity because not all variants are possible; only certain avenues are open (the thesis of Whyte 1965). Each trait is the work of a gene or gene combination, but for each genotype, or genetic configuration, there is only a finite universe of admissible modifications, and each choice precludes many other choices. The organism is an integrated unit with poorly understood tendencies to maintain its pattern or restore its integrity. Organs hold to their design; only pathological cancer cells flout the bounds.

Attractors in the Genome

The genome may be regarded as a set of integrated informational complexes, for which one may borrow a term from the study of dynamic systems and chaos theory: *attractors.* In mathematical terms, an attractor is in a multidimensional phase space; it is a set of permitted states of a system. The attractor is defined by the limits of mutations. It represents elements that cohere, alterations being usually unviable at some stage from cell division to the fitness of the phenotype—the less likely to be viable the earlier they affect morphogenesis. Deviations or rearrangements are allowed within usually narrow limits; the genomic attractors are tight, as indicated by the remarkable coherence of types.

This is equivalent to stating that the information transmitted by the genes is somehow packaged to form patterns, not a single grand blueprint but a set of many overlapping partial blueprints, just as the manufacture of an automobile is directed. In the making of the organism, the multiplying cells seem to be instructed as to their respective destinies, and they become permanently differentiated to compose organs. But the pattern or instructions for making the whole structure remain somehow latent. When a part is injured, some cells become undifferentiated to make new specialized tissues. Excise the lens of a newt's eye, and it makes a new lens. If the large claw of a snapper shrimp is removed, the smaller pincer claw is converted into a snapper, and a pincer grows in place of the snapper; the shrimp does not simply repair the member but restores the gestalt (Govind 1989, 468). If a planarian flatworm is cut in half, cells from the rear part proceed to rebuild the head, with brain, pharynx, and so forth (Pearse et al. 1987, 214–225). A small fraction of a ciliate protozoa, such as *Stentor,* can regrow into a whole new animal. If a hydra is cut into little pieces, each one can remake itself into a little hydra. Even much larger animals with more specialized tissues than the newt, not to speak of the hydra, have capacities of regeneration. A lizard can (imperfectly) regrow a discarded tail; the cut-off tip of a child's finger grows back.

There is thus a self-directed capacity for restoration of pattern, although the range of repair decreases with both maturity and complexity. A very early embryo, even of a human, if divided becomes twins, and a child can restore far more than an adult. A spider remakes amputated legs, but a human can restore skin only if it is not fully destroyed.

Most remarkable, an organ so complex as the human brain seems to have a sort of image of itself and to some extent can reassign functions to compensate for injuries. If innervation to an area is severed, especially but not only in the young, the idled area takes on new tasks, and greater use of a functional area causes it to enlarge (Gazzaniga 1988, 67). That the brain can reprogram itself to some extent is paradoxical; it seems there is pattern not only in terms of structures but of functions.

Faculties of repair can hardly be specifically acquired by mutation and selection. To account for them is like the basic problem of morphogenesis—how the instructions in the genes are converted into structures—but it is more. There must be a signal; the newt, or the newt's eye, somehow "knows" that it is missing a lens, and cells begin multiplying in the proper places and sequences to replace it. One might speculate that the organism retains something of the instructions that made possible its growth and differentiation in the first place, but to say this does not so much account for the faculty of repair as underline the mystery of morphogenesis.

The pattern is superior to the parts. It also seems somehow to stand above the specific organization of the genetic material. On one hand, a single new gene can put a species on a different track; on the other, many different genotypes can produce superficially identical phenotypes (Mayr 1963, 213, 220). The number of chromosomes and amount of DNA can differ by a hundredfold or more in rather similar forms (John and Miklos, 1988, 258). The organization of DNA into chromosomes similarly varies radically; the number of chromosomes within a single genus of deer (*Muntiacus*) differs from six to forty-six.

Most remarkable, species with this difference of number of chromosomes can be crossed—that is, their chromosomes can line up correctly—although the hybrid is not fertile (Miklos and John 1987, 265–267). Ants of the genus *Myrmecia* have from two to eighty-four chromosomes (Hölldobler and Wilson 1990, 195); the pattern must be superior to the division into chromosomes.

The essence of the genome is self-organizedness—elements that fit together and form informational units. Complexes of effective genes form coherent wholes, which vary within usually stable patterns. As Ernst Mayr put it, "Most genes are tied together in balanced complexes that resist change (Mayr 1984, 71). Single genes can produce proteins (under intricate controls), but the formation of structures—morphogenesis—requires a set or sets of coadapted structural and regulatory genes, all interacting and all subject to selection at different levels.

Differently stated, the genome is a system of linked attractors at all levels of genetic stability, from the basics of cell physiology to the body plan of the phylum to the characteristics of the species. There are weaker attractors within stronger ones—some more or less autonomous, others overlapping, in a fractal-like descending sequence. Any self-reinforcing pattern, or a self-ordering part within the self-ordering whole of the organism, such as the brain, a limb, or the liver, may be considered an attractor. Any taxonomic group, as far as it represents a set of organisms that belong together, represents an attractor at its level.

There is no known reason that organisms are classifiable into fairly coherent taxa, species, genera, families, and so forth other than that genomes are organized, attractor fashion, at each level. As far as traits cohere to form an attractor at any level, they may be regarded as units of selection. If the "selfish gene" may be deemed to seek self-multiplication because this lies in the nature of life, this must be no less true of complexes of genes that have a teamlike existence at whatever level one wishes to consider.

The attractor is the essence of self-organization. Just what constitutes it and how the organism shifts from one attractor to another is a task for genetics and microbiology to elucidate. The problem is difficult because regulatory factors cannot be identified by product proteins, and configurations are the outcome of unknowable history. Change of attractors might, however, be envisaged as rearrangement, transposition, or breaking of directive elements of heredity, in contrast to the substitutions of bases in DNA that constitute mutations of structural genes. Alterations of interrelationships are to be expected because elements of the cell are mobile, especially in mitosis. An attractor would consist of a relatively stable configuration, or perhaps a configuration that could sustain itself by reproductive success. Any viable rearrangement would place regulatory genes in new relationships to other regulator genes or factors and to structural genes, thereby advancing or retarding their effects.

If the interpretation of the genome as a set of attractors is difficult to apply concretely, it is nonetheless realistic. In a sense, the genome is an abstraction. It is not to be seen simply as an agglomeration of proteins, nucleic acids, and sundry structures but as their resultant. It is not so much a team of players as their game plan, which may remain the same although players are rotated in or out. The genome is a plan (or a combination of many plans) for building an organism; it is a pattern or a program.

The genome-attractor has something in common with the conventional metaphor of adaptive peaks from which species can depart only by descending into valleys of inferior adaptation. The peaks rise in a "fitness" dimension, although the meaning of "fitness" in this connection is vague. Most logically, and as originally used by Sewall Wright, it would refer to the ability of a genetically differentiated population to compete with other populations of the same species in the same environment. But more interesting is the ability of a population to move into a new niche or way of life, perhaps in competition with other species. In this situation, the new fitness, or the height of the new peak, is unrelated to the old peak. In other words, the landscape is changed by genetic change. To reach a new stability may be treated as a shift to a new attractor.

Less stable attractors, permitting variations like those of butterfly wings or shapes of oak leaves, may be regarded as the equivalent of species, with variability within genetic homeostasis. Taxa of higher levels may be considered attractors of higher degrees of stability, as though surrounded by walls of differing heights. Some, like those bounding classes and orders, seem impossible to surmount for many million years, in contrast to low and perhaps ambiguous boundaries between species. The genotype is integrated at many levels and cannot be reshaped ad lib like putty.

Stasis means the stability of attractors. This is close to what is often called canalization—the tendency of a species to evolve in a definite way despite genetic and environmental differences. The power of the attractor makes clearer why species may differ substantially in their genetics yet have similar structures (as in the case of many sibling species) or differ little in their genetics yet have very different structures (as do humans and chimpanzees): the higher attractor prevails. Similarly, a reordering may be envisaged as breaking away from one attractor (somewhat like an adaptive peak, in the conventional metaphor) to reach another. This is equivalent to the contention of the punctuationalists that evolutionary change results not from the gradual transformation of characters but from the emergence of new configurations, or new taxa (S. Gould 1980a, 126).

Attractors in the genomic complex may be of as many different levels as there are different levels in the structure and functioning of the organism governed by a hierarchy of regulators in the genome. The frequency or possibility of transitions must vary in accordance with the degree to which the attractor is embedded in the genome, and the possibility of a variant set of genes fissioning, so to speak, in such a way as to make a viable new integration. The deeper the change is, the more genes have to move together and the less probable the transition becomes. It is hence understandable that major innovations have been rare ever since life, early in the Cambrian, explored and learned to exploit the possibilities of multicellular organization. Innovations become fewer and more superficial as the fundamentals of structure are overlaid by new developments. To generate a new species requires not merely new genes but a new genetic cohesion (Mayr 1984, 79).

For example, the six-leggedness of insects is obviously strongly fixed, since it is constant for millions of species, but that attractor permits infinite variation within less inclusive attractors, such as that defining beetles (chiefly hard wing covers) or flies (two wings and complete metamorphosis). The four-leggedness of tetrapods is less firmly set; it is constant only in the fetus. Some amphibians (caecilians) have lost all limbs, as have members of five different families of lizards as well as snakes (Estes et al. 1988, 248). Of mammals, cetaceans and sirenians (but no burrowing mammals) have lost legs. The basic architecture of tetrapod limbs must also be considered embedded in a stable configuration, in terms of dynamic systematics, an attractor.

Taxonomic classes, insofar as they are real groups, can be considered attractors. For example, birds are characterized by a large number of traits—not only flight and feathers but endothermy, oxidative metabolism, horny beak, absence of teeth, very short tail, four-chambered heart, and so forth. These characteristics cannot all have been developed at once, but if traits varied independently, either randomly or adaptively, there would be a great deal of scattering. To the contrary, the major traits of birdness came together into the bird pattern, or the attractor, which imposed itself.

Small mutations, replacements by stochastic process of amino acids in proteins, occur continually at a slow rate, but it seems that breaking away from an attractor sets species on the course to new structures and ways of life, bringing irreversible changes. This implies that the most important competition is not among individuals and their lineages, as in the Darwinian view, but between new forms and old. The old must nearly always win, but the few newcomers that score an upset victory carry away the prize of the future.

A shift away from a stable attractor is, of course, highly improbable. It implies a small population because the change is at first doubtfully adaptive. In a large majority of cases, an incipient shift will fail. Even if it succeeds, at length reaching a new and better integration, it will leave scant record in the strata. Such metamorphoses imply rapidity of evolution because the transitional form is mutable. Major innovations may be thought of as transitions from a relatively integrated and stable pattern or attractor to another. This implies that during the transition, the genome is probably not well adapted; in other words, change is rapid and the population is small, perhaps restricted to an unusually hospitable environment. The more profound the shift is, the more rapid is the pace of change to be expected and the smaller are the initial numbers. In some cases, a big change takes much less time than a modest one. *Hyracotherium* was over 50 million years on the way from terrier size to horse size, developing teeth suitable for grazing and reducing its toes to a single hoof—no overwhelming transformation. Apparently less than 10 million years were required for a land animal to become a whale and a ground-living animal to become a full-fledged bat.

The newly inaugurated scheme, however, is not necessarily full fledged. A change need only set a trend (S. Gould 1980a, 127), and filling out the pattern may be many millions of years away. For example, *Archaeopteryx* had crossed the divide between land-bound and air-borne animals, but the complete bird pattern was still in the future. Perhaps *Archaeopteryx* was really a reptile with feathers, and the bird attractor took shape in a way concerning which we know nothing.

Convergent or parallel evolution is understandable in terms of attractors: different forms may share similar attractors and hence are able to develop in similar directions. A striking example is the barbed stinger, the use of which costs the life of its possessor; this odd trait appears in bees, wasps, and ants, somehow emerging from a genetic configuration shared by different taxa.

Such are the potentialities of the attractor. The idea is powerful for a general grasp of many phenomena of evolution.

Variation

What life has discovered is bewildering in its variety, complexity, and potency. Its fantasies are much more than can be attributed simply to organisms' fitting themselves, by a process of random variation, to the needs and opportunities of their world. There is not necessarily any adaptive meaning for patterns of butterfly wings, shapes of diatoms, songbird plumage, or the incredible variety of insect genitalia. The bizarrely varied flanges and projections of the many bats' noses, and also their faces and ears, cannot correspond to different needs of the many species making a living in about the same way. Twenty varieties of African antelopes have 20 different shapes of horns. One-celled radiolarians (marine protozoa) have hundreds of fantastic shapes, with all manner of facets and projections, doing nothing more complicated than catching microscopic prey adrift in the ocean (O. Anderson 1983, 266). It is difficult to suggest any adaptive functions for the multiform strange crests of duckbill dinosaurs (hadrosaurs), through which their breathing passage made a long detour (Fenton and Fenton 1989, 468). The hadrosaurs could not have been much

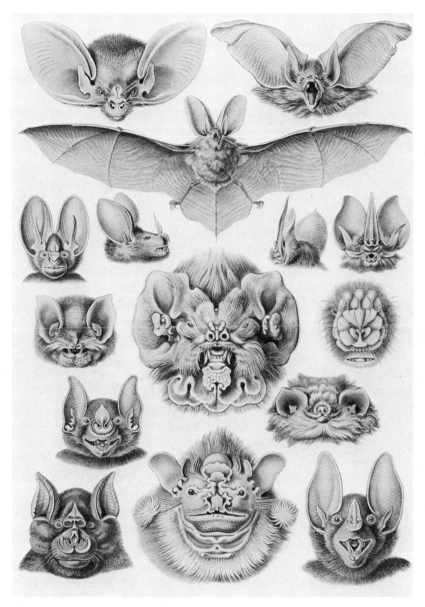

Bats are mostly much alike—except for their beautiful faces. (Illustration by Ernst Haeckel, *Art Forms in Nature*, © 1974 by Dover Publications, Inc. Originally published by the Verlag des Bibliographischen Instituts, Leipzig and Vienna, 1904.)

These are only a small sample of the fantastic variety of radiolarians. (Illustration by Ernst Haeckel, *Art Forms in Nature*, © 1974 by Dover Publications, Inc. Originally published by the Verlag des Bibliographischen Instituts, Leipzig and Vienna, 1904.)

handicapped, however, because they were by far the most numerous family of dinosaurs as the reign of that order neared its end (Strickberger 1990, 353).

Independence of environmental determinants is also evident in many morphs, or variants within a species. Especially among fish, amphibians, snails, and insects, less commonly among birds and other animals, individuals of different aspects share the same environment and interbreed; sometimes the differences seem to be related to camouflage and sometimes not. Countless details of living creatures, from the swirls of human fingerprints to the patterns of plumage of male ducks and social structures of ants, do not necessarily have any specific meaning or adaptive function.

Must there be any simple or adaptive reason that male lions have a mane, some birds have topknots, some squirrels have tufts on their ears, leopards have spots instead of the jaguar's mottling, robins have red vests, and the like? Or why some birds display tail feathers for mates and others dance? One ornithologist finds certain birds monogamous because they eat more insects and less fruit than polygynous cousins; another finds the reverse of another family of birds (Beehler 1989, 120–122). There may be some such concrete cause—or there may be none.

The life of plants is simpler than that of animals; with no complications of mobility, they have only to utilize solar radiation, water, and minerals drawn from the earth to build tissues and to scatter their seed. But their genomes have been shaping and reshaping themselves for as long as those of animals, and plants quite as much as animals have evolved tremendous diversity beyond apparent material needs. There is no reason to suppose that there exists any particular utility for a myriad of very different shapes and configurations of leaves, or tree bark, or the shape of stems fulfilling the same functions in the same environment.

This is something like a law of nature: within certain commonalities of pattern, there is a turbulent freedom of diversity. Snowflakes are hexagonal, and many of them tend to resemble one another approximately, but it is nearly impossible to find two that are identical. Much as ocean waves are similar but different, thousands of varieties of orchids fulfill the same broad pattern with designs that, if humanly made, would be called imaginative. Species, too, sometimes proliferate with abandon, the best example being the Hawaiian *Drosophila*. Within species also, there is a great deal of genetic diversity, contrary to the notion of population genetics that less favorable genes will be eliminated (Trewavar 1988, 3).

Not everything, of course, is variable; it seems that parts of the genome are relaxed while others are held tight. Orchids are fantastically varied in their flowers, but leaves, stems, and seed capsules are much alike. It is only necessary, however, that any particular characteristic be sufficiently functional to permit the species to survive. If there is an optimal shape of leaf for certain conditions of light and humidity, or of webs for snaring flies, and so forth, most species are far from it. If one best way could impose itself generally, there would be far fewer species. The existing number, guessed at between 10 and 50 million (May 1988, 1447), is much higher than can be accounted for by simple principles of natural selection. Although evolution creates new niches by coadaptation, the number of niches different enough to require adaptive specialization must be much less than the number of species of plants and animals (excluding parasites). In a single marine locale, one finds hundreds of species of creatures of totally diverse types—worms, tunicates, brachiopods, bivalves, sponges, corals, and others—in endless variety, all making a living by filtering the water.

It is not necessary to burden evolutionary theory with accounting for a host of differences. It would be extremely satisfying to find or surmise an adaptive reason for everything, but there may be no particular meaning in a million details of living nature, just as there is no significance in the details (within certain patterns) of snowflakes or the ocean breakers or the folds of mountain ranges.

If details are often, perhaps usually, insignificant, stochastic or chaotic, this may be true also of the inception of major differences. Most little variations lead nowhere; a very few are germs of greatness and point to new futures. Whatever behavior or structure started a dinosaur on the way to the birds was probably quite trivial—perhaps a detail of courtship or a flaring out of some scales. Innovations doubtless start—if they really have a starting point—with some improbable combination of internal directions and external circumstances.

To learn exactly how a species or any taxon came into being is as impossible as reconstructing the past of a breaking wave from knowledge of its present shape. Conceivably the behavior of a microbe in the pre-Cambrian, with effects escalating over the eons, may have led to multicellularity, the chordate phylum, and ultimately the human line. The genesis of life and its long course must have involved many highly unlikely and unforeseeable events, great and small, as dynamic systems produce things beyond imagination.

The Role of Chaos

If the imagination is overwhelmed by the variegated panorama of life, it is helpful to think in terms of chaos theory, an approach that has been found broadly applicable to living processes. Chaos appears in neural responses, cell physiology, and ion exchanges (Goldberger and West 1987, 7). Brain waves are chaotic in a controlled fashion, and their chaos may be essential to thinking. The healthy human heart is subject to chaotic variability; loss of chaos presages a heart attack (Goldberger, Rigney, and West 1990, 43–49). Chaotic dynamics possibly played a role in sequence diversity in prebiological polymers and so in the origins of life (Conrad 1987, 9, 11). Chaos has been found to be applicable to coding sequences in DNA (Ohno 1988, 4378–4385). The unpredictable fluttering of a butterfly, making it difficult to catch, may be based in a chaotic neural mechanism (Conrad 1987, 10).

Feedback is the heart of chaos. Waves in the ocean become chaotic because the wind, roughening the surface, increases resistance; the orbit of Pluto is chaotic because a deviation in approaching another planet makes the next deviation larger. Feedback being everywhere in living nature, living communities must be highly chaotic, the more so since reproduction is synonymous with self-compounding (Barryman and Millstein 1989, 26). In living processes, positive feedback multiplies molecular instructions until checked by negative feedback.

Within the self-regulation of organisms and ecosystems, dynamic interactions must lead to variation of genetic structures and population sizes. Even a small degree of instability in self-reproducing organisms' mixing their patterns must lead, as structures are piled up and multiplied, to ever-increasing complexity up to limits of the system.

According to Ford, "Evolution is chaos with feedback" (cited by Gleick 1988, 314). Being the long-term result of changes of the genome and population fluctuations, neither random nor predictable, evolution must be subject to chaotic development. In the interrelations of many organisms with variable genetic endowment and under pressures of competition and the environment, small, even unrecognizable differences can show indefinitely expanding effects. Living things have had billions of years to build up endlessly elaborate interlocking arrangements, in which positive and negative feedbacks play a dominant part, and small changes can cascade into big alterations (S. Gould 1983, 196). Changes are not additive: A, then B is not equivalent to B, then A. Chaos in the genome generates constrained randomness (Crutchfield, Farmer, and Packard 1986, 53).

Consequently, much of the infinite complexity of life—by which we understand the number and differentiation of elements of the system and their interactions—must be internally and chaotically generated. It has long been accepted that genetic drift contributes to evolutionary change and speciation by random neutralist or nearly neutralist changes of frequencies of genes (or alleles) in isolated populations. But chaos means not only stochastic mutation of structural genes but reorganization of elements of the genotype.

Growth of complexity accords with the principle of chaos: a dynamic system moves away from its initial conditions and becomes more difficult to describe. The wave pushed by the wind rises and becomes more turbulent up to the plateau where the energy lost to friction equals the energy supplied by the wind. In the interplay of structures, environment, and population, new combinations and new structures are always being formed, and they make possible ever more combinations.

Evolution is exploration of a vast unknowable space of possibilities, living forms advancing in one direction or another, like the pseudopods of an amoeba, usually stopped short, sometimes finding fertile expanses in which to proliferate. The directions it takes are entirely unforeseeable but are the outcome of previous conditions. Evolution is infused with deterministic chaos, which converts stochastic molecular events into evolutionary change. Under natural selection, this implies something that may be reasonably called progress.

Radical innovation becomes more understandable in terms of the ability of chaos to produce not only uncertainty—like the wavering of the stock market—but the unexpectable—like a crash, a catastrophe both mathematically and economically. Weather conditions that cause thunderstorms may give rise to tornadoes. A star, because of chaotic acceleration, can sometimes escape from the cluster that normally binds it. It is not necessary that there be any external cause for a readjustment of the system.

Feedback

The Ecosystem

Organisms evolve as part of a community, that is, an ecosystem, and evolutionists have recently given increasing attention to the entirety, which necessarily evolves together. One might better speak not of the origins of species but of the development of ecosystems. There are competing and cooperating populations, species, and ways of living, with individuals and groups on all levels striving for space, water, light, and nourishment, never finding equilibrium in the ceaseless turbulence of interaction. There is feedback between organisms and environment; some call the entirety of earth and biosphere "Gaia" and treat it practically as an organism (Lovelock 1988).

Chaos theory "could undermine much of the conceptual framework of contemporary ecology" (Schaffer and Kot 1986, 63). It has been applied most frequently to population studies, from the multiplication of amoebae to the rise and fall of numbers of lynx and hares. The simplest equations for population growth, with natality and mortality dependent on changing conditions, lead to chaotic cycling (May 1976, 459–467; May 1974, 645–647).

It is artificial, however, to pick out single aspects; the entire system is chaotic in principle. Adaptation usually depends less on fitness for physical conditions than on interactions with other species. The benefit of interaction may be one-sided; predators are not ordinarily considered to help their prey or parasites their hosts, but on balance it seems that most relations in a stable ecosystem are either innocuous or mutually advantageous. The community as a whole prospers because there is more room for cooperation than competition. Contrary to intuitive expectation, moderate predation usually increases the variety of prey species by reducing competition among them and checking any that are especially numerous.

Interactions are complicated by coevolution. For example, plants eaten by caterpillars develop toxins; the caterpillars evolve means of detoxification; the plants counter in various ways, such as covering the leaves with fuzz. The passionflower (*Passiflora*) makes poisonous hooks to catch the caterpillar's pseudopods; the insect responds by spinning a silk net on which to walk as it chews the margins of the leaf (Colinvaux 1986, 632).

Ecological studies concentrate on population numbers and densities, the apparent variability of which usually increases over time. Animal species (although not plants, which are less interactive) show periodic cycles and apparent chaos (Howlett 1990, 104). The evolution of the ecosystem is nothing more than changes of population numbers, which define the evolution of species.

Specialization

Cooperation is almost synonymous with specialization, which is the structure of the ecosystem. Species play their parts in a manner not wholly unlike the organs of a body. Specialization has had much to do with the dynamics of evolution; most organisms succeed by finding a role, or niche, in the web of life and fitting themselves for it better than their rivals. Where there are few species, they make use of a broad variety of resources; where they are many, they specialize more closely, as in the species-rich rain forest. Sibling species tend to find separate niches. Some warbler species go so far as to specialize in the part of the spruce tree on which they feed (Strickberger 1990, 496).

Specialization is of numerous kinds. Some species prosper by living where few others can, as in arctic or desert climates. Some take to particular habitats, from the branches of trees to underground burrows. Many mosquitoes specialize in sucking blood at certain hours. But most specializations of animals are nutritional, claiming the resource for which there is most competition: energy or food.

Specialization is self-reinforcing by feedback. It is a trap; it is easy to get into, but the more successful it is, the harder to escape. The panda found what must have been a more secure existence by eating bamboo, but it thereby lost the bear's ability to hunt, and the habits of the vegetarian make it impossible to recover the capabilities of the hunter. The koala has become so fixed on its diet that it will accept only a very few of the hundreds of varieties of eucalyptus and cannot live without them (Lee and Martin 1990, 37). The more defenses that prey species invent, the stronger is the pressure for those that eat them to concentrate on one or a few related species.

A specialization may entail either the acquisition of new structures appropriate for new ways, such as the long beak of the hummingbird, or the loss of structures no longer needed, like the eyes of the mole. The loss comes easier; new organs are difficult to achieve, but unused ones may fall away with unaccountable rapidity.

It may be because deviation of a population from a successful specialization nearly always fails that successful animals seldom give rise to new forms; those that dominate an era are replaced by a quite different group. The ruling amphibians did not engender the reptiles; the mammallike reptiles did not give birth to the dinosaurs, which succeeded them as lords of earth.

The evolution of our species was possible because in arboreal life primates were able to use a variety of food resources. But the primates entered, very slowly at first, into a peculiar specialization that amounted to a generalization—superior faculties of cognition and behavioral flexibility. Hominids carried this unique adaptation forward with great rapidity and success, thanks to feedback. Delayed maturity made possible the development of culture, which made delayed maturity possible and essential for the further growth of culture. The enlargement of the brain and culture made a more complex existence in which the brain became more useful and culture accelerated cultural growth. Humanity is now engaged in an especially far-reaching specialization of chaotic character, with innumerable feedbacks, carrying the species forward at an unprecedented rate on an irreversible course toward an unknowable destiny.

Symbiosis

Nature has been branded "red in tooth and claw," ruled by a merciless struggle for survival and procreation, wherein a few succeed and many fail. Yet "Nature green in root and flower" may also be valid (Boucher 1986, 22).

The most cooperative form of interaction is symbiosis, the mutual dependence of two probably unrelated species. Symbioses are everywhere. One may be partly responsible for the advance from bacterium to protist; it is believed, as Lynn Mar-

gulis proposes, that the eucaryotic cell became possible because
an overgrown bacterium, having gathered genetic material into
a central nucleus, reached a modus vivendi with smaller bacteria that it swallowed or that invaded it. Bacteria are frequent
symbionts inside eucaryotic cells. Organelles, mitochondria,
plastids or chloroplasts, and perhaps other parts such as micro-
tubules and cilia may have come about thus; they have something of a life of their own, reproducing independently and
making their own proteins (Margulis and Sagan 1986).

Countless animals and plants of different species are associated in all manner of relationships, which are called (although
the categories are not distinct) symbiosis if intimate and beneficial to both sides, mutualism if less intimate but reciprocally
beneficial, commensalism if beneficial to one and not harmful
to the other, and parasitism if beneficial for one at a cost to
the other. Termites depend on protozoa in their gut, and the
protozoa on termites. Farmer-ants rely on species of fungus
not known to exist independently. Some worms and fish make
themselves luminous by incorporating light-emitting bacteria
in special organs. Ratels (African honey-badgers) count on
honey guides to lead them to hives; honey guides count on
ratels (and humans) to open the hives.

An important cooperative arrangement is the little-noticed
growth of fungi in or on the roots of woody plants, a large
majority of which rely on these partners to increase their capacity to absorb moisture and minerals from the soil. Among the
commonest symbioses are the combinations of fungi and algae
in lichens. Fungi of a half-dozen different families have hit on
the useful arrangement, joining with more than 20 different
genera of algae of several widely separated classes. Lichens
cover trees in most tropical forests, but they usually are most
abundant in very harsh environments, such as bare rocks and
arctic tundra. The fungus and alga can be made to join only
when nutrients are scarce (Smith and Douglas 1987, 142).

In a symbiosis, species coevolve, developing in tandem. The
acacia provides stinging ants food and shelter, and the ants
drive away other insects and browsing animals and even
remove competing vegetation. The habit of rewarding ants for
protection must be easily reached because many plants exhibit
it in some variation. Even more common is the less conspicuous
rewarding of ants for carrying away seeds. These are provided
with a nutritious fatty lobe; ants take the seed home, eat the

edible part, and leave the hard seed to sprout in or near the ants' nest, where conditions are usually favorable for its growth. No fewer than 3,000 species of plants have this habit, evolved independently many times in at least 60 families. The relationship is loose, however; no ants depend on the seeds of particular plants, and no plants rely on a particular species of ant (Handel and Beattie 1990).

The lack of narrow specialization for seed dispersal contrasts with frequent close mutual dependence for pollination. A well-known example is that of the yucca (*Yucca*) and the yucca moth (*Tegeticula*). The yucca relies entirely on the moth, which not merely pollinates by brushing against the organs of different flowers while gathering nectar but by active transfer. Without even asking for nectar, the moth collects pollen from stamens, makes it into a ball with a special proboscis, flies to a different plant (not to another flower), and rubs it on the stigma. After laying a few eggs on the yucca ovary, she gathers another pollen ball and is off to another plant to repeat the process. The moth larvae eat only a few of the many developing seeds in payment for the service rendered (Perry 1983, 80).

The cooperation of tiny wasps with figs is equally complicated. The fig tree lines a receptacle that becomes the edible fruit with male flowers and two kinds of female flowers. A wasp of the species that serves the particular kind of fig collects pollen in sacs designed for the purpose, goes to a new fig, shakes the pollen over the stigmas of the kind of female flower in which it cannot lay, and then oviposits in the female flowers made for it. There may be still another kind of female flower to attract another species of wasp. In addition, each kind of wasp has its own parasitic wasp, and a cuckoo wasp may lay in the fig without contributing to pollination (Begon, Harper, and Townsend 1986, 470).

How symbioses have come about can only be guessed, but many of them must have arisen from a simple exploitation. A fungus attacking tree roots would gain by keeping them alive; if it supplied nutrients, the tree would find it advantageous to encourage the fungus. An organism adjusting to a parasite may come to depend on it. An amoeba (*Amoeba proteus*) after a few years could not live without a bacterium that originally sickened it (Smith and Douglas 1987, 210). The many origins of lichens, numerous kinds of fungus sheltering numerous kinds of algae, suggest that the symbiosis could originate simply if a fungus living on an algal colony protected it instead of consuming all of it and took only part of the carbohydrates it produced.

Many symbioses or mutualisms, however, involve unlikely instincts and peculiar structures. For example, some small fish, especially wrasse (*Labroides dimidiatus*), pick parasites off larger fish, even inside their mouths. Fish of many species go to established stations for this purpose and wait to be cleaned. It would be suicidal for the wrasse to approach the big fish if this did not check its appetite, and the big fish would have no reason to refrain from eating the wrasse unless it expected to be cleaned. Humans could establish such an arrangement only by negotiation and on the basis of trust. The situation is complicated by another small fish, a blenny (*Aspidontus taeniatus*), that mimics the wrasse closely. It is allowed to approach the big fish as though its intentions were good; but it uses this confidence to take a bite out of the big fish's tail fin. The way of life of the blenny is inherently marginal; if it multiplied very much, tolerance for the cleaner would come to an end. As is, older fish learn to distinguish between cleaner and mimic (Wickler 1968, 157–164).

Sometimes creatures cooperate in making a living. When an animal dies, there is a race to get at the feast of flesh, and the first to arrive and lay its eggs has a big head start. Carrion-loving flies frequently do so, and their larvae get the better of those of a carrion beetle (*Necrophorus*). But the beetle, arriving a little late, carries with it a few dozen mites. The mites disembark, hunt out fly eggs, and consume them, allowing the beetle larvae to flourish. When the beetle is ready to look for another corpse, some mites clamber aboard; others stay to accompany the beetle larvae when they mature (Keddy 1989, 13).

If organisms depend on one another, they form a unit of selection, whether totally integrated, like eucaryotic cells, or partially integrated, like siphonophores, or related individuals, like members of an ant colony, or interdependent species, like moth and yucca. The evolution of the symbiotic unit is more complex than that of an individual, and its ability to make difficult adaptations is enlarged.

Close cooperation of organisms has been central to the epochal advances in the history of life. It relieved the nucleus of the eucaryotic cell of housekeeping functions, which are carried on by organelles. The multicellular organism is a grand symbiosis of specialized cells, each kind expressing a different part of the genetic instructions of the entirety. Social insects are symbionts in fundamentally the same way, as the different castes take on allotted specialities. The evolutionary revolution represented by human culture must have taken its origin from the improved ability of an ape to cooperate for mutual benefit.

Symbiotic or cooperative relations have thus played a very large, and largely neglected, role in evolution; specialized abilities become more effective as they coevolve and cooperate. No one can say how much of evolution is competition or cooperation; they are inseparable.

Philosophers have liked to see the cooperativeness of nature as an example for humans. Peter Kropotkin, in his most important work, *Mutual Aid,* held that cooperation rather than the conflict implied by survival of the fittest is the key to evolution, and he tried to apply this principle to human affairs. Although his effort to redefine revolutionary theory was not successful, focusing on the fact that living creatures depend on one another gives a different perspective from the commoner emphasis on competition.

Parasitism

Parasitism easily becomes symbiosis because it is against the interests of the parasite to harm its host; a majority of parasites have little deleterious effect (Cheng 1970, 222). Bacterial infections, for example, tend to lose virulence. They may be helpful, like the intestinal flora that aids many animals, especially ruminants, to utilize vegetable material.

Parasitism may reach a happy ending in symbiosis, but it is a one-way road. Slight dependence, if successful, leads to loss of the capacity for independent existence. An extreme example is *Sacculina,* a crustacean related to barnacles that became a parasite on crabs, presumably because a barnacle that took to fastening itself to crabs found it easier to extract nutrition from the crab than to filter it from the water. A few cells of the free-swimming *Sacculina* larva penetrate the crab, lodge in its stomach, and send out rootlike projections. It comes to look more like a branching tumor than an arthropod. It is still capable of fairly complicated behavior, however. When the crab molts, the *Sacculina* protrudes its reproductive organs. It can do this, however, only if the crab is a female; so if the host is male, the

Sacculina destroys its testes, causing it to take on female shape (Schram 1986, 473–475).

Many more plants than is commonly realized depend on other plants, varying from countless epiphytes that use trees as platforms to those that draw fluids from the tree but photosynthesize for themselves (as mistletoe) and many that have become total parasites on roots. *Rafflesia* lives almost entirely inside the host like *Sacculina*. In compensation, it produces the world's largest flower, a thick mass a yard across.

Parasitism leads not only to what may be called degeneracy but also to bizarre ways, as in the seemingly unnecessarily involved cycles of worms, briefly noted in chapter 4. Going through several hosts, worms undertake metamorphic gymnastics, changing shape and means of reproduction. This is puzzling. If a helminth lives in a mammal's guts, the logical procedure would be to design its reproduction to get its offspring directly into other animals of the same species (as many parasites do), not to make a circuitous trip through other hosts, such as snails and ants. On the other hand, if the worm can enter snails and multiply in them, it would seem well advised to use its situation in a snail habitat to spread to other snails. The journey through plural hosts is contrary to the principle of specialization on a single host. It must be difficult genetically to develop something like plural genomes to be switched on in sequence, each adapted to quite different needs of penetration and existence inside very different animals. Yet it works. Schistosomiasis, or liver flukes, for which snails are the secondary host, is probably the most widespread of human infections, debilitating hundreds of millions of persons.

An especially interesting and complex parasitism is that of social hymenoptera. Some vespine wasps (yellow jackets and relatives) show all stages of parasitism, from facultative colony formation by a queen's entering a colony of her or a related species, to obligatory temporary parasitism, in which the usurper kills the host queen and produces her own workers, to full parasitism, in which the host queen is allowed to live and the parasite produces only sexual forms (Matsuura and Yamane 1984, 196–203). There are similarly social parasites among bees, especially bumblebees. For the most part, their parasitism is merely the practice of late emerging females taking over already established nests. Some (of the genus *Psithyrus*) have become wholly incapable of founding colonies; they have developed stronger stings, jaws, and protective armor than their hosts, and have lost the worker caste (Alford 1975, 93). In ants, such behavior must have begun scores of times. Over a hundred species of socially parasitic ants are known, and many genera are exclusively parasitic.

Two practices seem to have contributed to this dependence in ants. Colonies raid others of their own species; pupae brought back may be permitted to hatch instead of being eaten. And fertilized females after the nuptial flight sometimes return to their own colony or enter another colony of their species or a kindred species, where they take the place of the host queen. The two practices are often associated, and either may advance by feedback. A slight advantage either in raiding (because of greater aggressivity) or in colony formation by temporary parasitism (perhaps producing smaller and more ingratiating queens) could grow into a specialization.

In the first stage, plundering neighbors is irregular, or a species is only a facultative parasite, able to start colonies independently. Next, the parasitic queen, becoming adapted to forcing her way into a colony, ceases to be able to found a colony on her own. Many species have reached only this level of temporary parasitism.

In a third stage, the parasitic species raids colonies of the same kind that it originally parasitized, carrying away brood to hatch as slaves. This would seem optimal; although quite capable of carrying on their own affairs, the raiders add to their labor force. But the evolutionary sequence is for the raiders to become more specialized. Typically (as in *Polyergus* and *Strongylognathus*) they develop murderous mandibles. By corollary, the slave makers cease to forage on their own or to do household work. They may lose the ability to feed themselves, although some species (such as *Leptothorax duloticus*) can recover some practical abilities if deprived of slaves.

In a fourth stage, the parasite ceases to conduct slave raids, perhaps because it becomes better specialized for insinuating itself into host colonies. Instead the host queen is allowed to live and produce slave workers but is prevented from producing winged brood (reproductives) of her own. The parasitic species reduces its worker (fighter) caste, relying on its reproductives' entering host colonies. Finally, the useless nonreproductive caste disappears entirely, and only sexual forms of the parasite are produced. For example, one parasitic species (*Teleutomyrmex*) lives on the back of the host queen, relying on food regurgitated by host workers. Decadence becomes degeneracy; males lose wings and mate only with sisters in the nest. Such a sequence seems to have occurred, with variations, many times in rather different groups of ants (Starr 1981, 379–384; Hölldobler and Wilson 1990, 436–469).

These stages follow logically, but there is no apparent reason for loss of faculties. Slave makers could forage and care for brood when not engaged in raids, which occupy very little of their time. In one genus (*Kyidris*) the parasites make a feeble pretense of doing something useful, such as tending brood. But slaveholders, not having to work, seem to prefer their leisure and lose the habits, ultimately the capacity of useful labor. They slide into total dependency. Next, they renounce the mission of capturing slaves.

Each stage of adaptation to social parasitism is less common, that is, less successful, than the preceding one. Species that found colonies parasitically are less abundant than independent species in the same genera. Facultative slave makers are less common than temporary parasites, and obligatory slave makers are still less common, that is, less successful. Almost all full parasites are rare; many of them have been found only once (Hölldobler and Wilson 1990, 446).

It can hardly be that each step on the way was brought by reproductive success; there seems to have been a counteradaptive drive. Possibly the progression is related to dual feedback, as specialization promotes further specialization and a gain for the parasite represents a loss for the host species and a limitation on the success of the parasite.

There are many more (although rare) species of full parasites than either temporary parasites or slave makers. Hölldobler and Wilson list 13 temporary parasites (there are more), 12 slaveholders, and 68 that have nearly or entirely lost the worker caste (Hölldobler and Wilson 1990, 438–445). In many genera only full parasites are known. It seems that, once started on the course, a species fairly rapidly comes to the final stage and then continues to exist in small numbers that do not seriously affect the host species. A form must persist five or six times longer as a "degenerate" full parasite than as either a temporary or fighting parasite. It cannot evolve further.

Parasitism is thus usually, if not always, an irreversible specialization, as independent abilities are lost. For a generalized carnivore, breaking away from dependency on game is not impossible. Basically carnivorous bears, for example, are rather omnivorous carnivores. But parasites irretrievably relinquish attributes of self-sufficiency.

Sexual Selection

Like parasitism, sexual selection, with choice involving display, is self-reinforcing. It presumably had its inception as a particular signal came to trigger sexual reflexes. There would consequently be competition among males, as the sex bearing the smaller cost of reproduction, to present this signal. Its specific manifestation is presumably chaotic in origin—why would a female frigate bird have initially been charmed by a male who happened to have a red spot on his throat?—and it proceeds by positive feedback. It is independent of adaptation in the usual sense insofar as the organism is shaped not by needs of meeting the basic needs of life but by competition for a mate. The incredible, capricious, whimsical, elaborate, and often difficult practices of animal courtship and mating (for examples, see Hapgood 1979 and Kevles 1986) seem to represent chaos not only in the ordinary but also in the technical sense of the word.

In many groups, especially insects and arachnids, the shapes of sexual organs—penises, oviducts, pedipalps, spermatophores, and the like—are exaggeratedly varied and frequently quite different in each species. Courtship antics are similarly capriciously diverse. Such variants can have no direct adaptive significance because in related groups the corresponding organ or behavior is simple and fairly uniform (Eberhard 1985, 185).

Australian bowerbirds exemplify nonutilitarian sexual selection. Males of at least 14 species build bowers; these may be quite large, constructed of such materials as lichens, seeds, insect parts, and all manner of shiny or colorful objects, including trash of humans, with as many as 500 decorations. Some collect fresh flowers and replace them as they wilt; several kinds apply a paint of chewed fruit. One species makes a paintbrush of fibrous material. A large part of the male's energies go into bower building, but it is only for show. There is a trade-off: the more drab the bird's plumage is, the more ornate is the construction.

If a female stops by to admire his work, the male of McGregor's bowerbird hides coyly from her for a time and then rushes out; she flees but may—or may not—return to mate. After mating, he drives the female away, less concerned that she might take another lover than hopeful of luring another temporary mistress. The female nests some distance away, with no assistance from the male in housebuilding, incubation, or rearing the young. Instead, males raid one another's bowers (Pruett-Jones and Pruett-Jones 1983, 49–52).

Some members of the same genus build no bowers, and the sexes cooperate in brood rearing, a much more productive expenditure of energies. The more practical species, in which the male assists in rearing the family, raise a larger number of chicks, yet the impractical bowerbirds continue to prosper (Alcock 1988, 431).

In an effort to save the principle of adaptation in sexual selection, it is often argued that the ability of the male to waste substance and encumber himself proves his general fitness to the admiring female, who prefers his superior genes and so improves the species (as by Diamond 1990a). But it is odd to contend that an animal becomes more fit by becoming less fit. If the female chooses the male with the most cumbersome tail feathers, her offspring will carry genes for the encumbrance and metabolic cost of the exaggerated feathers. And so far as the male with longest feathers is favored, it becomes advantageous not only for the male to produce ever longer feathers but also for the female to produce sons with longer feathers. It would be more logical for the female to prefer the male that can demonstrate useful strength by pushing others aside. Yet in fish, amphibians, mammals, and insects, it is quite common that the female shows no prejudice against smaller or weaker males.

The costs of sexual display are serious. The exaggerated plumes that the African nightjar displays must be a serious impediment to flying (Welty 1982, 48). He bites off his elongated feathers after the courting season (Kirkpatrick 1987, 43–70). A male bird of paradise (*Paradisea raggiana*) dedicates himself to courting for half the year (Beehler 1989, 117). Claws of male fiddler crabs comprise as much as half of the weight of the animal, yet they are useless for feeding and nearly so for defense (Pearse et al. 1987, 498). Sexual selection may even threaten extinction. It certainly contributed to the passing of the Irish elk, whose antlers were up to 13 feet across.

Darwin paid much attention to sexual selection, although it is somewhat incongruous with his basic theory. Wallace did not like the idea. Darwinism explains a contest of strength among males fairly well, but rivalry by extravagant display is inconsistent with its general principles. It reduces fitness in competition with other species using similar resources.

It is difficult to understand how sexual selection by display can exist. Force can overrule show; when male birds gather to strut their charms, it should be easy for one with shorter feathers but stronger beak and more truculent disposition to drive others away. And the female, if she cannot have a helpmate, would do better to choose a strong fighter than the bearer of a gorgeous cape. Sexual selection by dueling carries little penalty; the qualities that give victory against a rival male are generally useful for the species. Yet only a few birds (such as ostriches) fight for the summum desideratum of possession of mates, although many are territorial.

Males come together in what are called leks, as though in agreement to flaunt their charms competitively. It is difficult to find evolutionary causation for this practice, but it is very common, not only in birds but in amphibians and many invertebrates. For example, ant males of various species (as *Pogonomyrmex*) congregate in large numbers in the same spot year after year—by some localized instinct?—to await virgin females (Hölldobler and Wilson 1990, 153).

There are less aesthetic sexual feedbacks. If the female mantis is hungry when approached by a mate to be, she may make him the entree of the wedding feast. The lower part of his body goes on mating quite competently after his head has disappeared into her belly (Thornhill and Alcock 1983, 240). In most spider species, the male (almost always the weaker sex) makes love at the risk of his life. There are consequently contrary selective pressures in relation to each sex: the parents want their sons to be adapted to get away and father another family, their daughters to be able to catch the lover and use his substance for the production of their eggs. The male's chief hope is a quick getaway, but various species have particular strategies: trussing up the female in silk, giving her a present, approaching her only when she already has had a good meal, or growing knobs to hold off her fangs. Sometimes one sex is victorious, sometimes the other. Males of many species, including the black widow spider (*Latrodectus*), often escape. In some, the male regularly pays for love with his flesh. For example, the female must catch the male of an orb-weaver spider (*Araneus pallidus*) in her jaws to hold him in position for mating, and she eats him as he fertilizes her (Foelix 1982, 197). Couples of a few species cohabit (Crompton 1950, 183–205).

Summary

The way of life is to try new combinations, tinkering with building blocks to explore what can be done with the materials—organs and enzymes—at hand. Life is enormously creative, with great frequency on a small scale and sometimes on a large scale, in ways that are unpredictable and are not dictated by adaptation although subject to the constraints of environment. Selection corrects deviations, but it is not necessary or possible to find clear-cut reasons for countless major or minor variations of plants and animals.

Thinking in terms of chaos and attractors not only relieves biology of attempting to account for countless unaccountable details; it also makes extraordinary adaptations more understandable. It is not necessary to assume that all stages of a development must have been advantageous; to the contrary, evolution can proceed without selection, as random variations are captured and enlarged (Collier 1986, 5). This is true not only of organs but also, and perhaps to a higher degree, of instincts. If tangible structures are subject to chaotic variability, this should be more strongly the case with behavioral patterns, which are more complex than organic traits and more subject to dynamic interactions with feedback. Their genetics are entirely obscure, but the chaotic aspects of the genome must reach their climax in the turbulence of a near-infinity of combinations of behavioral and perceptual patterns. On this basis, it becomes somewhat easier to appreciate how animals may have developed behaviors seemingly requiring thoughtful calculation.

The paths of evolution are contorted by multiple feedbacks. Organisms interact with one another in countless ways; the environment consists not only of physical nature but also, and perhaps more decisively, of other organisms, of the same as well as different species. Consequently evolution can go its way independently of direct requirements of the environment, and there is an interplay of adaptations divorced from obvious needs of the organism. Specialization, symbiosis, mutualism, and parasitism all involve complex and probably chaotic feedbacks. Something of the complexity of parasitism results from the contest between means of defense and means of offense and the contradiction between the imperative of the parasite to live on the host and its need to get to new hosts, with the qualification that the parasite's destroying the host is disastrous. The interaction of male and female has incredibly varied, often seemingly whimsical results, not least of which are the wasteful expenditures of sexual selection.

Evolution represents freedom within constraints, and the freedom owes much to chaos. Nothing in the material substance tells us that there should exist anything like chaos; its most obvious form, turbulence, has long been a major mystery of physics. It is a cosmic fact, like relativity and quantum mechanics. It makes possible new self-organization within a dissipative energy flow, thereby overcoming locally the increase of entropy and creating new order. It does not explain why a multitude of incredible adaptations have come about, but it makes them somewhat less incredible. Without chaos, there would be no creatures capable of musing about their origins.

Insofar as living nature is interactive and dynamic, as it is in countless ways, evolution must be nonmechanistic. Feedback and chaos can bring forth strange things, and evolution is a tale with many strange twists.

9

The Power of Attractors

Coherence of Pattern

Just as branches of the tree of life, as seen in the fossil record, are disconnected, organisms belong to groups. A century before Darwin, Carl Linneaus grouped organisms in a hierarchic system that has been retained to this day. The basic category of plants or animals considered a single kind is the species. Species that belong together form a genus; genera are joined to make families, and so forth through orders, classes, phyla, and kingdoms. The dog is species *familiaris* of genus *Canis*, family Canidae, order Carnivora, class Mammalia, phylum Chordata (subphylum Vertebrata), and the animal kingdom. There are differences of opinion, and taxonomists try to improve the scheme by inventing intermediate classifications, such as subfamilies and superfamilies, but only a few deviant forms mock the systematic scheme. For example, egg-laying mammals (monotremes of Australia) go against our sense of mammalness because they lay eggs, reptilian style, but they have mammalian-type organs as well as hair, warm-bloodedness, and milk. They are not in the mainstream of evolution, and species seeming to fall between major groups are quite exceptional.

There are shared patterns at all levels of generality, from the chemistry common to bacteria, palm trees, and hummingbirds to the particular marks of species or varieties within species. The deepest division between organisms is between bacteria (which are themselves deeply divided into separate kingdoms) and creatures with nucleated (eucaryotic) cells. The latter have many things in common, most importantly the elaborate mechanism for cell division, which is basically alike in all higher organisms, plants as well as animals. It is more surprising that the architecture of cilia, the whiplike projections of cells for moving them or fluids, is similar from protozoa to humans: nine double fibrils surround a pair of single fibrils and are enclosed in an extension of the cell membrane. Somehow the instructions for making them are deeply embedded in the pattern of heredity, although it would seem easy to alter such an external, unlike chemicals basic to metabolism, without disturbing the rest of the cell.

The next largest groups that obviously belong together are the phyla. To some extent the phyla can be grouped, especially on the basis of embryology, but the resemblances are much less impressive than the differences. What insects and mammals have in common, for example, does not strike the eye; the body plans are about as different as could be. Yet a series of *Drosophila* developmental genes (the so-called homeobox) have a 60-amino-acid domain that is strictly parallel to a similar series in mice. This system governing the designation of sections of the embryo has been conserved since near the beginning of metazoans; arthropods and vertebrates are on different sides of the basic embryological division of animals (Miklos and John 1987, 272; De Robertis, Oliver, and Wright 1990).

Animal phyla are sharply characterized. For example, the cnidarians (or coelentarates, including jellyfish, sea anemones, corals, and the like) have tentacles armed with specialized stinging cells, nematocysts, that shoot out a poisonous dart. No other animals have anything comparable, a coiled tube that can be instantly turned inside out to project a hypodermic needle tip and inject venom into a prey. Invention of this difficult device apparently gave rise to a large and quite successful phylum; the cnidarians dominate the oldest metazoan fossils and are still omnipresent in the oceans.

The phylum of echinoderms (starfish and relatives) is also clear-cut; it is characterized by a calcified skeleton with a unique structure and liquid-filled tube feet for locomotion and feeding. Even more striking is the almost invariable fivefold radial symmetry commonly seen in the starfish on the beach. Strangely, this symmetry characterizes only adults and is only external; there can be no adaptive need for it, and only echinoderms have it.

The arthropods (crabs, millipedes, spiders, insects, and so forth) are the most successful phylum in terms of number of species, comprising probably nine-tenths of them all. It is also the most diverse. Yet it is well marked by a horny external skeleton, jointed legs, a segmented bilaterally symmetrical body, and a circulatory body cavity.

Less inclusive groups are no less coherent. Our subphylum, the vertebrates, has a body organized around a spinal column carrying the main nerve channels; the head, with a skull, holds the brain and chief sense organs; the heart circulates blood through vessels. Vertebrates from jawless hagfish to humans share at least 90 percent of their genes (Loomis 1988, 216).

The stability of such patterns is not easily explained beyond stating the obvious: the basics are so embedded in the genetic structure that changes are either excluded or highly disadvantageous. Secondary traits are also persistent. Since the amphibians came onto the land, their descendants have kept the outlines of the tetrapod limb, at least in embryonic stages: a strong bone is joined to a girdle, front or back; there follow two parallel bones; to these are attached small bones, carpals or tarsals; the limb terminates in digits, primitively five or more. The joint to the girdle is ball-and-socket; the middle joint of the limb is a hinge. For no evident reason, the first digit (the thumb) has one less bone than the others. The pattern is the same for fore and hind limbs, like the repeated legs of arthropods. The pattern is conservative; the human hand is recognizably like the forefoot of a crocodile, although the common ancestor lived 300 million years ago. This is true despite the fact that extra or missing digits are common mutations.

Not only is the basic architecture of fore and rear tetrapod limbs the same; their modifications are genetically linked. In most tetrapods (flying and swimming animals excluded), front and back feet are much more alike than adaptation would require. This is easily observed in the fore and hind feet of dogs, elephants, squirrels, horses, seals, alligators, turtles, and others. There are no animals with single hooves in front and cloven hooves behind or with hooves behind and paws in front. The evolution of the front and hind feet of the horse, from several toes to one, was closely parallel. The panda has extended a wrist bone to make a sort of thumb for holding bamboo; the corresponding bone in the foot is also enlarged for no evident purpose (S. Gould 1980b, 24).

Many groups have their special mode. No other animal comes near the squatting-hopping habit of the anurans (frogs and toads). A large majority of them, unlike other vertebrates, have poisonous skins. They also have a metamorphosis (from tadpole to adult) unique among vertebrates. Turtles, cetaceans, and many other orders are equally distinctive. The snakes are not only legless but alike in being carnivorous and capable only of swallowing prey entire. They all have a single lung, an odor-sensitive tongue, unhinged jaws enabling them to engulf prey of larger diameter than themselves, fangs, broad belly scales, and an inordinate number of vertebrae. They have no outer or middle ears, and their inner ear receives vibrations from the lung, so their hearing is weak. Their glassy eyes without eyelids are unique among tetrapods (except for a few lizards). They focus not by changing the shape of the lens but by moving it, they have no fovea, they have little or no stereoscopic vision to judge the distance of a prey, and their vision is mediocre to poor (Klauber 1972, 385–386).

The remarkable fact about these characteristic traits of the snakes is that a number of them seem to be maladaptive. Hearing, good vision, and legs—not only for running but grasping prey—seem unequivocally useful for a land carnivore. There must be special compensating assets in the snake constitution, but it is not obvious what they are. One may be their venom. A few small mammals possess poisonous saliva and means of injecting it, with much less potency. So does the Gila monster (*Heloderma*) of the American Southwest; oddly, it has minimal need for poison because it subsists on bird eggs, chicks, and baby mammals (Alcock 1990, 42). No fish, amphibians, or birds have a poisonous bite.

Yet venom does not account for the success of the snakes, a majority of which get along quite well without it. The snakes presumably lost not only legs but hearing and good vision in subterranean living. Specializations requiring extensive adaptations are rarely reversible; some extraordinary change or changes enabled the burrowing animal to triumph in an open habitat.

Some of the traits setting our mammalian class apart are clearly adaptive, others less so. Hair is useful, if only for conservation of heat; mammals usually spend most of their energy keeping warm. Some mammals with no particular need for hair, such as whales, elephants, and armadillos, keep at least vestiges of it, while animals with much more need for it, such as sealions and humans, have largely lost it. Mammary glands help offspring to get a start in life. They are part of a broader characteristic, the possession of skin glands. Reptiles and birds have almost none, but mammals have a great variety, producing not only milk but sweat, fatty substances, and diverse odors and protective secretions.

Mammals are unique in sundry things, such as the three bones of the middle ear, a diaphragm separating thorax and abdomen, and the cortex enveloping the cerebrum. They have seven cervical vertebrae (with very few exceptions). Their jaws are different from those of reptiles, with a single bone instead of several, and they have unique multirooted, multicrowned teeth. Mammals never have proper scales, useful as these seem to be for reptiles. Armadillos, pangolins, and the like make do with toughened skin or fused hair. Mammals, presumably having lost coloration as nocturnal animals under the dominion of the dinosaurs, have not recovered it. Many vertebrates, fish, amphibians, reptiles, and birds are bright red, yellow, green, or blue, and such decoration would serve some mammals. With a few exceptions, such as the bright faces and purple buttocks of some baboons and other monkeys, they manage only dull red or yellow.

Birds are less structurally diverse than mammals. They all have horny beaks, although this has nothing to do with flying (or with living in a shell, as in the case of the turtles). They lack teeth or movable lips, which mammals find serviceable. Their brains are quite different from those of reptiles and mammals. Unlike many reptiles, all birds lay eggs, although it would well suit some, especially ground-nesting or flightless species, to give birth to developed young. Their eggs have calcified shells and big yolks. A remarkable specialty of birds is the system of air sacs for through-put respiration. Feathers seem to have changed very little at least since the time of *Archaeopteryx,* 150 million years ago, and they remain fairly stable in all birds. There are no featherless birds like nearly hairless mammals.

Within the class of Aves, families are homogeneous. Thousands of species of songbirds are indeed birds of a feather. All penguins are much alike, with no intermediaries to other families (Müller-Schwarze 1984, 5). All hummingbirds are small, feed mostly on nectar, have long bills with tubular tongues, and display more or less iridescent plumage (Skutch 1973, 43–48).

Plant families also share traits beyond the dictates of environmental needs. The fronds of true ferns (and hardly any other plants), in all their variety of shapes and habitats, grow by uncurling a crozier. Many ferns (and all tree ferns) have the feathery fractal fronds commonly associated with their class and not found elsewhere. Beginning life with a single leaf, or cotyledon, marks a group diverse as palms, grasses, and orchids, and they typically have three-part flowers, no cambium layer, and parallel venation in the leaves. The flowers of some 25,000 species of orchids are fantastically varied in their flowers, but all are perennial, with simple, rather similar leaves and stems; most are tropical epiphytes.

Palms are notably restricted. Their very large leaves are confined to palmate or pinnate design. They live only in mild to hot climates. With one primitive exception, they have a single unbranching trunk; that is, they have lost the ability to branch. Their trunk cannot grow stronger as the plant grows, so it is unnecessarily thick when short. The palm dies if its single apical bud is killed. It cannot enlarge roots as needed to support the growing plant but can only make more of them (Corner 1966, 84–86; B. Wilson 1984, 11). Evolution has strikingly reduced the options of the palms, yet palms are a prominent part of tropical flora.

To say that groups are well defined is equivalent to saying that intermediates are missing. There exist a few species that are suggestive of melanges, like the monotremes already mentioned. *Peripatus* looks somewhat like a cross between an annelid worm and an arthropod, with jointless and clawless legs and a soft integument, but it is believed to be a separate offshoot of a wormlike animal, not an arthropod ancestor (Manton 1977, 44). The lungfish has heart and lungs like an amphibian but is a legless fish. The earth is not populated by the mixtures of amphibian and reptile, reptile and mammal, and so forth that would result if traits were developed independently because of their utility.

It follows that any group sharing inherited traits may be regarded as a unit of selection. If the selfish gene may be deemed to seek self-multiplication because this lies in the nature of life, this must be no less true of complexes of genes that have a teamlike existence. Individuals compete within species, species compete to some extent with one another and, less immediately, over the ages, genera, even classes compete. For example, reptiles displaced amphibians, and birds took the place of pterosaurs. Dinosaurs kept mammals in check as long as they lasted. In recent times, mammalian predators have driven predatory ground-dwelling birds and big lizards almost entirely to extinction. The Komodo monitor (perhaps the original monster of mythology) can hold out on its island only because there are no mammal carnivores. Natural selection operates not only on genes and individuals but on an indefinite set of groups, the less clearly as the group is broad.

Limits of Change

In a few thousand years, mostly in the last few centuries, the wolf has given rise to hundreds of breeds of dogs, from whippets to pekinese, many of them superficially much more distinct than wolves and jackals. It is easy to conclude, as Darwin did, that such changes can go on indefinitely and that nature has thus, over millions of years, evolved the different genera, families, classes, and so forth that we observe.

Animal breeding, however, also shows the limitations of the variation-selection principle, and Darwin's contemporaries did not find it a very convincing explanation of the origin of species (Ruse 1982, 48). Different as domesticated breeds may be, they do not become new species; dogs remain dogs, interbreeding freely so far as difference of size permits and readily hybridizing with wolves. Cats, pigeons, and so forth also belong to the species of their ancestors.

Breeders can work only with the reservoir of variation in the species. Variability is larger in dogs than in pigeons, pigs, or horses, but it is limited. The behaviors for which dogs can be bred are intensifications of behaviors of wolves: territorial possessiveness makes the guard dog; fighting, the pit bull; stalking, the pointer; bringing food back for the pups, the retriever; hunting, the sheep dog (Fox 1978, 255). A dog obeys its master as a wolf submits to the alpha wolf of the pack; cats are not so respectful of the top cat. If one starts out with a generalized dog, one can breed for changes in many directions; if one starts with a specialized dog, one makes little further progress in the specialization. Cats are much less modifiable than dogs. Despite 5,000 years of domestication, they have hardly changed in morphology from the wild type, differing mostly in their coats.

Breeders come up against limits. As different as toy dogs are from great danes, dogs cannot be grown small as mice or big as horses. Neither can they produce tusks, make curly tails, or sprout bristles like pigs. The speed of greyhounds has reached a plateau. The financial rewards of racing have strongly motivated horse breeders, but for eighty years, despite improvements of care and training, veterinary medicine, and knowledge of genetics, they have been unable to reduce track times appreciably. Breeding faster horses would presumably require making changes not accessible by gradual improvements, perhaps changes that would lower speed until perfected and combined with other changes.

It is easier to manipulate plants genetically, but there has been no great advancement in the quality of fruits and vegetables for decades. Cultivated peaches are much sweeter and larger than their wild ancestors, but they have not been bred to be seedless or the size of melons. To make further progress after some generations of selection, one may hope for a new departure by awaiting a rare fortunate mutation or by artificially modifying the genome by radiation or a mutagenic chemical. Or one may develop new combinations by hybridization, which works better with plants than animals. The modern agronomist also resorts to genetic splicing, but desired mutations may be unattainable. American chestnut trees have not developed immunity to the imported blight that killed billions of them; scientists were unable, even with x-rays and mutagens, to evoke the needed mutation—the more surprisingly because closely related species are resistant (Burnham 1988, 478–487).

It is a general rule that selective breeding for a trait reaches a plateau in 30 to 50 generations in plants and animals; after many more generations, there is sometimes a rise to another plateau (Endler 1986, 229). The intensively bred organism may also become unviable as negative side effects appear. Breeding a dog with the optimum muzzle for show purposes often entails defective hips or weak kidneys.

Despite their superficial diversity, domesticated animals have little genetic diversity; the reverse is true of the wild forms. And artificially selected traits are unstable. Domestic breeds, if allowed to reproduce without selection, revert in not many generations more or less to the wild type, as Darwin noted (C. Darwin, 1960, 157). Feral dogs lose the exquisite features of various breeds and tend to a nondescript brown or blackish medium-sized mongrel type that is about the same the world around. Cats go feral much more readily than dogs and live much as their ancestors did (Serpell 1988, 151–152). Domestic pigs also easily go wild, growing more bristles, recovering tusks, and having stripes when young. Fanciers have bred more than a hundred varieties of goldfish, many rather grotesque, with growths around the head, bulging eyes, and the like. Placed in a pond and left to themselves, they soon revert to their ancestral shape.

Fruit flies mutate to monstrosities, but they are still *Drosophila*. Laboratory selection for a high number of abdominal bristles brought a rise at first, but the number reached a plateau after 20 generations with about half again as many as when the selection began. Selection for a low number brought a moderate decrease, but the strain died out after 35 generations (Mayr 1963, 285–286). A few generations of selection can make *Drosophila* either very resistant to DDT or supersensitive to it. Without selection, both the resistant and the supersensitive strains soon return to normal (Ehrlich 1986, 65). Loss of immunity is attributed to the cost of producing the necessary enzyme, but this cannot cause selection pressure comparable to the death of nonresistant organisms. The rapid loss of supersensitivity is even more difficult to explain.

Drosophila were selected through 20 generations for preference for dark or light (negative or positive phototaxis), and widely divergent strains were produced. But when the two strains were no longer subjected to selection, both reverted to type as rapidly as they had diverged under very strong selection (Dobzhansky 1970, 207). It would seem that without selection the attraction of the old pattern was as powerful as the intervention of the experimenter. *Drosophila* sometimes mutates to eyeless, and a homozygous eyeless strain (without recessive genes for eyes) can be bred. But if allowed to breed freely, it will recover almost normal vision in 8 to 10 generations (J. Huxley 1942, 43–44).

The making of new species seems to require breaking away from an attractor.

Parallel Evolution

Lineages often show convergence: starting from diverse bases, they become more alike. For example, bats believed to be descended from insectivores look like bats believed to be descended from primatelike ancestors. A slightly broader phenomenon is parallelism, following similar tracks despite different origins. The Tasmanian wolf (*Thylacinus*) evolved in parallel to the placental wolf.

The coherence of patterns, from the biochemical essentials of life down to single species, represents a sort of parallelism. The fact that groups are clearly definable means that in important ways their species have evolved much more similarly than demanded by needs of adaptation. Birds, mammals, spiders, snakes, and so forth have kept basic traits as though unalterable; the chief reason we consider certain traits basic is that they remain changeless despite great diversity of habits and needs. Many traits of birds are irrelevant to their mode of life, whether they catch insects on the wing, fish under water, eat seed on the ground, or swoop down to clutch small animals. The pattern is retained, although in any individual species elements of it must be maladaptive.

It is another kind of parallelism that many members of a group develop a particular capacity in various ways. The facility of snakes for venom has been mentioned: the poisons made and the means of injecting them are quite varied, but it is a strong specialty of the order. All spiders use silk, but in different ways, to make an egg case, to line a nest, to sail with the wind, or to entangle prey. Their silk is especially strong, or elastic, or thin, or sticky as required. The makers of great orb webs have at least six kinds. Some advanced orb weavers make flowerlike designs on their webs with special silk that reflects ultraviolet light, which most insects perceive; they land in a trap instead of a flower (Craig and Bernard 1990). Yet geometric orb webs are believed to have been originated independently, using different kinds of silk, in different families (Foelix 1982, 144–147; Shear 1986, 369).

Different forms responding to similar opportunities sometimes become more alike in ways not ascribable to adaptive needs. For example, fruit bats (Megachiroptera) and insectivorous bats (Microchiroptera) are classified as a single order, Chiroptera, because they are much alike, have membrane wings framed by grossly extended fingers, hang from their hind feet, and so forth. But the facts that the architecture of the wing is approximately the same in the two suborders and that they both hang from their feet instead of perching are probably not because of common ancestry but because of convergence. The two suborders of bats are apparently no more closely related by ancestry than monkeys and shrews. There are some consistent differences: fruit bats never hibernate, whereas insectivorous bats outside the tropics usually do so; the former have two claws, the latter one claw on the wing; the former are almost all larger than the largest of the latter, as the scientific names suggest. For the microchiropterans, echolocation is a primary adaptation. The earliest fossil bats of nearly 60 million years ago, which are clearly microchiropteran, are believed to have had well-developed echolocation. The megachiropterans, the earliest known fossils of which date to only 35 million years back, almost entirely lack it. They could hardly have lost a device so useful for a night flyer. Even the few microchiropterans that have taken to eating fruit have retained it. A few megachiropterans have a faculty of echolocation, but it is differently evolved from that of microchiropterans—of lower pitch and less accurate.

It is strong evidence of separate origins that the dentition of megachiropterans is quite different from that of microchiropterans and could not have been derived from their insectivore-like teeth. The brain and central nervous system and the circulatory and reproductive systems of the two suborders also differ importantly; in these respects the megachiropterans are more like primates (Hall and Stuart 1984, 29–37; Pettigrew 1986, 1304–1306). The eyes and ears of megachiropterans are also consistently different from those of microchiropterans (Griffin 1986, 9). It would seem that in matters related to flying, the two groups have assumed traits that come easily to mammals taking to the air (although not necessarily dictated by needs of flight) and so acquired a similar general aspect. In most structures not related to flying, they remain dissimilar.

Entering an environment vacated by the extinction of marine reptiles along with the dinosaurs brought about the convergence of cetaceans (whales and porpoises) and sirenians (sea cows, manatees, and dugongs); they somehow hit on the same plan of adaptation to the ocean, a plan very different from that of the pinnipeds (the single ancestry of which is questionable).

The two orders took to the water independently, the whale having arisen from carnivorelike stock and the sirenians having a probable affinity with the elephants (Hartman 1978, 132–

134). The life-styles of the two orders are quite different; the whales have always been carnivores, the sirenians herbivores. The former live in the open ocean, the latter in shallow waters rich in plant life. Yet both have acquired fairly similar tapered bodies. They have also made the many adaptations necessary for giving birth and suckling in the water, a useful ability that has eluded the pinnipeds (seals and relatives) for many million years but is a necessity for marine animals with no means of locomotion on land. Both orders are practically hairless, although hairlessness is not necessarily advantageous for marine mammals; fur seals, polar bears, and sea otters, having fur, do not need such a thick layer of fat to maintain body heat in cold water. In addition, cetaceans and sirenians have made the apparently useless change of moving the tail mostly verti-cally—uniquely among the many tetrapods that have taken to the ocean.

It is most interesting that they discarded hind legs entirely, although the many other orders of tetrapods entering the sea have made the obvious adaptation of converting all four limbs into paddes or flippers. This seems somehow to have been genetically forbidden for the whale. The earliest fossil whale, *Basilosaurus isis,* of 40 to 42 million years ago, which must have been rather abundant because numerous skeletons have been found, had enigmatically tiny and seemingly useless hind limbs. They were far too small to support the highly elongated animal and were not at all adapted for swimming (Gingerich, Smith, and Simons 1990). The only other conceivable use, as acces-sories to mating, seems unlikely; legs are not necessary for the purpose. The modification of hind limbs to paddles must have been incompatible with the attractor that shaped the whale (like sirenians) for aquatic life.

The parallels of even-toed ungulates (artiodactyls—deer, antelope, sheep, cattle, and relatives) and odd-toed ungulates (perissodactyls—horses, rhinoceroses, tapirs) are less striking but also evolutionarily significant. Their limbs are structurally similar, and both groups have turned claws into hooves (Thom-son 1988, 137).

A potential for paired horns seems to have developed in the ancestry of artiodactyls after separation from the perissoda-tyls. Only artiodactyls have paired horns, but their horns are made in several different ways and have evolved independently as many as seven times (Janis 1982, 278); something in the genome facilitates the formation of paired horns without specifying how they are to be made. The utility of horns is diminished in many species by their being turned backward.

There is also some equivalence between horns and tusks, different as these are in their formation. Some artiodactyls—members of the pig family—have overgrown teeth, or tusks, that fulfill similar functions as horns and are also likely to curve backward. Very small deer (especially *Muntiacus* and *Moschus*), with small horns or none at all, have tusks. Perissodactyls have no tusks or horns, except for the rhinoceros's central horn or horns made of hair.

In these examples, two groups have been involved. In numerous other cases, unlikely patterns have emerged independently a dozen times or more. The bedbugs' practice of injecting sperm into the body cavity has apparently arisen many times in the family (Hinton 1964, 95). The habit of at least nine species of cichlid fish of eating scales snatched from larger fish seems strange because approaching the larger fish is likely to be dangerous, and scales are difficult to digest and not very nutritious. Living on them requires specially adapted teeth and guts. Yet several scale-eating species seem to have developed the practice independently (Fryer and Iles 1972, 86–87).

An eccentric termite defense is also believed to be a multiple invention. Most termite soldiers have cannon in the head. A gland, which may take up much of the body, secretes a poisonous or sticky substance that the blind termite can shoot through a nozzle for a considerable distance and rather accurately. Although a number of animals, ants, beetles, skunks, and so forth engage in chemical warfare from their rear end, and spitting cobras from the mouth, no other animal has anything resembling the termites' snout. It requires extensive bodily changes and includes loss of functional mandibles. Yet it is reported to have originated at least nine times in as many genera (E. Wilson 1971, 188).

There are striking parallels of various Australian marsupials with placental mammals. The placental mole, flying and nonflying squirrels, jumping mice, ordinary mice, cats, and wolves all have marsupial counterparts. The resemblance of the (extinct) Tasmanian wolf to the northern wolf is especially close, down to the shape of teeth, claws, ears, and skull. That their convergences are beyond the needs of adaptation for their respective ways of life suggests that certain patterns are

The variety of horns of antelopes is only a small subclass of the horns of artiodactyls. (Illustration by Ernst Haeckel, *Art Forms in Nature*, © 1974 by Dover Publications, Inc. Originally published by the Verlag des Bibliographischen Instituts, Leipzig and Vienna, 1904.)

most compatible with the mammalian plan, that is the basic mammalian attractor. The way genes can be fitted together must be similar to marsupials and placentals, although their common ancestry lies at least 100 million years in the past. It is somehow inherent in the mammalian pattern that, in suitable environments, primitive mammals could radiate in similar directions, evolving similar burrowing insectivores, coursing carnivores, and so forth. There are, of course, marsupial specialties, such as kangaroos which fill the place of ungulates. So far as is known, no other animal bigger than a mouse has ever used the kangaroo mode of locomotion.

There are many parallelisms in plants, some quite as remarkable as those cited here in animals. For example, many bushes in New Zealand and almost none elsewhere show an odd style of growth: twigs branch at nearly right angles from the stem at rather long intervals with quite small leaves, giving the plant an outer protection of interlaced twigs. This peculiar mode, known as divarication, is found in about 50 species in 21 families, usually in only 1 or 2 species in a genus. This mode seems to have evolved many times independently as protection against browsing by the now extinct moa (J. Diamond 1990b, 769). It is remarkable, however, that many plant families had the capacity for this adaptation. Similarly, the eucalyptus-type leaf, elongated, tapered at both ends, bifacial, smooth, and rather hard, appearing only in the adult plant, occurs in many families and in very different environments in Australia—and only in Australia (Went 1971, 198–205).

There is parallelism of instincts as well as structures. Hunting wasps (*Pompilus* and others) paralyze spiders by stinging and haul them away to their nests to nourish their young. Wasps of related genera (*Evagetes* or *Ceropales*) hang around the nesting places. When a *Pompilus* brings in a spider, *Evagetes* sneaks in and replaces the *Pompilus* egg. *Ceropales* waits until the *Pompilus* has left the spider, then lays an egg that proceeds to hatch before that of the rightful owner and eat the latter along with its spider. Such stealing of another animal's food supply, called cleptoparasitism, has arisen with many variations dozens of times among wasps and bees (Evans and Eberhard, 1970, 215–216).

Organized animal societies have been evolved many times. Specialization of workers and reproductives has originated at least a dozen times, probably many more, in wasps, bees, and ants (Michener 1974, 236). It also appears in the very distantly related termites, with rather similar "institutions." It is less well known that group living has evolved independently in all major orders of carnivores, from mongooses to cats (Gittleman 1985, 200). Brood parasitism in birds has evolved at least seven times (F. Gill 1990, 388).

Social parasitism in ants has arisen independently many times, the number being impossible to fix because there is no way to determine at what stage related species took separate evolutionary roads. The parallels between slave makers in quite different divisions of the ant family are quite close, however, with similar raiding habits and potent jaws. The full parasites, those that have quite lost the worker cast and rely entirely on winning entry into a host colony, show more parallels. They tend to much reduced size, in some cases being smaller than host workers. The male partly or entirely loses wings and may never really emerge from the pupal state; mating is probably restricted to the home nest. Oddly, the new queens do not scatter far, a fact that certainly contributes to rarity. The antennae are reduced, and the exoskeleton has become relatively thin and smooth (Hölldobler and Wilson 1990, 467–469).

The tropical American army ants and African driver ants, which are similar in general appearance and habits, have sacrificed useful traits in acquiring the abilities that make them among the most numerous of animals. Both groups (along with an Asian counterpart) have very large to immense colonies, up to 20 million or so in Africa. They are periodically nomadic and have polymorphic workers, large winged males, and very large wingless queens. They swarm across jungle or savanna, capturing and killing all small animals that cannot flee. Yet it is believed on various grounds, mostly morphological, that African, Asian, and American mass-predatory ants have separate, possibly rather recent, origins (Gotwald 1979, 462–476).

Some of their similarities can be related to the life-style of mass predation. Because they rely on numbers, colonies cannot be founded by lone queens, in the manner of most ants, but must be started by fission or swarming, in the way of honeybees. Hence queens are wingless. It is unclear, however, why colonies should have to be so exaggeratedly immense, up to millions, such that a column several feet across may take hours to pass. A hundred thousand could ravage a considerable territory and overwhelm any likely prey. The extreme size of colonies means that they quickly exhaust the game within marching distance and are forced into continual nomadism, which greatly complicates their existence.

The huge numbers of the army ant colony require correspondingly large egg production, yet they never have more than one queen, and dependence on a single mother is dangerous because the colony is at an end if she is lost outside the brief season for raising reproductives. Yet five legionary ant groups have separately evolved the same type of queen, not known among other ants, so distinctive that she has been awarded her own awkward technical title, dichthadiigyne (Hölldobler and Wilson 1990, 305–306). It also seems wasteful that they have large males. Most ants have relatively small males, as is appropriate for mere sperm carriers. A far greater disadvantage of the army-driver ants is that they are nearly or quite blind and have weak stings, although their presumed ancestors had both good vision and strong stings (Ranks 1979, 142–144; Hölldobler and Wilson 1990, 574–595). Surely the army ants would be far more formidable if they could chase their prey by sight and dispatch it by venom. But even with their handicaps, they are, in terms of numbers and ecological impact, among the most successful creatures on earth.

Loss of the power to sting is very common; many bees and ants have suffered it, the loss bearing no obvious relation to habitat or way of life. Some stingless bees (*Trigona*) compensate by squirting an acid secretion when they bite or by placing sticky resinous pellets on an attacker (Mayr 1960, 375)—a means of defense seemingly less efficient than the abandoned sting. In at least nine cases, the stingless bee acknowledges inferiority by mimicking a bee or a wasp that possesses a powerful sting (Michener 1974, 217).

Ants have followed a similar course. Primitive forms, like their wasp ancestors, have stings that are a menace for the collector. Most of the highly evolved ants have reduced or lost the sting, and this bears no obvious relation to way of life. Predators such as army ants have lost the power; some seed eaters have retained it. Many stingless ants, like stingless bees, have substituted a noxious spray.

The barbed stinger of the honeybee cannot be withdrawn after insertion but is pulled out of the abdomen, poison sac attached; the honeybee can sting only once. This is an odd adaptation, the utility of which is not clear. Hornets sting fiercely without such sacrifice. However, several genera of social wasps (*Polybia* and others) have independently developed barbed stingers like the honeybee (Evans and Eberhard 1970, 12). So do a few ants, such as *Pogonomyrmex* (Hölldobler and Wilson, 179).

Something in the hymenopteran makeup must favor a barbed stinger. In sum, it often appears that certain attributes go with broad patterns, for reasons unclear, and this fact must have been of great, perhaps decisive, importance in evolution.

The Meaning of Parallelism

Many traits may be viewed as more or less convergent or parallel. For example, both scorpions and spiders, rather different arachnids, sometimes eat their mates (and scorpions occasionally eat some of their young), but this may have little meaning beyond the expression of rapacious instincts. It may be more or less accidental that the pincer claws of scorpions and crabs are much alike. Is the resemblance of the cones of magnolia (an ancient tree) to pine cones significant?

Convergences may simply indicate that there are a limited number of ways, mechanical or physiological, to accomplish objectives. One attributes no deep significance to the similar shapes of sharks, extinct marine reptiles, and porpoises beyond admiring the ability of the animal to become beautifully streamlined. One also assumes that the octopus eye resembles the vertebrate eye because there are not many ways to make a cameralike apparatus. Birds and mammals (and also crocodilians) revamped the reptilian circulatory system in approximately the same way, completing the division of the heart and separating blood flow to the lungs. This is not surprising; one wonders that lizards and turtles have not done the same. Birds and mammals also improved reptilian metabolism similarly, and perhaps there was only one way to accomplish it.

Various fish and amphibians nourish embryonic young by something like the mammalian placenta. Pigeons (males as well as females) feed their chicks a sort of milk (Welty 1982, 399). The pigeon's "milk" is stimulated by the same hormone, prolactin, that governs lactation in mammals (Kevles 1986, 136). The number of enzymes is limited, and they may serve related purposes. The hormone that produces brooding pouches on the backs of frogs is used in the mammalian uterus (Tyler 1983, 134).

If the same enzymes play a role in very different organisms, this means that the same or nearly the same genes must be at work. By extension, it is possible that some of the genes that dictate the formation of the molluscan eye may be nearly or quite the same as those responsible for a vertebrate eye. The combinations of proteins available to organisms to direct the formation of an eye are not infinite.

This means that it is difficult to separate analogy from homology. Analogy implies unrelated origins of outwardly similar organs. An insect wing is analogous to a bird's but is entirely different in derivation. Homology, to the contrary, is similarity of structure attributable to common ancestry. The bird wing is homologous to the human arm because both are derived from and keep the basic pattern of the amphibian forelimb. Other examples are not so clear. The wing of the bat is both analogous and homologous to that of the bird. The heat receptors above the nostrils of pit vipers are extraordinary infrared-sensitive cameras. Yet boas and pythons have analogous or homologous heat-sensitive organs along their lips.

A group of related animals has sometimes seemed to share a capacity for going in one direction. For example, several groups of mammal-like reptiles evolved in somewhat parallel fashion in toward the mammal pattern, different groups being closer to mammals in different ways. Possibly several branches of mammals evolved separately from reptiles; in fact, it is probable that the egg-laying mammals of today (monotremes) represent a separate line of descent from a reptilian ancestor (Colbert 1980, 246–260).

A quite different generalized ability is that of electric fish. This strange faculty is known to exist only in fish, but it has originated independently at least six times, and the faculty of electroreception has originated independently three times. There had to be a predisposition for use of electricity in the ancient fish genome. It is not a specific but a rather amorphous faculty. The current is most commonly head positive, tail negative, but it is often the reverse, even within a single genus. *Stenarchus* generates alternating current, which has no apparent advantage but represents a real trick for a batterylike apparatus. The larval fish frequently have electric organs in different parts of the body from those of the adult. The generating cells are mostly derived from trunk muscles, but in one case (*Astroscopus*) from muscles around the eyeballs and in another (*Apteronotus*) from nerve tissue (Bass 1986, 20–32).

All innovations have to be made within the framework of a successful genome. Directions are set by the potentials given by the past, as well as by the selective pressures of the present (Thomson 1988, 131). Differences often owe much more to ancestry than adaptation, even within closely related groups (as observed of lizards, Schwenk 1988, 593). According to John H. Campbell, "Not only the mechanisms but also the goals of evolution are defined by the genetic message" (J. Campbell 1987, 284). Parallel evolution shows that organisms may be predisposed to evolutionary tracks that are definable more in terms of function than of structures. Certain lines are open, wherein species may develop toward similar purposes in physiologically and organically different ways.

The strength of parallelism has yet other consequences for evolutionary theory. It is commonly taken for granted that if two or more groups share a trait, this trait must have been present in their common ancestor, but it can reasonably be assumed only that a common ancestor had a protentiality for the trait.

If different families follow parallel courses beyond the adaptive requirements, it shows a shared potentiality. For example, several kinds of not closely related birds have developed nest parasitism. This generally implies that the parasite chick kills or crowds out host nestlings; this faculty seems to be related to the inclination of chicks of many species to maltreat nestmates (Alvarez, de Reyna, and Segura 1976, 915). Similarly, siblicide, the first chick's killing of younger chicks, has evolved independently at least eight times (Anderson 1990, 334), although this would seem to be an unlikely trait or at least a mode contrary to the good order of families. A propensity toward siblicide and nest parasitism is evidently embedded in the genome of many diverse birds.

If seemingly improbable traits, such as the bedbug mode of sex, the termite squirt cannon, heterostylous flowers (one species having flowers with differently arranged anthers and stamens to assure cross-pollination), and the like, arise independently several or many times, they cannot be so very improbable. They are not to be dismissed as the results of rare accidents but must be regarded as the attainable—in some sense likely—product of an ancestral genotype.

The idea that the direction of evolution is in any way predetermined is disliked because it seems to hint divine purpose. But evolution is history, and historical outcomes are partly, if not largely, determined by their past. It would be strange if a genotype did not favor mutation in certain directions, or if certain sequences of mutations were not much more probable than others, within the confines of a class or family. The development in different groups of similar improbable organs and instincts confirms this coherence, especially when the parallel traits are of doubtful utility.

Scientists, finding confusion in abundant fossil remains of various groups of fishes and reptiles, asked, "Do the animals look alike because there are unknown structural constraints that mandate their size and shape?" (Bürgin et al. 1989, 81). The answer must be positive. Some things are possible within the plan, and others are not. Emerging groups acquire not only new traits but capacities for new directions and new traits. It may be, for example, that an ape, adopting upright stance and making correlated changes, acquired the ability to enlarge the brain.

Evolutionary Inertia

There are clear-cut long-term tendencies in evolution, as one sees in the development of the human from a monkeylike primate. Tendencies of development imply something like momentum, strengthening traits beyond needs of adaptation or fitness; animals seemingly find it easy to continue on a course, even to exaggeration.

The huge size of some dinosaurs and whales may be partly ascribable to something like inertia of development; evolutionary increase of size is so common that it has been given a name, Cope's law. Decrease of size is rare. There may be feedback, making the specialization irreversible; and very large animals are likely to become extinct. Their smaller cousins last longer and come through extinctions better (Benton 1990, 150–151). Giantism may be adaptive along the way, but it is ultimately unadaptive.

It is not that evolution proceeds toward a goal but that it carries on in a direction that has been adaptive in the past.

There must be advantages of being a millipede with many segments, but the two hundred or more legs of some species seem exaggerated. Many millipedes are burrowers, and a multitude of legs may help push the animals through soil (Manton 1977, 328), but they would not appear well designed for this purpose. The simple fact seems to be that millipedes have a genetic propensity for leggyness. Hummingbirds have usefully elongated the bill to reach nectar deep in flowers; one kind (*Ensifera*) has a bill twice as long as its body.

Many ants, including all members of the genus *Camponotus*, have a noxious defensive spray. A Malaysian species, *C. saundersi*, has carried this faculty to a self-destructive extreme. It has a huge poison-filled sac extending the length of the body; if attacked, the ant virtually explodes (Hölldobler and Wilson 1990, 179–180). This self-sacrifice, like that of the bees, wasps, and ants with barbed stingers, is hardly necessary for making itself unpleasant.

Although the height of the giraffe is obviously useful, it seems to represent not simply an adaptation but the fulfillment of a potential. The evolving giraffe line left no middling branches on the way, and there is nothing, living or fossil, between the moderate neck of the okapi and the greatly elongated giraffe. The several varieties of giraffe are all about the same height. There are a number of fossil giraffids with more or less the shape of the okapi; it would seem that one of them rather suddenly took off and grew to the practical limits of a giraffe.

Many fossil sequences show something that looks like orthogenesis. One that has been much remarked is that of the ammonites. From the origins of the group around 400 million years ago, the septa partitioning the interior of the coiled shell became rather rumpled, and for some 150 million years, they grew more and more elaborate, with intricate patterns, ultimately with designs not unlike frost on a windowpane. The fancier types died out, but the survivors had ever more crumpled and patterned septa, only to revert to simpler types before dying out along with the dinosaurs (Fenton and Fenton 1989, 281–283). This is the more notable because the partitions inside the shell could have little adaptive significance. Flat septa would give maximum strength with minimum cost of materials.

The course from the rather nondescript little *Hyracotherium* to a horse similarly may owe much to what may be called internal direction. It was allowed by natural selection, but we do not know how much credit goes to natural selection. The single hoof seems appropriate for life on the grasslands, but the rhinoceros, elephant, ostrich, wild dog, and so forth do not find it necessary. On the other hand, in a remarkable bit of parallel evolution, an extinct South American ungulate, the litoptern *Thoatherium,* had a head, body, and legs very similar in aspect to horses that arose tens of millions of years later in North America. The architecture of the feet was practically the same, and the side toes were even more completely reduced. Paradoxically, however, although the evolution of horses is conventionally associated with a switch from browsing to grazing, *Thoatherium* had teeth suitable for browsing (Simpson 1980, 98–100). One-toedness must come readily to perissodactyls.

Such a built-in evolutionary tendency may be surmised to have helped shape *H. sapiens.* The growth of the brain seems somehow to come easily to mammals. From the beginnings of the class, very early in the age of reptiles, mammals had much larger brains than reptiles. After the passing of the dinosaurs, mammals rapidly increased brain size. The primates developed exceptionally large brains for reasons that cannot be clearly attributed to their way of life. This trend was continued and exaggerated in the human line, and again the role of natural selection is unclear.

It is broadly understandable that a successful genome carries traits beyond the dictates of natural selection. Selection for certain mutations is at the same time selection for the corresponding mutability, the combination of enzymes and proteins, transposons, repressors, and others that made the useful mutations possible. Patterns and trends become self-reinforcing as species evolve in certain ways (Campbell 1985, 145–146). The genetic conditions favoring increased number of legs or enlargement of the brain could continue even if there were no particular benefit from additional legs or development of the brain. The interdependence of elements of the genome may entail momentum of development within limits of viability.

Some directions are easy, others difficult or impossible. This implies something like orthogenesis—not a mystic impulse but the result of characteristics of the genome. It is logically to be expected, moreover, and is explicable in genetic terms (J. Campbell 1985) that if selection favors a trait, this may mean favoring the configuration that made possible mutation for that trait. Change would continue on the same track.

This notion may be broadened speculatively: selection for any trait may carry with it visible or invisible correlated changes in the genome. Indeed, it would seem that traits that may be adopted more or less independently (there are probably no traits without side effects) are evolutionarily unimportant, like details of dogs' coats. If changes were unrelated, evolution probably could go nowhere.

10

Evolutionary Change

Speciation

Darwin's problem was the origin of species, although he did
not tackle it directly. It is still unsolved. The idea of species
belongs to the common culture, a classification that is intuitively, although imprecisely, obvious. We are aware of cats,
hardly of felids, carnivores, mammals, and so forth. The first
task of evolutionary theory was therefore to deal with the
familiar category, the species, which were traditionally
regarded as separately created and changeless. Moreover, it
has been the natural but unsound assumption that understanding of the development of species is equivalent to the understanding of evolution.

The species classification is natural; New Guinea natives have
names for 137 of the 138 species of birds distinguished by
ornithologists (Strickberger 1990, 485). Yet there is no satisfactory definition. Dachshunds and greyhounds are capable of
interbreeding, although in practice they hardly do so, and are
regarded as belonging to the same species, like Eskimos and
Sudanese. Taxonomists prefer a morphological definition,
which means that species are populations of plants and animals
that resemble each other and are visibly distinct from other
populations—taking into account accidental differences, sex,
and so forth. But there is no agreement on how close the
resemblance must be or when a variety becomes a species. For
the biologist, it is more scientifically satisfactory to think of
species as groups of organisms that form an interbreeding
community, isolated from other such communities, a definition
advanced especially by Ernst Mayr.

Species have evolved many barriers to hybridization, such as
differences of courtship or season of reproduction. If mixed

matings are sterile or the offspring are sterile, one assumes that the parents belong to different species. This definition is more theoretical than practical, however; the reproductive isolation of only an infinitesimal fraction of the millions of species has been studied, and in practice species are almost always defined by their morphology, sometimes by their behavior.

The definition by reproductive isolation is often difficult to apply. When populations are geographically isolated—on islands or mountain tops—it may be difficult to decide whether they constitute separate species. When a population varies continually over a large range, if variation is continuous, it is regarded as a single species. But the extremes may be incapable of interbreeding. Leopard frogs in the United States are an example (Mayr 1963, 24). On the other hand, some populations regarded as different species, such as dogs and wolves, interbreed freely if they are allowed to.

Neighboring different populations are regarded as distinct species despite hybridization if the large majority of the populations clearly belongs to one kind or the other. Many quite unlike species of oaks, for example, produce fertile hybrids in nature. No fewer than 20 are known in California, and they are vigorous and fertile (Tucker 1990, 13–18). But where the ranges of two species meet, the zone of hybridization is usually narrow; nature does not seem to favor the mixing of two separately successful types. Population outcrossing ordinarily makes for vigor, species crossing more commonly for weakness (Mayr 1984, 73).

Many similar species are distinguished chiefly by some habit—their food, breeding locale, time of activity, or the like—but it is usually possible to find slight structural differences on close examination. Detailed study sometimes shows that what was thought to be a single species actually is several; for example, the European malarial mosquito (*Anopheles*) was found to be difficult to combat because it is really five species (Mayr 1963, 35). An extreme case is the protozoan *Paramecium aurelia*, which has 28 known mating types in 14 classes, which could be called distinct species, although they are indistinguishable, because they do not interbreed (Sleigh 1989, 84).

Similar but noninterbreeding populations are called sibling species, but there is no clear-cut delimitation between clearly distinct species and sibling species, or between species and subspecies, varieties, and ecological and geographic races; taxonomists often differ in their judgment. Plants are more variable than animals, and many botanists are skeptical of the species concept.

Morphs, or variants within interbreeding populations, such as different colorations of birds or fishes, are common. A single species may be very diverse. A small Puerto Rican frog (*Elentherodactylus coqui*) comes in more than 20 different patterns (Pough, Heiser, and McFarland 1989, 21). One vegetable species, *Brassica oleracea,* includes not only cabbages of sundry shapes and colors but kale, brussel sprouts, broccoli, and cauliflower. Morphs are not easily accounted for by principles of natural selection; one would suppose that a single type would be best adapted in a given habitat. They often increase or decrease for no apparent reason (Mayr 1963, 244). They may conceivably be incipient species.

It is intellectually most appealing to define a species as a genetically homeostatic group, that is, one with tendencies to revert to type despite secondary differences. The dog is a good example. Genetic homeostasis is difficult to test, but the definition is realistic. It underlines the fact that species seem generally to have settled into stability; the fossil record usually shows them changing little over long periods, although the paleontologist, who sees only a skeleton, has to regard a species differently from a field naturalist, who sees differences mostly in nonfossilizable parts. The hundreds of species of African cichlids would be scores if only their bones remained.

During their geological lifetimes, species usually vary only within rather narrow limits and without any overall direction or trend pointing to a successor species (Katz 1987, 287). They cannot merely have become very well adapted to the environment; if this were the case, they would vary continuously with slightly changing environmental conditions. To the contrary, a species represents an effectively integrated genetic combination, or attractor, that is likely to persist with little reflection of environmental change.

An established species has tried out and assimilated or rejected the mutations that ordinarily occur in it. To reach a superior set of traits, it might well have to pass through a condition of inferior adaptation to a new condition of stability. It is disputed to what extent this requires geographic separation—being marooned on an island, for example, where a small, detached population could change, drift, or experiment randomly, possibly becoming less well adapted. Mayr would practically deny the possibility of a new species' appearing without physical separation. The numbers should be small, perhaps originating with a tiny founder population (Mayr's term).

The fewer individuals that are involved, the greater is the likelihood for changes to survive by sheer chance or for mutations to become established; possibly a single pair could set the process of speciation in motion. Probably a number of mutations must come together, in ways possible only in a small population (Bates 1989, 229). "Selection" is, after all, mostly accident: a seed is eaten, another falls into moist earth; one baby fish is swallowed, another is overlooked, regardless of the qualities that might have appeared in the adult. Speciation becomes a possibly random transition between shallow attractors. In a small population, there will be proportionally fewer mutations; speciation can be favored only if there are many small populations.

Archipelagos, with numerous partially isolated territories, such as the Hawaiian or Galapagos Islands, show the importance of geographic separation. In the Hawaiian Islands, the oldest of which counts only about 6 million years (although older islands have disappeared) and the youngest under a million, there may be over 800 species of fruit fly, probably all descended in 10 million years or so from a single fortunate colonization (Carson and Kaneshiro 1976, 311–345). They comprise more than a third of all members of the genus *Drosophila* worldwide and show more variation than all the others together. Many of them are adapted to different ways of making a living, not only herbivorous but nectivorous, scavenging, parasitic, and predatory, including niches not occupied elsewhere in the world by *Drosophila* or even by flies. Some species, on the other hand, have different structures with similar habits (Dominey 1984, 240).

Almost all species are endemic to a single island. The large number of islands plus the existence of many patches of vegetation isolated by lava flows must have made the extraordinary speciation possible, although they cannot entirely account for it. The Hawaiian Islands have many other species swarms, both animals (such as snails) and plants (especially a composite, silvershields), presumably descendants of a single colonization.

The diversification of Hawaiian honeycreepers far exceeds that of the finches that Darwin made famous.

The Galapagos, 13 sizable islands and many little ones, have developed in a few million years 15 varieties of tortoises with the same diet but slightly different shells, 14 of Darwin's finches, 66 of snails, and so forth, most of them limited to one or a few islands, all in practically the same arid environment unconducive to exuberance of life (Lack 1947, 154). The finches have been most studied, because of Darwin's attention to them, but it remains unclear to what extent their differences, especially of beaks, are the result of dietary specialization or of independent drift on the separate islands, most of which harbor a single species. For the most part, the finches on the smaller islands have simply gone their independent way. Such diversification is the more significant because comparable continental species inhabit a wide range of environments. Ironically, the fauna of the Galapagos supports the fact of evolution but not the means Darwin proposed.

A combination of adaptation and random drift, leading, for example, to the great variety of fruit flies in the Hawaiian Islands, would represent evolutionary change, but this can hardly be the whole story of speciation, perhaps not even the major part of it. Geographical isolation followed by reunion is not frequent, and most species must have arisen within the territory of older species. This was the case, for example, of each successive species of *Homo* (Groves 1989, 320). Separation does not have to be geographic; isolation is the key. An important factor may be entry into a new environment.

There is no reason that morphs, without a clear-cut spatial isolation, might not give rise to species. It would only be necessary that, in a variable population, like should cleave unto like, with assortative mating leading under favorable circumstances to reproductive isolation. Many species of birds, fish, snails, butterflies, beetles, and so forth show widely different colorations, perhaps so unlike as to have been classified as separate species. Frequencies of morphs vary in different territories, which in the case of less mobile animals may be only a mile or so apart (Owen 1966, 91–95). The morphs that naturalists note are almost exclusively of the coat, but one may feel sure that many less conspicuous traits vary in similar fashion. If, then, a morph related to sexual preference became associated with some difference of habit or environment, a new species would be on the way. In this view, speciation should be most rapid when adults and offspring tend to remain together, are patchily distributed, and are territorial.

Social dynamics may thus be a major factor in evolutionary change, at least of higher animals. The formation of new species depends on many things and often, perhaps usually, seems to be qualitatively different from microevolutionary change, probably because speciation involves not single genes but combinations of genes (Williamson 1981, 442).

Molecular Change: Microevolution

Within a species, such as the human, individuals (except identical twins) differ in a multitude of ways; we suppose that anyone is easily identifiable by physiognomy. Naturalists also keep track of individual giraffes or whales by noting distinctive markings. Typically, 10 to 30 percent of genes in animals have different alleles. Proteins always differ slightly. To the discomfiture of organ transplant patients, the body rejects alien tissues, the vigor of rejection depending on the degree of difference of proteins. There are hundreds of variants of the human hemoglobin molecule, most without biological significance (Fitch 1988, 37).

It is a corollary that proteins are gradually changing or evolving. The number of variations in the amino acid sequence of basic proteins in different species is rather closely correlated with degree of separation as judged from morphology. For example, the hemoglobin molecule differs unimportantly from humans to the nearest ape, much more from humans to dog, and so on to lizard or fish. The human-chimpanzee difference is 1 percent; human-carp, 50 percent.

Despite some anomalies, differences of proteins correlate rather well with length of time since common ancestry as indicated by fossils. But the divergence of such proteins as cytochrome-c (important for metabolism) bears little relation to anatomical difference. There is no sequence from jawless fish to bony fish to amphibians, reptiles, and mammals; egg-laying mammals are as far from reptiles in their proteins as we are. The biochemical difference between a bony fish, such as carp, and a jawless fish, as the lamprey, is about the same as that between human and lamprey (Maynard Smith 1986, 51). All members of any group have approximately the same degree of difference from all members of a different group.

Such changes are not the prime materials of evolution. Changes of proteins and evolution of morphology go on at virtually independent rates (Raff et al. 1987, 117): "Structural [protein-making] genes, as shown by alterations in protein sequences, do not seem to play a major role in evolution" (Langridge 1987, 253). To change an animal it is probably unnecessary to change proteins or "structural" genes at all. The Hawaiian *Drosophila* are extremely diverse in morphology but very close in proteins. Numerous species of African cichlids have almost identical proteins (Pough, Heiser, and McFarland 1989, 30).

This difference appears by comparison of anurans (frogs and toads) with mammals. The anurans arose over three times as long ago as the postdinosaurian mammals, and in proteins they are much more divergent than mammals. But in anatomy they are much less so. Frogs are classified as a single order, while mammals fall into 16; one frog genus, *Xenopus*, has been around with little change for 90 million years (A. Wilson 1985, 170). They have not been exempt from mutation but have varied in regulatory genes much less than mammals. Consequently, frog species are much more hybridizable than mammals (A. Wilson 1975, 122–126). Why frogs should be so conservative in morphology is unexplained; their diet of small animals certainly permits a great variety of adaptations. Their odd squatting habit hardly seems relevant. While some must find it very useful to jump, others mostly swim or climb; good runners they are not. In care of young, frogs are far more inventive than mammals or birds.

In similar fashion, there is a high degree of similarity in proteins of humans and chimpanzees (99 percent in forty-four proteins tested) despite substantial differences between the two animals. The two species have the same cell types; the differences are of relative growth of parts. That is, they must be primarily, perhaps almost entirely, ascribable to regulatory genes and differences or relations of groups of genes. This is indicated by the fact that chromosomes of the two species are separated by at least ten fusions, inversions, and tranlocations (John and Miklos 1988, 293).

The anurans all have basically the same pattern but endless variations of detail. (Illustration by Ernst Haeckel, *Art Forms in Nature*, © 1974 by Dover Publications, Inc. Originally published by the Verlag des Bibliographischen Instituts, Leipzig and Vienna, 1904.)

Structural Change: Neoteny

Small typos that happen to make sense can give new twists, but they cannot make a new story while keeping the work always readable. One can build a bigger wall by adding a brick at a time, or add flourishes ad lib, but one cannot make an arch or a dome by bits. Nature has an abundance of elegant arches and domes. There are obviously different kinds of evolution, although it is unclear whether there is a sharp difference between them or, as seems more likely, they are extremes of a continuum.

How new organs and types of tissues evolve is obscure, but much differentiation, even as significant as that which made an ape into a human, can be interpreted simply as changes of rates and times of growth. In simple form, this means allometric growth, or change of relative size of parts. A great deal of development from embryo to adult is allometric; the newborn baby's legs are programmed to grow about five times more than its eyes. The human cranium continues to grow after birth much more than that of the chimpanzee, whose jaws grow much more than those of the human.

It is supposed that much of the difference of the giant panda from its bear progenitor consists in the better development of the fore part of the body, particularly the chewing mechanism, at the cost of weaker development of the rear portion (Stanley 1981, 129). A clearer case is the enlargement of deer antlers more than proportionally to the body; those of a large deer are not only absolutely but relatively larger (Grant 1985, 365), the prime example of the overgrowth of antlers being the Irish elk. Very small deer of several Asian genera, under 20 inches in height, have very short antlers or none at all. In compensation, they have tusks (Gross 1983, 6–27). The primitive condition is the possession of tusks for fighting purposes; as deer grew larger, the antlers increased disproportionately, despite the cost of annually renewing them, and the tusks disappeared.

Allometric growth comes naturally because parts have to grow at different rates in morphogenesis. If the animal becomes larger evolutionarily, that is, continues growth longer, the different rates have greater effect. On the other hand, it is possible for certain parts or aspects of the organism to retain more or less juvenile form although increasing in size. This is juvenilization, or neoteny in the usual term. It must rest on mutations of regulator genes—perhaps only a few being needed to alter the pattern.

Neoteny is conducive to innovation. It permits a new beginning and relatively rapid change as the organism backs up evolutionarily to get a better start. Adult specializations are left behind, and the species can proceed to develop in new directions. It is a related fact that smaller animals, usually having less developed structures, are more likely to give rise to new forms (Benton 1990, 151).

Some traits are more subject to neotenic change than others. The reproductive system, of course, has to mature while other organs are held back. From our point of view, the most interesting aspect of neoteny is the relative growth of the head. In vertebrates from fish up, the head takes a lead in the embryo and is relatively large at birth, as one observes in baby cats, pigs, and humans, and neotenic vertebrates are usually more or less big headed.

Neoteny may be responsible for the origin of insects from myriapods, which are born with six legs, and for the origin of our chordate phylum; larval tunicates have the fundamental traits leading to our lineage. Several groups of salamanders in neotenic fashion retain larval gills when fully grown unless given a thyroid hormone to cause maturation (Young 1981, 271). Lampreys have a disposition to neotenic near-repression of the adult. Ordinary lampreys live as larvae in brooks, undergo metamorphosis, and go down to the ocean (or lake) where they fasten onto fish and continue to grow for years. But many lamprey species in various parts of the world have independently developed sister species that become sexually mature, spawn, and die immediately after metamorphosing (Young 1981, 103–104). Neoteny also helps island birds to become flightless; the adults are like overgrown chicks and retain chicklike features (D. Anderson 1987, 147).

Toy dogs are neotenic; the pekingese is like a puppy, with rounded skull, relatively large cranium, short face, big eyes, short legs, curly tail, and soft fur. Other breeds with large puppylike heads and less developed legs, including boxers and bulldogs, are neotenic. Many domestic animals show such a tendency; pigs, cattle, sheep, goats, and to a degree cats have shorter skulls than their wild ancestors (Zeuner 1963, 67–70). This may be because of selective breeding for early reproduction. Humans are the most striking example; neoteny has long been believed to explain something, perhaps much, of human peculiarities.

Neoteny implies that many genes are repressed if not eliminated, making way for the expression of new or previously repressed genes. It is difficult for mutations to be useful unless they are slight, maintaining existing physiology and structures; neoteny is important, if not essential for much evolutionary change because it sets aside adult structures. Large innovations in metazoans probably entail changes early in the embryological sequence (Thomson 1988, 122, 132). The fact that the genome is a set of instructions, not a blueprint, is essential. Neoteny simplifies the instructions, making it easier for a small change to alter the line of development.

Moreover, it may not be necessary to carry out a large alteration immediately, only to take a new direction. In *Archaeopteryx* most of the alterations carrying to birdness remained to be made—not only adaptation of the skeleton but traits irrelevant to flight, such as stub tail, toothlessness, and probably the calcified eggshell. That the innovation is only a turn of direction would relieve the problem of how the mutant—the traditional "hopeful monster"—finds a mate, and its innovation would have as good chances of becoming established as an ordinary mutation.

It is not surprising that large evolutionary innovations are not well understood. None has ever been observed, and we have no idea whether any may be in progress. There is no good fossil record of any. Because they are difficult, evolution has occupied billions, not hundreds of thousands of years.

Fixity

Species with low fecundity and under slight selective pressure maintain very complex organ systems apparently indefinitely. This fact may be related to the ability of many plants and animals to continue as though changeless over vast stretches of time.

Species usually stand still, as has been often noted. One study of foraminifera (planktonic protozoa) of the past 10 million years is ambiguous; most species seemed to be static and then to change relatively rapidly but not explosively (Malgren, Berggen, and Lohmann 1984, 317–319). But it is exceptional that the record shows considerable change in a lineage. For example, a detailed investigation of Cenozoic snails and bivalves found no gradual change in any lineage but sudden replacement of one by another, usually under the stress of climatic change (Williams 1981, 437–443). British beetles have hardly varied in the past million years, and not greatly for at least 5.7 million years, while mammals have been evolving rapidly. There have been no important innovations and few extinctions, although ranges have shifted with the advances and retreats of ice sheets (Coope 1979, 247–267)—despite the fact that beetles (coleoptera) must have had a special faculty for innovation. Only one of 25 orders of insects, they comprise more species than all noninsects together (except mites).

Over a longer span, ants preserved in amber 25 million years old look very like those of today; some species are difficult to distinguish from modern descendants (Hölldobler and Wilson 1990, 24). There are many vastly older examples of rigidity, the "fossil species." Lungfish go back 350 million years, and horseshoe crabs (*Limulus*) have changed little, at least in their skeleton, for that time or longer. Some brachiopods (*Lingula*) are apparently unchanged, at least in their shells, for 450 million years. Other holdouts of bygone ages include crocodiles, some turtles, and a variety of bony fish and sharks. A number of plants, such as ginkgoes, cycads, horsetails, and clubmosses, are at least 100 million years old. One leafless plant without true roots, *Psilotum*, seems to be much like the first vascular plants that appeared on land about 400 million years ago. Dragonflies and cockroaches not unlike the contemporary species have been flitting around for 300 million years. The bony-finned coelacanth, thought to be long extinct but rediscovered in 1938, has been approximately static some 450 million years (Avers 1989, 317). First prize may go to the peripatus, a soft-bodied legged animal between an annelid and an arthropod; a creature at least superficially very like it was found in the Burgess Shales, about 570 million years old (S. Gould 1989b, 168).

Such changeless creatures must have achieved an optimum within the range of changes accessible to them. Some of them, such as opossums, horseshoe crabs, dragonflies, and cockroaches, are abundant today. But the antique forms typically have little diversity. There is only one extant species of ginkgo and (so far as known) of coelacanth. There are few species of such antiques as horsetails, cycads, clubmosses, lungfish, and horseshoe crabs. The many varieties of dragonflies are much alike.

The nearly timeless species are not exempt from the changes of proteins that go on in all living beings, and they could surely vary in many ways without loss of adaptiveness, but their patterns have become somehow frozen. The fact that modern ginkgo leaves are indistinguishable from those of dinosauran days does not mean that they are especially well adapted but that they are fixed in the ginkgo pattern. Fan shaped, bifacial, parallel veined, and two-lobed (usually), they are unique, unlike all the myriad shapes that other plants have found satisfactory. Yet leaves of modern ginkgoes are somewhat varied. They must have been slightly changeable for 100 million years without breaking out of the mold.

From the point of view of conventional evolutionary theory, long-term stasis is hard to explain. Rapid evolution is comprehensible as species adapt to new conditions or opportunities; but it is incongruous that species remain unchanged through changing conditions over many million years (Sheldon 1990, 114). Even without adaptive pressures, one would expect drift to alter appearances. Yet there is logic in fixity. Minor changes may be admissible (as the mating habits of the dragonfly are varied), but larger changes are probably harmful. No change being particularly required, the soundest procedure is to stand fast. The record often shows that rapid change slows and ceases (for some fish families, Young 1981, 221). In other words, the cohesion of the genotype seems to grow stronger as it becomes settled over a long period (Mayr 1984, 77), or attractors in the genome become more stable as variants are discarded, and inevitably occurring mutations have no cumulative effect. Because stability, overall or in certain areas, has survival value, there must be selection for configurations maximizing stability. Some species in effect learn how to evolve; others learn how to avoid evolving.

This harmonizes well with the fact that the big branching of metazoan life occurred very early, the phyla—and very many classes—appearing in startling variety 570 million years ago. There were relatively many phyla—many more than now—when there were few species. For the phyla to result from a large number of species, changes would have required many more species than there were, appearing and replacing one another over a long period, a process that would surely have left an abundant record (Valentine and Erwin 1987, 97). It has been suggested that the metazoans underwent the phenomenal radiation preserved in the Burgess Shales because they were new in an unoccupied landscape, but this is contradicted by the fact that the radiation was not of species but of phyla and classes. Moreover, the environment prior to the appearance of metazoa was relatively uniform, with few niches; most niches, after all, are made by living creatures.

The phyla grew old. The foundations, overlaid by a multitude of secondary modifications, long ago became unalterable. When animals emerged on the land, no new phyla arose in the totally changed environment; the old basic patterns sufficed (McMenamin and McMenamin 1989, 169). Classes also became solidified. For example, the trilobites burst on the scene in the early Cambrian and were the dominant animals for some 200 million years. But after their radiation, forms changed little for 300 million years; in the last 200 million years of their history, no important new group appeared (Eldredge 1987, 75). In due course, subphyla, classes, orders, families, and genera have tended to settle down. The mammals are exceptional in continuing rapid evolution of some orders, such as carnivores, ungulates, and primates, into recent times.

Radiation

From time to time the stability of living forms has given way to variability, generating the almost inconceivable variety of life. When the integrity of a pattern—its successful cohesiveness—is abandoned, the way is opened to multiple changes in different directions, perhaps amounting to what is called an evolutionary radiation. Variability is variable, in both multiplicity and depth, from superficial externals to deeper reorganizations.

Humans, for example, are a rather new animal and a sharp departure from the great apes; it is hence not surprising that the tribe is versatile. Fossil species of *Homo* are decidedly diverse (Groves 1989, 317), to the considerable confusion of paleoanthropologists. One wonders whether they may not have been, in nonfossilizable traits, even more varied than the single surviving species. If the same criteria were applied to humans as naturalists apply to animals, a dozen persons might well be placed in a dozen different species on grounds of build (from pygmies to lanky Nilotic herdsmen and squat Eskimoes), distribution of fat (buttocks, bellies, breasts), hair color (black to red to flaxen), hair shape (from tightly kinky to straight), different body hair, skin color, eye color, lips, noses, and so forth—traits for the most part seemingly without adaptive significance.

Variability may be concentrated in a single area. Although bats are remarkably similar in their general aspect, habits, and mode of flying, their facial features differ phenomenally. One bat has a hood it can pull up from under its chin to cover the face, with transparent spots like goggles for the eyes; others have eccentric ears or a huge larynx (Voelker 1986, 47), and they would be leading contestants in an ugliness contest.

Orchids show an extreme of speciation. The family is relatively recent, but it is the most species rich of all plant families, with some 25,000 species. There is little variation in leaves, stems, or seed capsules, and the greater number have a single habit, tropical epiphytes. But the flowers are a marvel of invention, a kaleidoscope of shapes and colors. Many of them are especially adapted to particular insect pollinizers, but the adaptation is entirely on the part of the orchids; the insects have not coevolved. The obvious explanation might be a strong need for outbreeding, hence specialization to attract particular instincts. This does not fully explain the creativity of the orchids, however, because they hybridize freely and do not seem to suffer inbreeding depression (D. Gill 1989, 463–466). Perhaps the best that can be said is that in the orchid family the attractor governing flower shape is unstable, and as variegated blossoms proved attractive to certain pollinators, there was a feedback rewarding and strengthening traits attractive to those particular wasps or flies or beetles. This hypothesis, however, is negated by the fact that many orchids deceive the pollinators by declining to reward them with nectar.

Variability may be shallow. For example, cuckoos can match the eggs of their hosts because they have the faculty of varying the patterns of their own eggs; eggs of nonparasitic relatives of the cuckoo are also quite varied (Skutch 1987, 31). Birds of paradise, with generally similar ways of life and bodies, are creative in plumage; of forty-two species, thirty-three are gaudy to spectacular, displaying splendid crests, coils, and capes, with feathers long and short, thick and thin, even reduced to near-threads flaring at the ends, projecting from the head, the tail, or the wings (Beehler 1989, 118). Different genes must dictate narrow green feathers ornamenting the head and long blue feathers on the tail, but a set of genes dealing with feathers must have been set loose to vary. In other words, there must have been selection for the ability to produce ornamental plumage.

Variability can be increased rather easily. Thus, selection for mutations can increase the rate of mutation in bacteria (J. Campbell 1985, 137, 146). Wild canines are no more varied than wild felines, but from ancient times, dogs have been bred much more than cats for different purposes, perhaps because dogs are not only larger but also, as social animals, more trainable and responsive to their masters. Cats are rather ornamental, valued for elegant appearances. Hence it is less surprising that dogs have undergone a varietal radiation while cats have stayed close to the wild type.

If a species has the good fortune to move into a vacant environment, variability becomes very advantageous; the colonization of a new region puts a premium on change (Mayr 1984, 78). The explosions of species of fruit flies, snails, and other species in the Hawaiian Islands show such a propensity for variation and experimentation, a propensity that seems to have its inertia and carries forward beyond environmental needs. Having filled the available niches, the fruit flies have continued to diversify in such fashion that numerous species overlap and share a territory that elsewhere would have a single species.

Interconnected lakes are somewhat like islands; well over 500 species of cichlid fish have arisen in seven East African lakes in less than a million years, many of them in less than 10,000 years. Of 126 species in Lake Tanganyika, none is found elsewhere. They display every possible kind of family life, including monogamy, polygamy, maternal, paternal, or two-parent rearing of young, plus mouthbrooding. Their feeding habits have become phenomenally specialized. Although many species are omnivorous, there are forms adapted morphologically for at least the following: (1) catching floating algae by means of mucus in the mouth; (2) picking up vegetable material on the bottom; (3) scraping algae from rocks; (4) eating algae growing on larger plants; (5) grazing on plants; (6) eating snails and mussels, either crushing and swallowing the molluscs or pulling the animal from its shell; (7) catching water insects and crustaceans, with subspecializations on different kinds; (8) digging in bottom sand and filtering contents through the gills or picking out edibles in the mouth; (9) taking zooplankton in open waters; (10) picking off scales of other fish; (11) snatching pieces of fin or biting out the eyes of other fish; (12) catching other fish; (13) eating eggs and/or larvae of other fish, chiefly cichlids. One species grasps a mouthbrooder by the jaw and forces it to release its young to be scooped up (P. Greenwood 1974, 5, 119; Fryer and Iles 1972, 61–104).

Like Hawaiian *Drosophila,* African cichlids not only have numerous species competing in a single habitat without interbreeding but have entered many niches that the family does not occupy elsewhere (Dominey 1984, 236, 240). Not surprisingly, the African cichlids are very diverse in traits related to diet, jaws, teeth, and so forth but are similar in general body shape and in proteins. The cichlids also have great diversity of reproductive strategies.

How fluid the cichlid pattern may be is shown by a Mexican cichlid, *Cichlasoma,* which takes advantage of a variety of edibles—snails, algae, fish, and arthropods—by producing within a single brood individuals with different teeth and digestive apparatus, even when all are raised on soft food in the laboratory (Sage and Selander 1975, 4669–4673). It should not be supposed, however, that the cichlid family is unique; there are various other, although less spectacular, species swarms of fish as in the Philippines, Mexico, and Lake Baikal.

A microcosm of microevolution occurred when house sparrows were introduced into the United States in 1850. Spreading into a fairly open field, the human-modified habitat, the sparrow diversified into numerous races of differing size and plumage (Ehrlich, Dobkin, and Wheye 1988, 634). A more striking example occurred when fish (*Bairdiella*) were introduced into the newly formed Salton Sea in southern California early in this century. In the unoccupied habitat, up to a quarter of the fish became markedly deviant, some with what might be called gross deformities. They did not have sufficient space and time, however, to engender new species, and the deviants were eliminated as the population reached carrying capacity (Williamson 1987, 135–136). Sometimes evolutionary exuberance seems capricious. For example, there are 290 species of amphipods, crustaceans like tiny shrimp, in chilly Lake Baikal, with very diverse feeding habits (Dominey 1984, 239).

New species can appear in a few hundred years. Polynesians brought bananas to Hawaii probably not more than a thousand years ago, and there are no native members of the banana family in the area. Yet at least five closely related species of moths (of the genus *Hedylepta*), all with different territories, feed obligatorily on bananas of the strains that have gone wild on the islands. They seem to be descended from a moth that subsists on palms (Zimmerman 1960, 137–138).

Extravagances of evolution such as the African cichlids and the Hawaiian fruit flies are intuitively understandable, but questions remain unresolved. Fish of various genera entered the newly formed African lakes; only the cichlids exploded, becoming dominant in both numbers and species. On the other hand, cichlids have not radiated extraordinarily where they live from India to South America (Greenwood 1984, 141). It is conjectured that discriminating sexual selection, with extensive courtship, has been an important factor in diversificiation of species flocks (Dominey 1984, 235–236). But several investigators suggest "a need for a new driving force for this sort of evolution" (Sage, Loiselle, and Wilson 1984, 189).

That change facilitates more change not only implies new external forms but internal remodeling. Big alterations of design—fish to amphibian to reptile, reptile to bird, and so forth—involved not only changes directly connected with the new way of life but many transformations of organs. Birds, mammals, and reptiles differ quite as much in soft as hard parts. Reptiles are unlike amphibians in dozens of facts of the anatomy, from claws and scales to—perhaps most significant— the invention of an egg much improved over what the amphibians had (or have produced to this day), with membranes to protect the embryo, means of breathing, and shell. Supposedly because of their multiform improvements, reptiles enjoyed a grand radiation into many orders much like that of mammals 160 million years later (Paul 1988, 164).

It seems that feathers were the key development of the birds. But feathers, however remarkable, gave no quick success. Birds were long very scarce, to judge from the fewness of fossils of birds or birdlike creatures for some 120 million years after protoavis. All known specimens of *Archaeopteryx* came from a single deposit. Rather suddenly in the early Cretaceous, water birds (especially *Hesperornis*) close to modern type appear in some abundance. Their success could not have come from feathers and the power of flight, which had been around many million years, but perhaps from such advances as a high rate of metabolism, a four-chambered heart, and a set of air sacs and ducts to pump air continuously through tubes in the lungs (much more efficient than the reptilian and mammalian respiratory system). The bird pattern is so effective that many large species, of which ostriches are a contemporary example, have been able to compete with mammals on the ground, despite the lack of forelegs. In fact, there were flightless birds almost as soon as birds proper appeared (Paul 1988, 68).

A radiation of variants occurs because of the relaxation of natural selection, a new freedom. This may come about in several ways. One is an invention inviting the proliferation of new forms, such as cell differentiation to make multicellular organisms, or jaws, or limbs, or the reptilian egg (giving independence of water for reproduction). It is guessed that the radiation of the flowering plants in the Cretaceous occurred because of the development of insect pollination (Stanley 1981, 90). Or the opportunity may be entry into a new environment, as emergence onto the land or taking to the air. Or survivors may profit from a massive extinction of competitors.

The most notable case of seizure of a vacant stage occurred in the mammals' succession to the dinosaurs. The story of the mammals is strange. The mammal-like reptiles of 250 million years ago were the dominant order of the time, well before the appearance of the dinosaurs. They disappeared, however, and dinosaurs became rulers of the land as no other tribe before or after; there was no nondinosaur larger than a poodle during the dinosaur reign. Our ancestors, far from profiting by the superiority which we naturally attribute to them, had shrunk in size and number by 180 million years ago. For over 100 million years, one might have reasonably assumed that mammals were not the promise of the future but a relic of the past. One is reminded of the approximately 80 million years during which birdlike reptiles were too few to leave more than two small fossil finds.

When the dominant reptiles vanished, the mammals came into their own, although the most prominent group, the multituberculata, disappeared not long afterward. More than 30 orders appeared in a relatively brief time. This differentiation and expansion must have included many changes of internal anatomy and physiology, as well as of the skeleton. Only extraordinary genetic freedom could have made possible the transformation of quadrupeds into whales and bats in 10 million years or less after the demise of the dinosaurs. Some 16 families of ungulates differentiated within 2 or 3 million years from the condylarths (Carroll 1988, 578).

The modern orders of birds also appeared shortly after the departure of the dinosaurs. Although birds had progressed more than mammals in the Mesozoic, it is possible that only one species survived the catastrophe that overtook the dinosaurs and many other animals, both marine and terrestrial. The fortunate bird or birds gave rise to an exuberant radiation parallel to that of mammals (Wyles, Kimbal, and Wilson 1983, 4395), including giant flightless birds where mammalian predators were absent, as on Madagascar and New Zealand. More surprising, reptiles that survived the catastrophe grew huge, competing with the big new mammals. A turtle, *Colossochelys,* was over several meters long and must have weighed well in the elephant range. A land-going crocodile (Rhamphosuchus) reached 15 meters. Into recent times, an Australian lizard of the clan of the Komodo monitor but about fifteen times larger (*Megalania*) terrorized the kangaroos (Bellairs 1970, 29; Colbert 1980, 233; Grzimek 1974, 6:40, 327). The triumph of mammals was hardly to be taken for granted.

Extinction

The great extinction at the end of the Cretaceous, clearing the stage for modern animal orders, was only one of many die-offs of animal (not plant) families and genera, a sequence of eight major and many minor extinction episodes, extending back as far as the fossil record goes (Donovan 1989, xii). The first of which we have knowledge probably occurred about 650 million years ago and did away with the Ediacaran-Vendian soft fauna, making way for the Burgess Shale fauna with fossilizable hard parts (McMenamin 1990, 180). The greatest was at the end of the Permian, about 250 million years back, when possibly 95 percent of species perished, compared with up to 75 percent that accompanied the dinosaurs into oblivion (Clarkson 1986, 58–59). The trilobites that had swarmed in the seas for 300 million years then came to an end. Planktonic foraminifera (protozoa) have repeatedly been cut back by extinction events and then reradiated, as have echinoderms and other animals (Eldredge 1987, 170).

There are many guesses why great numbers of families and orders die out, including changes of sea level, climate, volcanism, disease, and the wounding of the earth by a comet or asteroid. But none makes clear why all of some classes—all dinosaurs and all pterosaurs, and so forth—should succumb, while other classes survived, some of them apparently undiminished.

The most studied extinction, about 65 million years ago, closed the age of reptiles. The dinosaurs had come through previous lesser extinctions, at the end of the Jurassic, 144 million years ago, and at the end of the Triassic, 208 million years ago, and recovered to new total dominance, only to be swept away at the end of the Cretaceous. They were far from alone. Not only land but marine animals were removed, not only marine reptiles but many others; foraminifera, molluscs, and other marine animals were severely reduced. Mammals and birds came through, of course, but a large majority of marsupials perished (Carroll 1988, 328). Crocodiles, lizards, and turtles escaped.

In the logic of natural selection, one group of animals might replace another gradually, as the new class developed superior adaptations and outcompeted the dominant form. But in all probability, some of the most vigorous and best or specially adapted or isolated species of the outcompeted group would remain in their niches or areas, like marsupials in Australia. But the dinosaurs disappeared as though accursed. This would seem impossible as long as the earth remained more or less habitable. The dinosaurs were very numerous and diverse; as the only sizable land animals, they occupied all manner of niches, from beyond the arctic circle south. If some of them were wiped out, it should have been an opportunity for others to take their places.

Credence has consequently been given to the Alvarez hypothesis (Alvarez 1983) that an asteroid or comet striking the earth raised a global dust cloud, chilling the land and halting plant growth. Persuasive evidence for this is the iridium-rich layer marking the boundary between Cretaceous and Tertiary, the Mesozoic and Cenozoic strata, which is easily explained only by the disintegration of an extraterrestrial body (meteorites usually contain much more iridium than earth rocks). Shock-metamorphosed quartz particles of the proper age have also been found in many places (Dixon 1988, 6–9). Most plants would survive the months of darkness by means of rootstocks or seeds; large animals needing fodder would die, as would their predators. Small animals that could hibernate or scrounge amid the debris would survive. Insects, however, most of which eat plants or plant-eaters, would suffer severely. The fossil record of insects is rather good, but it shows no evidence of abrupt change at the boundary (Whalley 1987, 123).

Enough asteroids cross the earth's orbit that a globe-shaking collision every few million years is probable, and the evidence of some such impact is strong (Jablonski 1990, 165, 170). But most paleontologists doubt that such a catastrophe could have been the chief cause of the extinction of a large majority of animal species. The dinosaurs had been on the wane for some millions of years, failing to replace genera that died out, as had occurred in better times. Nine orders of dinosaurs and the pterosaurs stopped at the boundary, but 16 orders and the last of the marine reptiles (plesiosaurs) came to an end in the preceding 10 million years (Carroll 1988, 287). Alternative explanations of the extinction are feeble, however (Glen 1990, 354–370).

Although the oceanic environment is more stable than the terrestrial, there have been 12 major reductions of marine life in the past 250 million years since late Permian, at intervals of 5 million to 34 million years (Carroll 1988, 589). Most extinctions show no evidence of an extraterrestrial shock, while there are iridium-rich layers unrelated to an extinction. The most devastating of extinctions, at the end of the Permian, cannot be attributed to an asteroid. It is especially puzzling because it had little effect on vertebrates, terrestrial or marine, although it wiped out the bulk of marine invertebrates. It was altogether different from the Cretaceous-Tertiary extinction of 65 million years ago (Erwin 1990, 193).

The size and variety of the earth and the ability to compensate for disturbances make it difficult to envision how any such change could completely destroy widespread and vigorous forms. A climatic change ordinarily causes zones of warmth and moisture to move north or south, and animals move with them. If the climate became adverse in some areas, it would probably remain or become favorable in others; if numbers of a large group of animals were much reduced, probably many would survive in refuges. When the bad conditions ceased in a few hundred thousand or million years, the survivors would recover and reoccupy their old territories.

The great ice ages of recent times caused no major extinction; lions, horses, and mammoths grew thick coats. The loss of many great mammals in North America and Eurasia about 12,000 years ago is usually laid to human hunters. Marine life may suffer from reduction of shallow seas, but there must always have been continental shelves. Volcanoes have been much more active at some times than others, but they could increase tenfold over their present level without representing a serious threat of mass extinction. An enormous upsurge of eruptions is difficult to postulate because their ultimate source, heat from radioactivity within the earth, is utterly steady.

The extinctions may or may not have been accidental or independent in causation from the dynamics of life. They may have been like death from lightning, or death from chilling, to which the elderly are sensitive. One may have the impression that the dinosaurs, like the trilobites and many groups between, eventually grew tired of life. It may be relevant that dominant forms seldom give rise to new dominant classes. The extinctions, contingent or not, give a different twist to the story of life on earth, quite out of harmony with the traditional picture of gradual change, expansion, and improvement over the ages.

However the extinctions were caused, they may have been essential for evolution toward more complex and effective organisms. Without them, living forms would probably vary, as the dinosaurs did, without breaking out of broad patterns. Wiping away most species, successful as well as less successful, invites new starts, and removal of large species gives latitude to more inventive smaller species, permitting experiment and innovation, new patterns that could not have gotten started in competition with the old rulers of the earth.

In the new freedom, the genome invents new, perhaps extravagant, shapes. There is a protean shifting of life's patterns and shapes. There is intense, perhaps chaotic, competition among lineages and species. In time, forms settle down, and change becomes less rewarding; natural selection becomes conservative, serving to eliminate deviants. Most species remain more or less static, until or unless somehow the gates are opened to a new radiation.

Why extinctions should make open spaces for new life from time to time hardly enters evolutionary theory. But they take an honorable place among the mysteries and show how far we are from understanding evolution.

Summary

It was the sense of "descent by modification" (Darwin's term) that animals were shaped bit by bit, much as a worm made of modeling clay might be squeezed a little here and built up elsewhere to turn it into a fish, then into a salamander, a more upright mammallike reptile, and eventually into a monkey and ultimately a biologist. The means of change proposed by Lamarck was mostly use and disuse; Darwin had the better idea of accumulation of those little variations that favored the production of more offspring than others. But many facts indicate that the story is far more complicated and underline the importance of the integrated organization that makes possible such a complex entity as an animal.

Of such facts, the most commented is the discontinuity of the fossil record. Only to a minor degree do forms change gradually into successor species; there are smaller leaps, as between species or genera, and larger leaps, as to new classes. This corresponds to the fact that existent taxa have much integrity. Only a few odd forms, such as the platypus, are suggestive of transition.

The power of the organizing principles of the genome is also apparent in something so banally commonplace, yet so little understood, as the ability of the genome to direct a single cell to make a body of trillions of cells organized in thousands of organs. This is the more notable for its flexibility: it is not a rigid blueprint but is subject to feedback. Divide an early embryo, or an adult flatworm, and the cells make two perfect embryos or worms. Similarly, in higher and less flexible animals, the body retains capabilities of self-restoration: in the crab or newt, limbs regrow; in the mammal, bones heal, and many organs can remake themselves after partial destruction.

A less familiar fact is the superiority of the whole: the plan of the building prevails over the details of the bricks going into it. This is seen in sibling species that have different genetic components but very similar bodies. In a sibling species swarm of ciliate protozoa (*Tetrahymena*), different proteins are used in major structures and membranes, yet the animals are quite similar (Frankel 1983, 281).

This means that there are distinct kinds of evolution, of proteins and of structures. Changes of molecular structure are going on continually at a slow rate, and comparison of basic proteins in different forms gives a fair idea of the time since their common ancestry. For example, American species of *Drosophila* are rather similar in appearance and habits but, having been around for many million years, are genetically rather different. Yet visible evolution may be rapid. Hawaiian species are quite diverse in appearance though genetically close (Stebbins and Ayala 1985, 79). Frogs have been on the scene far longer than modern mammals and are much more dissimilar in proteins, but they are much less so in morphology. Humans and chimpanzees are practically sibling species in regard to their proteins but differ in general aspect more than wolves and cheetahs, which belong to separate families. Almost every bone can be distinguished, and the pelvises are much more unlike than the brains. The chimpanzee has changed organically little from the apelike ancestor, the human quite markedly (King and Wilson 1985, 107–112).

The genome, on the one hand, is constrained; on the other, it holds latent potentialities. Domesticated animals, highly modified by artificial selection, tend to revert to something like a norm for the species. Buried traits may surface. Horses, for example, sometimes show something of the toes of their ancestors. It is difficult to believe that the hoatzin's claws were developed because of some special need for climbing. To the contrary, they must be part of the morphogenetic sequence because they regress after a few days (F. Gill 1990, 545).

Parallel evolution shows something of the capacity of the genome for the realization of certain possibilities. This is exemplified by the suborders of bats apparently of quite different ancestries. The bat style, with a membrane stretched from enormously elongated fingers, seems to be the way a mammal develops a usable airfoil. Something like this may even be in the vertebrate way of things. It would seem far easier to extend the forearm than fingers, but the pterosaurs, like the bats, took the seemingly farfetched course of flying with fingers.

The self-directedness of the genotype is shown by both fixity and plasticity. Living fossils are subject to random mutations like all other species, yet they are held to virtually changeless patterns through the changing environments of the ages. They are in the time dimension like the species that make themselves at home on five continents. The importance of the pattern is equally shown when it is relaxed, permitting a swarm of species to differentiate into a multitude of habitats and ways of life, not only forming species to fill niches but a multitude of species beyond environmental needs.

This dominance of pattern should not be surprising. The construction of organs is a very different problem from the production of proteins, and much more difficult to master. Doubtless for this reason, four-fifths of the history of life has been unicellular organisms. Evolution has been a process of discovering ways in which organisms can be put together, and it is to be expected that the genome must be a complexly integrated entity, like the animal that it can organize. It is not a jumble of traits but a unitary self-regulating and self-organizing system. In order for the organism to become established and to subsist, it must have an almost incomprehensible inherent stability.

11

Positive Adaptation

The Persistence of Lamarckism

From its beginnings, evolutionary theory has wrestled with the question of how animals fit themselves to their environment and way of life. The answer given by Jean-Baptiste Lamarck a half-century before Darwin was that the animal's striving was the key, especially the use and disuse of organs. This did not satisfy the scientifically minded because it suggested no means whereby the effects of striving—as the giraffe's reaching for higher branches—could become hereditary. Darwin supplied a simpler answer: by the accidents of heredity, some proto-giraffes were born with longer necks and forelegs; they were better nourished by being able to browse higher than other animals, hence had more offspring, and so on, until the species reached a practical limit in the modern giraffe.

This theory excludes anything purposive; progress comes only by the elimination of the less fit in a randomly varying population. The randomness of variation is the key point; the nonexistence of direct influence of the environment on the genes is central in the neo-Darwinist synthesis.

Darwin himself, however, conceded the possibility or probability of inheritance of characteristics acquired by the organism in coping with its environment, the kind of inheritance that Lamarck had earlier made the basis of his theory. As he wrote, "Something, but how much we do not know, may be attributed to the definite action of the conditions of life. Some, perhaps a great, effect, may be attributed to the increased use or disuse of parts," although selection "seems to have been the predominant power" (Darwin 1964, 108). He gave many examples of the supposed inheritance of acquired traits, such as stronger gums from hard chewing, myopia from close work, and larger lungs (of Indians in the Andes) from breathing rarefied air (Darwin 1871, 118–120).

Mendel's discovery of well-defined units of heredity—genes, they were named—made this compromise seem unnecessary. In the standard doctrine put together in the first part of this century, the mechanistic view prevailed: inheritable variations occurred without relation to actions or needs of the organism. The germ plasm, Weismann had earlier contended, was divorced from the outside world. Evolutionary changes toward greater fitness—meaning greater capacity to reproduce—were the result of fortuitous mistakes that happened to prosper in the competition of life.

This proposition was reinforced by molecular biology and the discovery in the 1950s of the role of nucleic acid, placing Mendelian genetics on a solid material basis. Molecular biology showed a means of transferral of information from genes to proteins but gave no indication—until recently—of any transfer in the opposite direction. The inference that there could be none became what Francis Crick called the central dogma of molecular biology: "Once 'information' has passed into proteins, it cannot get out again" (Crick 1988, 109). This was prima facie unlikely, considering the multiform activity of proteins, but the belief was affirmed that information goes only from genes to the body.

Traits or capacities acquired by conditions or exertions could have no effect on the genes. How could the bird's need for a longer bill to extract insects from tree trunks change the bill-making genes? How could the use or disuse of faculties help or hurt the next generation? The sons of a professional wrestler would be no stronger by virtue of his exercising, and those of a wheelchair-bound victim of muscular dystrophy have a normal chance of becoming strongmen if they lift weights. It is supposed that cave animals are eyeless not directly because of disuse of eyes but because if they are not needed, animals without them reproduce more abundantly because of the metabolic saving. Similarly, once whales had lost legs, the animal was supposedly better off without a pelvis. Why vestiges of eyes remain and whales have rudiments of pelvises and sometimes of hind limbs is another question.

A classic experiment of August Weismann a century ago was taken as proving the noninheritance of acquired characteristics. He cut off the tails of 22 generations of mice and found that the next generation was born with normal tails. It was in any case unlikely that mice would be born tailless, if only because the continuation of the spinal column in the tail is probably resistant to genetic change. Lamarck's theory did not apply because the mice were not striving to lose their tails. The sacrifice of hundreds of mouse tails only proved that this class of mutilation in mice was not inherited.

It would have been more conclusive to breed three strains: one selecting for small tails, one selecting for small tails but amputating tails, and one amputating tails without selection. But in the general conviction that Lamarckism was impossible, Weismann's experiment has been repeatedly cited, even nearly a century later (for an example, Barash 1982, 21).

Many evolutionists have always been uncomfortable, however, with the idea that progress is simply a matter of selection of the best mistakes. Selection eliminates the unfit and stabilizes the species, but many adaptations seem to require something less banal.

If the environment guides heredity only by granting or denying the privilege of having descendants, it is like a market in which sellers have no contact with buyers except the knowledge that certain articles are bought and others rejected. The producers of furniture would use a roulette wheel to decide on modifications (mutations) in their offerings; then they would drop styles rejected by the market (the environment) and make larger runs of those that sold well. If the furniture business operated in this manner, furniture styles would remain acceptable, but not many elaborate and imaginative new models would appear. Yet in the marketplace of nature, there are many very well-designed innovative products.

Furniture makers try very hard, of course, to inform themselves about the market, and the idea that environment influences heredity more directly than by simply rejecting the less fit refuses to go away (Saunders 1985, 148–161). It has always been stronger in France and some other countries than in the United States, and it has appealed to nonpracticing biologists.

Philosophically, the holistic approach—the notion of the organism adapting to its surrounding—is attractive. It seems natural. The body, we well know, has remarkable and poorly understood abilities to adapt itself to use and suffers from disuse; the athlete in training gets stronger muscles, the couch potato's muscles weaken; the brain improves with exercise;

hikers' feet become callused; bones and tendons get stronger, within limits, as they are stressed. Most damaged organs can repair themselves; the liver or bladder will regrow even after most of their tissue is excised, recovering approximately their original shape. If one kidney is removed, the other grows. If a young person moves to high altitude, the number of blood corpuscles increases. If one loses blood, the marrow makes new cells at a far greater than normal rate. The skin thickens where it is worn, lateral vessels enlarge when a vein or artery is blocked, and parts of the brain take over functions of an area damaged by a stroke. That such responses can be programmed, changing the expression of the genes of the cells involved, suggests the possibility that organisms can be geared to respond genetically to external signals.

The enduring strength of Darwinism testifies to its scientific usefulness and validity; in like manner, the persistence of Lamarckism, despite the strong distaste of most biologists, indicates that it may have some substance. Lamarck, who was a pioneer of evolutionary thought many decades before Darwin and whose ideas were no more farfetched than those circulating in his day, has uniquely been regarded as an enemy—a fact that suggests that his ideas cannot be so easily abandoned as phlogiston or the ether.

A number of difficult questions suggest that Lamarck's approach—not his specific theory but the idea of interaction between organism and environment more active than the culling of the less fit—is to be considered seriously. "Use it or lose it" is particularly plausible. Is natural selection for metabolic economy an adequate explanation for reduction of unused organs? May disuse conceivably facilitate the inactivation of genes? Unless generations of sloth bring enfeeblement, why are humans only about a third as strong as chimpanzees of similar weight? Can there have been selection for feeble musculature among premodern humans?

Some recent ideas have edged toward a compromise between natural selection and inheritance of acquired characteristics. For example, Allan C. Wilson suggests that "cultural evolution" is dominant in birds and mammals: traits become established not only by the relative multiplication of their genetic possessors but by imitation; the "cultural" change leads to genetic fixation; therefore the rate of change in these intelligent classes is high and increases with the capacity of the brain (A. Wilson 1983, 172).

In this interpretation, an important part of the adaptation of the giraffe would have been protogiraffes' copying one another in stretching toward the higher leaves, and this would promote the selective process favoring longer-necked mutants. This still leaves a lot for natural selection to explain. The protogiraffe had not only to lengthen neck vertebrae (fixed at seven in mammals) but to make many concurrent modifications: the head, difficult to sustain atop the long neck, became relatively smaller; the circulatory system had to develop pressure to send blood higher; valves were needed to prevent overpressure when the animal lowered its head to drink; big lungs were necessary to compensate for breathing through a tube 10 feet long; many muscles, tendons, and bones had to be modified harmoniously; the forelegs were lengthened with corresponding restructuring of the frame; and many reflexes had to be reshaped. All these things had to be accomplished in step, and they must have been done rapidly because no record has been found of most of the transition. That it could all have come about by synchronized random mutations strains the definition of random. The most critical question, however, is how the original impetus to giraffeness—and a million other adaptations—got started and acquired sufficient utility to have selective value (John and Miklos 1988, 236).

Many more or less fabulous organs and instincts noted in previous pages are much more difficult to account for in terms of Darwinian selection than the giraffe's neck. The observer must be often tempted to suppose that organisms have responded to their conditions and needs more purposefully than strict Darwinian theory can allow. To admit more interaction—perhaps quite subtle—among the environment of an organism, its activity, and its genome means muddying evolutionary theory and admitting grave complications. But possibly such complications have made evolution possible.

The Responsive Genome

Some biologists have studied how an organism might adapt genetically to external conditions. A generation ago, C. H. Waddington theorized (Waddington 1975) that environmentally promoted changes could become hereditarily fixed or "assimilated." He tried to square this with Darwinism but was not wholly successful. More recently, Mae-Wan Ho, Peter T. Saunders, Sidney W. Fox, and others have questioned the adequacy of the conventional denial of the possibility of environmental influence on heredity. Molecular biology, moreover, has partially reversed its outlook; as the complexities of the apparatus of replication have become better known, it has raised doubts that information can flow in only one direction.

The impossibility of input from environment to the genome has never been proved. In view of the inventiveness of life, it would be surprising if a law of nature absolutely prohibited it. From the beginnings of life, there must have been selection not only for particular abilities but for the capacity to make useful mutations. A prime means of metazoan evolution is the improvement of the ability of cells to alter their hereditary characteristics to make possible specialization of tissues and increasingly complex multicellular organisms.

If billions of years of interaction have brought some means of adaptive response, not only of the individual but of its descendants, this would amount only to an extension of the ceaseless feedback to genetic programs. Within the cell, genes are continually being switched on and off, as certain proteins are needed or not; in metazoans, signals from other parts of the organism guide cells in their genetic programs. If one breaks up a hydra into separate cells and then allows them to rejoin, they specialize as stinging cells (nematocysts), muscular, digestive, or reproductive organs, somehow receiving information from one another to tell how many should turn into each type in the proper place. Slime mold cells (*Dictyostelium*) wander singly, amoebalike, while food is abundant; when nourishment runs out and it is necessary to colonize a new site, they send out signals causing cells to congregate in a slug, grow a stalk, and produce spores to float away and hatch a new clone of amoebas (Bonner 1982, 3–9). Even bacteria have some such responsiveness; cells in a chain specialize for particular functions, especially reproduction (Shapiro 1988, 82–89).

In the morphogenesis of higher animals, some kind of signal—apparently the diffusion of inducer substances—tells cells which genes are to be activated to make up to 150 or 200 different types in a large number of organs. The signals must be conditional for type and location of cells. For example, a thyroid hormone (thyroxin) that dictates the metamorphosis of tadpoles has very different effects on different parts, causing muscles in the tail to atrophy and those in the legs to grow (Pough, Heiser, and McFarland 1989, 405). Differentiation is inheritable; embryonic cells soon lose the capacity to form a whole new individual and become irreversibly liver or skin cells. Cultivated in a test tube, they continue to reproduce as liver or skin cells.

That protists required at least 500 million years to achieve multicellular differentiation shows that it must have been difficult to work out. Cells have to respond with great accuracy and sensitivity to instructions to become just what is needed at the right time in the correct relation to other specializing cells (Valentine 1977, 43). For instance, in the making of a limb, the bones, muscles, tendons, and so forth have to be precisely and complexly directed according to distance from the center.

Mechanisms of specialization must have been imperfect at first. Precambrian metazoans were simple, mostly flat, with little diversity of tissues. The less ancient Burgess Shales fauna shows much more complexity and differentiation, not only elaborate shapes but organs of perception and muscles. Subsequent evolution doubtless rested on the ever-improving ability of animals to direct the formation of different tissues to make more different and more elaborate organs.

Not only do cells receive a set of instructions telling them how to build the new organism; throughout life, instructions change. Genes take effect in their season, as parents become aware when their children reach puberty. The refashioning of caterpillar to butterfly shows a new set of genes coming dramatically into play. Worms transforming themselves through several hosts must have as many sets of genes under a master plan.

It is no great leap for cells to react genetically to signals from the environment. Dysentery-causing amoebae change from benign to malignant at least partly because of diet (Hart 1989, 216–217). In the immune system, an extremely elaborate web of orchestrated genetic interactions, lymphocytes, in reaction to an indefinite variety of antigens, use a limited amount of genetic information to produce appropriate immunoglobulins. A gene-activated enzyme causes recombinations of many genes, which are somehow selected for usefulness (Oettinger et al. 1990). In a somewhat similar manner, trypanosomes, the protozoa causing African sleeping sickness, alter their protein coat as rapidly as their host develops antibodies, going through as many as 20 different serotypes. They revert to the basic type on passage into a new host (Hoare 1972, 116–117). Cancer cells unfortunately raise their resistance to chemotherapy by gene amplification, that is, they increase the production of appropriate enzymes by multiplying the corresponding genes (Schimke 1982, 3).

Whole animals may be guided by signals analogous to those that dictate the formation of organs. Siphonophores, colonial relatives of the jellyfish, like the Portuguese man-of-war (*Physalia*), look like organisms with complex parts but consist of colonies of individual specialized organisms. The seven or eight different kinds of component animals serve for digestion, defense, reproduction, flotation, propulsion, and capture of prey (Lutz 1986, 130). Their dependence on the colony is as complete as though they were simply organs, but they have no nervous coordination. In each one, the genome is expressed in different fashion according to instructions somehow given by the colony.

The members of social insect colonies are much less totally specialized than the siphonophore zooids; they are obviously individuals, sometimes with a bit of apparent willfulness of their own. But the genetically dictated differences between castes in wasps, bees, and ants are conditioned by nutrition, pheromones, temperature, and/or behavior, in ways for the most part poorly understood (Hölldobler and Wilson 1990, 348–354). The specialization may be slight, as in most wasps; but queen ants can be as much as 4,000 times as large as the smallest workers, and workers may differ not only a hundred-fold or more in size but also in shape, especially with big-headed or big-jawed soldiers (Hölldobler and Wilson 1990, 566; Brian 1981, 157–172). Termites have queens hundreds or thousands of times larger than workers and soldiers quite different from ordinary workers.

There are also environmentally induced modifications. Aphids are extraordinarily plastic in their genetic expression. Depending on season, abundance of food, kind of plant on which they live, and even the attentions of ants, they may reproduce sexually or asexually, lay eggs or give birth to large, mobile young, be winged or wingless, and vary in size and shape. A single species may have as many as twenty different forms (Pearse et al. 1987, 597). Barnacles, bryozoa (moss animals), *Daphnia* (water fleas), protozoa, and rotifers develop spines, protective plates, or other defenses in the presence of predators (Dodson 1989, 447–452).

Plants have great capacities to respond flexibly to external stimuli. The embryo in the seed begins differentiation autonomously in somewhat the same way as animals, but external influences take over. Very little is known about how plants register environmental conditions and direct cells to activate the necessary genes. But cells specialize and continually make new organs according to conditions of light, day length, temperature, moisture, and wind, to grow stems, roots, flowers, and other parts. A cutting placed in water sprouts roots; flower buds respond to the season. Leaves of the water crowfoot are broad if out of water, threadlike if submerged. Trees grow low in windy places, and many plants become thicker and shorter if frequently bent; some genes are regulated in response to rain, touching, and injury (Braam and Davis 1990, 362). Changes of instructions may be hereditary. For example, some plants, such as ivy and creeping fig, produce differently shaped leaves when they come to flower; if a mature shoot is cut off and propagated, it continues to make leaves in the modified form (A. Huxley 1974, 64).

Some of the progress of plants may be seen as an improved ability to produce different kinds of tissues or modify development. For example, such ancient types as horsetails (*Equisetum*) and clubmoss (*Lycopodium*) are not much differentiated and have little capacity to vary form; they can produce shoots and roots only from the rhizome. Conifers are more plastic, but they are typically unable to send out new growth from a stump; they are hard to graft, and cuttings are hard to root. They are all woody perennials; none is herbaceous. Greater genetic flexibility is supposed to have been important in the rise of the dominant modern plants, the flowering plants or angiosperms; new reproductive structures were accompanied by several kinds of new tissues (Cronquist 1988, 131, 134). Dicotyledons (most ordinary plants) are more plastic than monocotyledons (palms, grasses). The former often cause leaves to become quite specialized, for example, as petals, tendrils, thorns, or even water pitchers; the latter generally have only one kind of leaf.

A much studied example of how cells can respond to needs is the *lac* operon of *Escherichia coli,* large numbers of which we all carry around. The *lac* operon causes the bacteria to turn on genes to enable it to digest the sugar lactose and to facilitate its transmission through the cell wall when and only when lactose is present. A repressor combines with lactose and is hence immobilized by its presence; lacking lactose, the repressor binds to DNA (Rothwell 1983, 482–483). This is in principle fairly simple, but enough attendant complexities were known twenty years ago to be made the subject of a book of 437 pages (Bechants and Zipses 1970). There are many other operons or operator systems to enable bacteria to take advantage of various nutrients in their medium.

Such modifications are, of course, genetically programmed in all their complexity. But if cells can respond genetically to make things as different as threadlike neurons, hard tooth enamel, phagocytes, and glands and if the genome can react to external signals by making structures as different as the digestive organ and the stinging tentacles of the Portuguese man-of-war, it is difficult to set limits in principle to genetic modification in response to external stimuli.

The differentiation of cells in the body or of insects in the colony means the activation of one or another part of the whole genome shared by all cells or by the varied members of the colony, not the formation of new genes. However, the faculty of giving expression to certain genes and not to others in response to external cues is not very different from a faculty for changing heredity. Only a small fraction of DNA, perhaps as little as 1 percent in the human, is actually used; the large amount of unused DNA presumably represents an accumulation of phased-out genes, as well as a great deal of duplication; some of it may well constitute a reservoir of potential changes. There is a huge storehouse of potential changes on which the body could draw without producing any new structural genes.

Adaptive Variation

Animals show a considerable range of responses to environmental conditions, responses that may be difficult to separate from hereditary adaptation. A camel is born with callosities on its knees; can this hereditary trait be detached from the ability of the skin to thicken in response to friction? Some rotifers program their offspring to grow smaller or larger according to the food supply (Pearse et al. 1987, 307–308). A cichlid fish

(*Astatoreochromis alluandi*) has equipment to crush snails where snails are abundant but not where snails are few; if offspring of snail-crushing fish are raised on a soft diet, they fail to grow the hard parts needed to handle snails, although they regard them as food (Fryer and Iles 1972, 103). According to Harvey Hoch, a fungus can change genetically quite rapidly to overcome the chemical defenses of the bean plant (Weisburg 1987, 214).

Animals seem to lose unneeded structures and instincts much more rapidly than can be accounted for by any selective advantage of not having them. Parasitic ants lose work habits that could be useful, although they have ceased to be necessary, much as human aristocrats shrink from manual labor. Insects, crustaceans, salamanders, and so forth lose eyes in caves, in some cases rather quickly. Mexican cave fish have lost vision in about 10,000 years (Culver 1982, 66). Many burrowing reptiles (legless lizards and snakes) and mammals have also become partially or entirely blind. Some fish with electrolocation (*Gymnarchus* and others) have eyes capable of little but distinguishing night and day. The trivial economy of reducing eyes can hardly be the sole reason for their rapid loss (Culver 1982, 73). Their metabolic cost is trivial compared even to a small tail.

Conceivably, disuse somehow relaxes the controls maintaining the integrity of the complex organ. The well-functioning eye is not deeply embedded in the vertebrate genome, although "eyeless" animals usually keep degenerate eyes covered by skin. Loss of vision is less of a problem than the maintenance of such a complex organ in the face of continual slightly deleterious mutations.

The shrinking of human dentition is also curious. The teeth of our ancestors began a steady decline about 100,000 years ago. Decreasing about 1 percent every 2,000 years, and recently 1 percent every 1,000 years, they have shrunk to about half their previous size. The shrinkage is associated with the coming of fire and cooking, the grinding of food, and the invention of pots in which to stew dinner. Strong teeth have become less necessary (Wilford 1988, B5). It would hardly seem, however, that there would be any substantial selection pressure in favor of small teeth. To the contrary, unless tastes have changed, fine teeth might be advantageous in sexual competition for both males and females (B. Campbell 1985, 300). It would not be surprising that disuse would lead to decline of an organ or instinct, unless it is (like six-leggedness of insects) strongly embedded; if there is nothing to maintain a complex dynamic structure, it may be expected to lapse.

In the 1920s an Austrian biologist, Paul Kammerer, claimed to demonstrate hereditary effects of use and disuse. His experiments were described, with their tragic outcome, by Arthur Koestler. Kammerer reported that the grippers used by the male toad (called "midwife" because he carries the eggs around) to clasp his mate in water, so-called nuptial pads, were absent in a land variety. He forced the toads to breed in water, and after a few generations the pads returned. He also found that salamanders hereditarily changed coloration according to the background on which they were raised (Koestler 1971, 39–47). So, he asserted, Lamarckism was proved. It was learned, however, that someone had injected ink into nuptial pads. An unfriendly evolutionist ridiculed the experiment, and Kammerer shot himself—such was the excitement generated by the issue of heredity and environment. The ink was hardly relevant, but Stephen J. Gould held that the restoration of the pads could be accounted for in strictly selectionist terms because of the forced readaptation to water, as they "may be tied to the set of genes that confers success in water" (S. Gould 1980b, 82). However, there was only the briefest possibility of selection for success in water, the adaptation was very rapid, and there is no evidence for the existence of a set of linked genes making for success in any particular environment. If there were, it would be an important amendment to Darwinist theory with Lamarckian implications.

Such matters being complicated in higher animals, biologists have endeavored to clarify them in bacteria. They have found considerable evidence of genetic adaptation. If bacteria are placed in a medium with nutrient molecules too large to pass through the pores of its membrane, they mutate to make the membrane more permeable (Benson 1988, 21–22). Bacteria grown in a salty medium become better able to survive and reproduce in seawater (Munro, Launod, and Gauthier 1987, 121–124). It has been learned that a bacterial transposon ("jumping gene") moves in the presence of the antibiotic erythromycin in such a way as to generate resistance; that is, "evolution is influenced by the environment not only through natural selection but by directed mutations" (Endler and McLellan 1989, 413).

The principal business of bacteria being to digest things, many experimenters have investigated their ability to use substances for which they were not adapted. John Cairns and colleagues found that a strain of the bacteria *Escherichia coli* unable to digest lactose became able to do so when they had only lactose as a nutrient. They did not do this by restoring the lost gene but by bringing forth another one; the presence of lactose, in Cairns's opinion, somehow triggered a new set of mutations (Cairns, Overbaugh, and Miller 1988, 142–145). No mutations to revert to the capacity to use lactose seem to occur as long as other nutrients are available, but they appear when the medium contains only lactose (Opadia-Kadima 1987, 27–35). This is only a step beyond the well-known *lac* operon, which activates a gene to make an enzyme to utilize lactose, and other operons work similarly to produce a "mutation on demand."

As observed by Barry G. Hall, when *E. coli* was given little other nourishment in a solution of salicin (a relative of aspirin), it underwent two otherwise rare mutations together at a rate thousands of times higher than in normal growth, enabling itself to metabolize the salicin. An unselected mutation was not affected; that is, the bacteria did not merely become more disposed to mutation (Hall 1988, 887–897). In a similar experiment, he started with a strain of *E. coli* unable to synthesize the essential amino acid tryptophan. He gave them tryptophan for a few days and then deprived them of it. The number mutating to produce tryptophan jumped as much as thirtyfold. He showed that the mutation did not arise until it was needed and did not occur in other bacteria starved of other amino acids. He claimed, "I can document [purposeful mutations] any day, every day, in the laboratory" (Stolzenburg 1990, 391).

It is axiomatic that an organism can respond genetically to external signals only to the extent and in the ways that it is keyed to do so. But since the beginnings of life, bacteria have had to find means to utilize different substances. Probably the first lifelike entity utilized only a single compound much like its own substance out of the many organic compounds in the primordial soup. As this became scarce, the proto-organism had to modify itself, or mutate, to utilize related compounds. This faculty of shifting capacities must have expanded during the more than 3.5 billion years of bacterial existence, during which bacteria are believed to have been constantly changing their chemistry (Loomis 1988, 162). This is less extraordinary than the trypanosome's altering its coat in response to the host's antibodies.

How bacteria modify themselves is complex; they are very complex systems. According to Cairns, "We now know that, in the processing of biological information, almost anything is possible. Sequences [of DNA] are spliced, rearranged, cast aside, resurrected, and to a limited extent may even be invented when the need arises, and so it should not be difficult for an organism to devise a way of testing phenotypes before adopting any new genotype" (Moffat 1989, 226).

Robert Mortlock, who has done a great deal of work on the utilization of sugars that bacteria do not encounter in nature, has a fairly clear idea how they manage, by stepwise changes, involving duplication of genes and alteration of regulatory genes (Mortlock 1982, 257). Enzymes are "borrowed" from established pathways; at first they doubtless work poorly, but the cell increases mutations under stress, modifies structural genes, perhaps activates repressed or "cryptic" genes, and becomes able both to transport the new nutrient through the cell wall and to metabolize it normally.

A bacterium cannot "perceive" that a novel substance is potentially digestible unless molecules can diffuse through the cell wall in at least small numbers and can be acted upon, albeit to a minimal degree, by enzymes already present—that is, unless the substance can be slightly utilized. This is possible for a range of substances because there are many potential pathways, any pathway includes numerous enzymes, enzymes may have a low degree of activity for substrates different from those for which are adapted, and uninduced pathways are not shut off absolutely but perhaps to the extent of 99.9 percent. In view of such leakages and the large numbers involved—about 10 billion cells per cc.—mutation to a new pathway may be rapid.

It may also be slow. If bacterial (*Klebsiella*) cells are placed in a medium with only L-xylose, a synthetic sugar, to work on, there is no observable growth for four to seven weeks, a great age in terms of bacterial generations. But they always adapt eventually, and multiplication becomes rapid (Mortlock 1990). This is remarkable. The bacteria obviously do not become able to use L-xylose by virtue of mutants already present in the culture when it was placed in the new medium; if there were such mutants, they would rapidly spread. On the other hand, there can hardly be random new mutations because the cells are completely or almost completely unable to grow and divide.

Bacteria normally have mutation rates 10 to 1,000 times lower than those of multicellular animals (Grant 1985, 47), but rates can, perhaps must, increase greatly under stress. Not only do correction mechanisms become less effective; it is to be expected that bacteria are normally geared to increase mutations when, as in near starvation, they are direly needed. In the case of the hungry *Klebsiella,* any cell with an ability, however slight, to admit and use the strange substance would be able to adopt and reproduce any mutation improving that ability, however slightly. Any mutation in a useful direction would serve as a basis for further mutations in the loosened genome. By cascading changes, the bacteria would become increasingly able to make the new nutrient fully available (Hartley 1984, 51).

The ability of bacteria to consume new substances has enormous and growing practical importance. To promote the biodegradation of polychlorinated biphenyls (PCBs), for example, an experimenter searches in dumps for strains that may have acquired a taste for various wastes. Bacteria are found with surprising ease capable of digesting hydrocarbons, toxic herbicides, and many other synthetic compounds, organic and inorganic. The ability is selective; bacteria that can handle one of the many varieties of PCB cannot cope with others (Browne 1989, B12; Lindow, Panopolous, and McFarland 1989, 1301–1305). "Biomediation," as it is called, is a booming business (Barnaby 1990).

Because of bacterial defenses against substances designed to kill them, pharmacologists have to be perennially discovering new antibiotics. The rapidity with which resistance is increased as much as 100,000-fold cannot be explained by random mutation and selection in the traditional sense (J. Campbell 1987, 285). James Shapiro points out that the genome has a "fantastic system for proofreading, editing, and repairing itself"; consequently, "the genome is smart; it can respond to selective conditions" (Moffat 1989, 226).

It is obviously much easier for a cell to respond genetically to the medium with which it is in direct contact than for a multicellular animal to convey information from the environment to its reproductive cells. But if one-celled ancestors of metazoan animals possessed a capacity for useful mutation, it would be extremely advantageous to continue and develop it, doubtless in more complex ways in the more complex organism. It appears at least that in animals, as in bacteria, stress increases variation in the gene pool, perhaps generating evolutionary novelties; "the transposition rate of at least one major category of mobile genetic units may be inducible by environmental stress" (McDonald et al. 1987, 240, 258).

In one respect, animals have shown a capability like that of bacteria. The disconcerting rapidity with which pests, from rats to insects, acquire resistance to poisons is hard to account for. All insects that humans try to kill fight back genetically. Some acquire resistance in a few seasons or even a single one. The Colorado potato beetle (*Leptinotarsa*) seems to be especially adept. It has acquired resistance to some pesticides the very season when they were introduced; it is known to resist at least 15.

Some 500 species of insects and mites have defeated one or another pesticide. At least 17 have developed resistance to pesticides of all major categories, either producing enzymes to destroy the chemical or altering the insect's physiology. They even learn to overcome insect hormones or mimics. Over a thousand species of plant pathogens, 55 of weeds, and 5 of rodents have also developed defenses against pesticides (Pedigo 1989, 492–498; Georghiu 1986, 16). A nematode parasite of sheep became completely resistant to a therapeutic drug in just three generations (Holmes 1983, 162).

It seems natural that plant-eating insects should be able to acquire resistance because they have to overcome the chemical defenses that plants are perpetually developing. But the unfortunate bugs encounter man-made substances designed especially to thwart their defenses. If they had to wait for the right protein to be produced by sheer accident, out of the infinity of all possible proteins, it would take vastly longer than half a year to develop immunity. Contrary to the assumption of Darwinist gradualism, resistance is almost always due to a single gene. "Therefore standard evolutionary theory does not seem to apply to the development of resistance" (Plapp 1986, 75). But the ability of bacteria and animals rapidly to acquire resistance to chemicals is much less than the ability of the immune system to develop antibodies to almost any of an infinite variety of molecules.

Many experiments have shown complex relations between environmental and genetic change. Plants under stress or placed in altered conditions may be genetically altered. For example, flax (*Linum usitatissimum*) undergoes considerable variation in size, weight, details of the seed capsule, various chemical products, and amount of nucleic acid in response to fertilization or lack of it and environmental pressures, especially of temperature. The variations, which are not primarily adaptive, are indefinitely inherited. Tobacco (*Nicotiana*) also shows genetic effects after a single generation in a changed environment (Cullis 1988, 53–57; Cullis 1987, 62–73).

C. H. Waddington found that when pupae of *Drosophila* were incubated at abnormally high temperatures, modifications of wing veins appeared. When strains were bred selectively for particular modifications, after a few generations the peculiar venation became hereditary and appeared with pupation at normal temperatures (Waddington 1975, 51). The process works in reverse, of course; the effect of temperature on wing venation can be reduced by selection (Thomson 1988, 95).

Waddington tried to reconcile this phenomenon, which he called "genetic assimilation," with the conventional theory by attributing it to selection. But there was no question of new genes' appearing appropriately at random to confirm the acquired trait. Mutations are too rare, and the probability that just the right mutation should occur in a small population (or even a large one) by chance would be slight. Waddington suggested that many genes, or near copies of a gene, might under selection strengthen the trait that was already a potentiality of the genome. That this idea is realistic is indicated by the fact that genetic assimilation proved limited to a fixed range (like the limits of variation of domestic animals), and it failed for inbred strains (Waddington 1975, 69, 179).

Genetic assimilation is thus evidently a matter of reassortment of genes or a change in modes of their expression, not of mutations of structural, protein-producing genes. But it indicates that environmental factors can somehow influence the way genes behave under selection—not directly, it seems, by bringing about adaptive change, but by increasing the probability of recombinations inducible by environmental conditions. For a genetic response to environmental pressures, it would suffice to modify the activity of the control system, bringing different combinations of already existent genes to the fore

and permitting mutations in a given area. Some genetic structures apparently serve to increase the capacity to change genetically (J. Campbell 1985, 137, 146).

It would have very great survival value for the species to be
able to open the way to genetic change in certain areas, and
there is no evidence that it is impossible. The hereditary mechanism itself is subject to evolutionary change, and it is clearly
possible for the organism to be tuned to alter its effective
genetic instructions in response to signals from without. It
would be especially beneficial—and favored by selection—if a
species were able to relax controls in ways to improve its ability
to move into new environments, use new food resources, or
assimilate learned behavior patterns. It would be remarkable
if starving bacteria were not able to increase mutations of genes
dealing with metabolic pathways. It seems improbable, in view
of the genome's great capacity for self-repair, that it should be
unable to permit increased mutation in a certain area, possibly
to increase the frequency of those potentially useful.

There is a contradiction between the easy assumption that
frequent random mutations suffice to produce resistance to
pesticides in short order and the contrary assumption that
extremely complicated organs—one thinks first of the eye—
are naturally stable. If it is true that mutations are much more
frequent where they are needed than where they are virtually
certain to be harmful, they cannot be held to be random.
Animals, in any event, do many things that are quite as incredible as having some—certainly rather limited—means of favoring useful mutations. So far as one finds two or more structures
or instincts that cannot be useful unless fully developed and
coordinated with each other, one is tempted to suspect that
some facility for mutation in functional areas has combined
with selection to bring them about.

There may have been a long-term selection of taxa for ability
to adapt to environment and realize new potentialities; that is,
animals may have become more capable of functional change
over the ages. This implies that evolution has accelerated, a
conclusion easily reached although not easily proved. There
was a great extravagance of shapes in the early Cambrian when
the phyla were born, but this probably had little adaptive meaning because ways of life were few. Much of the diversity of
form was probably like that of orchids, with many and various
shapes serving mostly similar purposes.

Subsequently, there has been increased adaptive change, although there have appeared no new phyla. This is most obvious in the post-Cretaceous radiation of birds and mammals. The profusion of specialized birds left the pterosaurs far behind. The diversity of orders of mammals—moles, bats, anteaters, ungulates, elephants, kangaroos, and many others—has far exceeded the reptiles' ability to find new ways of living. The superior adaptiveness of mammals, when the stage opened for them, can hardly be due to simple errors of replication. There must have been a basic adaptation; in some way, the mammal genome became better able to generate new forms through what has been called "evolutionary facilitation" (Wills 1989, 6–7).

To think in terms of a relationship between environment and genome more positive than the failure of the less fit does not imply a broad purposiveness. It is axiomatic that the organism can respond only in the ways for which it has become programmed and only to signals for which it is prepared. The *Daphnia* is tuned to grow spines when it senses predators; the crop-eating insect, to take countermeasures against poisons. There is no general tendency for mutation in directions dictated by needs of the organism, and the environment does not of itself induce suitable mutations. Organisms fail to develop many useful traits that would seem readily accessible. It is obviously more complicated for a multicellular animal than for a single cell to facilitate genetic change in response to environmental signals; but it may be, as Christopher Wills speculates (Wills 1989, 280–284), that the genome evolves the ability to evolve and becomes more expert in making useful mutations.

The question is not whether organisms can respond genetically to external signals but the range and kind of signals to which they can respond. Morphogenesis is a continuous process of feedback between cells and their environment in the broadest sense, in which the whole system exerts control, and adaptation is not merely of genes but of the entirety (Ho 1986, 426).

Behavioral Adaptation

If bacteria respond biochemically to their environment, higher animals mostly do so behaviorally. This means learning, that is, modifying behavior because of reward and/or punishment, making choices, and altering ways to meet opportunities. In the ever more complex interaction of animal and environment, behavior and structure inevitably change together. It is possible that structural or somatic change leads an animal into new habits and a new way of life (John and Miklos 1988, 243), but this seems less likely than that new habits and entry into a new way of life lead to structural change.

A philosopher of science, Karl Popper, seeing all organisms as problem solvers exploring the environment, emphasized the animal's initiative. In his view, the woodpecker must have arisen from a bird's specializing in finding insects in crevices of the bark, developing skill in searching, better neck muscles, and a longer, sharper beak, and by feedback strengthening the specialization of woodpecking (K. Popper 1988, 150). Popper would rehabilitate Lamarck's derided example of the giraffe's neck by suggesting that the critical initiative was indeed an antelope's stretching for the higher buds. He did not suppose that this action directly lengthened the neck of its offspring but that it started a train of development.

A more qualified evolutionist, Ernst Mayr, underlines the precedence of behavior: "In animals, almost invariably, a change in behavior is the crucial factor in initiating evolutionary innovation . . . the pacemaker of evolution" (Mayr 1988, 408). This is an easy conclusion for a student of nature to reach. But if behavior leads the way, the prime mover in innovation is not an inheritable random variation, as Darwin postulated, but the animal's choice.

This is logical because learning is purposeful, whereas random variation is almost certain not to be. Behavior is flexible, tentative, and experimental; it can be tried and discarded, selected by experience. Somatic modification is an irrevocable commitment. Moreover, it is unlikely, perhaps impossible, for a random mutation to give rise to a complex behavioral pattern. "No gene has been demonstrated to direct a complicated series of motor responses that produce an integrated behavior pattern" (Fuller and Thompson 1978, 208). This is as true today as it was in 1978.

Pressure to modify behavior is not lacking. Hunger drives to seeking new ways of procuring food. Pear thrips, insects about a millimeter in length, discovered in the later 1980s that sugar maples are as tasty as the fruit trees that they have always lived on. They had a population explosion and defoliated trees on millions of acres of the Northeast (Gold 1988, 1). Pigs on certain South Pacific islands have developed the habit of wandering widely over coral reefs during low tide to forage on small molluscs, fish, and so forth; they have become excellent swimmers and spend much of their time with heads submerged. Along the coasts of Scotland, sheep have learned to graze algae in the intertidal zone. They even swim short distances to reach a new patch of kelp (Riedman 1990, 50). The question remains why pigs or sheep or other animals in other lands under similar conditions have not ventured into the water. Why are there not a host of semiaquatic species at all stages of using marine resources?

Behavior that has become habitual is apparently translated into instinct. Garter snakes of coastal California live primarily on slugs; those of the drier inland where there are few slugs shy away from the slimy creatures (Alcock 1989, 77–78). A cichlid fish, *Haplochromis acidens,* with all the morphology of a piscivore, eats primarily entirely plants, apparently because in its lake (Victoria) there is a surfeit of piscivorous fish and a lack of vegetarian competition.

Most monkeys and apes are omnivorous; the gorilla has gone over to a vegetarian diet with only minor changes of its body, particularly a larger stomach and longer intestines, although still short for a herbivore (S. Gould 1987b, 22). Although it has grown too heavy for life in the branches and dwells almost entirely on the ground, the gorilla, with powerful arms and awkward gait, is better framed for an arboreal existence.

The panda is a vegetarian bear. In giving up the eating of meat, the panda, like the gorilla, has become in a sense maladapted. Morphology follows at a distance. The panda, like the gorilla, has a simple stomach and relatively short intestines; only its teeth are tolerably adapted to grinding bamboo (O'Brien 1987, 103–104). It has no capacity to utilize the chief components of vegetable matter, cellulose and lignin. Fortunately for the panda, bamboo is more nutritious than most other grasses, and it grows in great dense stands, but the panda needs an enormous amount of fodder and must spend most of the day eating. No convinced vegetarian, it loves meat when offered, and panda traps are baited with roast goat (Schaller et al. 1989, 218–235).

Behavior has apparently started change in many smaller creatures as well. The habit of mass predation must have led to the characteristic adaptations of army ants: very large colonies, large wingless queens, nomadism, and polymorphic workers. Termites are evidently descended from wood-eating cockroaches. Social parasitism in wasps, bees, and ants seems to have originated in the habit of colony formation by entry into a host nest.

Behavioral changes must have led into social organization, in insects as in birds and mammals. The subordination of inferiors among social animals, seemingly having begun through dominance like that of the pecking order, is hardened as it is made more structural. In the more advanced societies, the queen controls the workers by one or more pheromones, and they lose, or largely lose, the ability to reproduce. This occurs not only in all classes of social insects—wasps, bees, ants, and termites—but also in the African mole-rat, whose dominant female affirms superiority by a chemical in her urine and prevents maturation of subordinate females (J. Jarvis 1981, 571–573).

Behavioral change—perhaps with structural modifications—sometimes apparently evolves more readily than seemingly straightforward organic adaptation. For example, vampire bats have acquired the instinct for regurgitation for the benefit of hungry colony members instead of the very simple stratagem of fat storage—an instinct the more difficult to develop because it is altruistic. A species of Darwin's finch in the Galapagos uses thorns—probably broken from a cactus—to poke out insects from crevices, although the tool held in the beak must be a weak instrument; a longer and sharper bill would seem to be easily attainable by natural selection. Bills of closely related species are quite variable and, of course, almost all birds that search for insects under bark, woodpecker-style, have developed their built-in pecking tool.

It is even more striking that a much more primitive animal, a pelagic Antarctic amphipod, *Hyperiella dilatata,* has a semi-behavioral mode of defense against the fishes that consume it in large numbers. It has somehow become able to recognize a small planktonic mollusc, *Clione limacina,* that fish do not eat, to clasp the mollusc, and to put it on the amphipod's back, like a rider on horseback. Fish find the mollusc distasteful, so they leave the couple alone. The defense has some cost, because having the rider reduces the amphipod's mobility; it cannot be nearly so effective as a structural or chemical defense; and it would seem infinitely harder to acquire by natural selection.

A good imagination is necessary to envision how such adaptations came about. Biologists are tempted to think in terms of learning, even though they do not say so. Thus Waddington applied genetic assimilation in a decidedly Lamarckian sense: if an animal took to burrowing, the new activity would strengthen muscles and bones in a way that could, through selection, become hereditary (Waddington 1975, 88). A fish inflated itself defensively by taking in water; the habit became permanent, and after many generations, there was a puffer fish (E. Wilson 1980, 10). It appears that wasps, laying an egg in a prey, came to accompany it with a paralyzing secretion; the ovipositor became a sting, with poison gland and apparatus for drilling into the enemy (Cloudsley-Thompson 1988, 124).

The sonar system of the bat could hardly get started without an element of learning, as the bat found that echoes from squeaks helped in locating prey or avoiding obstacles in the dark. It may be thus that many animals have invented echolocation. As soon as the adaptation is begun, the behavior that is learned would be favored by selection, and new ways give utility to new or modified organs.

Behavior as well as morphology separates species. Field crickets are distinguished mostly by their songs. Different species of a genus of termites (*Apicotermes*) are visibly characterized only by the form of their nests (E. Wilson 1978, 232). Some species of wasps and bees are practically indistinguishable except for different styles of the nest of combs; diversification of hornets has been much more behavioral than morphological (Evans and Eberhard 1970, 164).

Higher taxa are also characterized by their habits. Termites live only on wood and related materials. Although many reptiles are herbivorous, snakes consume only living prey. Spiders, despite great variety of forms, are restricted to eating captured animals, which they kill by venom and digest externally, and they all use silk. One suborder of bats eats fruit, the other (almost exclusively) insects. Many other groups of mammals, from carnivores to anteaters, are characterized about as clearly by their diets as their skeletons, and it may be surmised that their divergence from primitive mammal stock was originally behavioral. This may well be the case of most or all evolutionary departures of higher animals.

Learning

All but the most primitive animals can learn. Even when guided by instinct, they respond flexibly within a range of situations and search for ways of satisfying needs. The almost brainless earthworm (*Lumbricus terrestris*) stretches out of its burrow, explores the vicinity, recognizes a leaf or other edible material, finds a suitable spot for grasping it, and drags it in. It also shows what Darwin called intelligence in using stones to plug the burrow (C. Darwin 1902, 60–65). Earthworms can be taught by stimulation with light, electrical shock, or an unpleasant surface to turn in one direction or another, and they retain the lesson for two or three days or even if the first five segments with the cerebral ganglia (or "brain") are amputated (Edwards and Lofty 1972, 75–76).

Learning must be within a genetically given framework. For example, a fly can associate wet feet with sugar. It is programmed to make this kind of connection; flies use their feet to sense edibles. They cannot be taught to associate food with color or shock (Nelson 1976, 353–368). This is like the limitation of learning in rats: they very easily associate a taste with nausea, even much later, but they cannot associate a sound or light with nausea (Barnett 1975, 201). The connection, which is good sense in the rat's world, is prefigured.

Within the animal's framework of response, however, it can be trained; the experimenter has only to find means of rewarding it. And animals train themselves, that is, learn. They move about and seek novelty, especially in unfamiliar circumstances. To explore means to make choices; if choices entail rewards or punishments, learning results. Ethologists write constantly of choice, search, avoidance, and strategy.

Mammals and, less often, birds learn by being playful and curious. Play tests capacities, develops social relations, and explores the environment. It is close to curiosity, which is sometimes so compelling as to lead animals to endanger themselves. Wildebeest and other grazing animals are inquisitive about lions and hyenas. Thompson gazelles run up to look at cheetahs and leopards, although the predator may turn and catch one (Kruuk 1972, 286–287).

Play can be regarded as preparation for adult life, as wolf cubs frolic in actions like hunting and fighting. That such practice should be necessary itself suggests plasticity of behavior; the new generation learns for itself, sharpening patterns that are not entirely instinctual. But adults, too, play. Mature whales and dolphins toss objects about and leap out of the water in ways not apparently related to biological needs. Hyenas frolic, sometimes going into the water they normally avoid, pretend to fight, and seem to take pleasure in baiting rhinos they cannot attack (Kruuk 1972, 249–250). Sea otters spend much of their time frolicking. Young baboons are extraordinarily curious, attentively examining novel objects. The chimpanzee is exceptionally varied and imaginative in its play (E. Wilson 1980, 86; Goodall 1986, passim). Play and curiosity have the same usefulness as commonly attributed to sex: the facilitation of innovation.

Elaborate instincts are no bar to learning within the animal's universe. Honeybees learn rapidly to visit feeding stations at different places and hours; they recognize flowers visually, remember when they open and close and where they are abundant; they seemingly have to learn the relation between the sun's position and the hour of day (Pearse et al. 1987, 617). They can remember a color associated with a source of nectar for at least two weeks; their waggle dance is a means of communication just short of language. If a feeder is moved daily, say 100 feet east, the bees anticipate the next move and come to the new locale even if there is no feeder there (Gould and Gould 1988, 221 and passim). They normally go through a preordained cycle, beginning as nurses and ending as foragers, but if there is a shortage of nurses, foragers revert to nursing, reviving the glands to secrete food for larvae. Although many wasps specialize in preying on one kind of insect or spider, others can hunt a wide variety but come by experience to prefer particular species (Evans and Eberhard 1970, 45–47, 52).

Cockroaches instinctively clean their antenna with their front legs. If the front legs are amputated, the roaches are at first frustrated, but they soon learn to use the middle legs (Barnard 1983, 132). An octopus that has never encountered a glass tube will discover in a few minutes that it cannot capture a shrimp enclosed in one. It can also learn not to attack a crab accompanied by an experimenter's warning signal, distinguishing, for example, between a square and a rectangle, and remember for weeks. Old octopi learn better than young ones (Wells 1962, 43, 52; Pearse et al. 1987, 378).

The seemingly learned behaviors of birds only hint at the capacities of the small-brained creatures. The black-headed gull carefully disposes of eggshells, the presence of which would give away to predators the location of the nest. Egyptian plovers flip sand over their chicks when danger looms (Alcock 1989, 77–78, 334). Ravens drop stones on persons threatening their nest (Griffin 1984, 120), an action we would commend as intelligent in a chimpanzee. A nutcracker, having cached pine seeds, will dig them up and remove them if it spies a potential thief watching (Wall and Balda 1983, 63). Crows break open whelks by dropping them from a suitable height onto a hard surface, repeatedly if necessary (Griffin 1984, 61–62). One can only guess whether a bird somehow figured that this was the way to an otherwise inaccessible meal or happened to be carrying around a whelk, accidentally dropped it, came down to retrieve it, and learned the strategy. Egyptian vultures have the opposite tactic of dropping stones on ostrich eggs to break the hard shell—no easy target for the bird to hit—a skill the young have to learn from their elders (Riopelle 1972, 387–388).

Such actions may be called intelligent, although the gull that cleans up around its nest is doing no better than ants (*Formica*) that carry away cocoon fragments after having been subjected to a raid by a more aggressive species (Hölldobler and Wilson 1990, 362). Birds, however, do things of which no insects are capable. A fisherman returning to check a baited line may find that a crow has pulled it up and eaten the fish caught on the hook (Welty 1982, 312). A green-backed heron has been observed to drop bait in the water to lure fish; when the bait drifted away, the bird retrieved it (F. Gill 1990, 142). Baboons climb trees to escape lions; when a human hunter approaches, they climb down and run away (Moss 1975, 222). British tits discovered that they could peck through the metal tops of milk bottles left at doorsteps; thousands were soon doing so, a dozen other species copied the act, and it spread to the Continent (Griffin 1984, 67; Marler 1972, 43). On some islands of the Galapagos, Darwin's finches draw blood by pecking boobies; on other islands, they have not learned this stratagem (Griffin 1984, 64). Darwin's finches probably got started using twigs as probes by playing with them (McFarland 1985, 511). Some jays do the same.

Mice eat darkling beetles, which have a noxious discharge from the rear end; some clever mouse must have learned the trick of pushing the beetle's abdomen into the ground and eating it from head down (Alcock 1989, 298). Hyenas kill by tearing out the guts of an animal; lions, leopards, and cheetahs more sophisticatedly cover the muzzle with their mouth or hold the throat until it suffocates (Schaller 1972b, 265–266), a tactic that is not obvious.

Ants (*Formica*) can learn a maze, albeit two to five times more slowly than rats; they memorize landmarks and recall them a week later. If they find food after having meandered for an hour or so, they make a beeline for the nest. They adjust their path to the movement of the sun, but they have to learn to do this. The learning process is basically the same as in mammals despite the minuteness of the ant brain (E. Wilson 1971, 208–209, 211, 216, 217, 227). Ants provide a multitude of examples of instinct akin to intelligence: the habit of protecting the bush that gives nourishment and shelter, the use of larvae's silkmaking to bind leaves into a nest, and careful protection of aphids are a few. But ants are probably not outstandingly intelligent among arthropods; they do not have exceptionally large brains.

It is easier to account for many such seemingly clever ways if behavior—that is learning—led the way. Detailed study of the cichlid fishes convinced ichthyologists that "behavior is a prime determinant of structure." The incredible radiation of that family may have had something to do with the fact that the cichlids much more than most other fish show awareness, alertness, and curiosity; it is not surprising that they are all assiduous parents (Fryer and Iles 1972, 50, 99).

Instinct

There is no clear distinction between organic response to stimuli—as when shortening days cause subarctic animals to grow fat for hibernation and birds to prepare for migration—and instinct, such as the bird's program for nest building. There is also no sharp line between inborn compulsive, more or less reflexive actions and inherited predilections, which merge into learned behaviors. Some songbirds are born knowing what to sing; others vary their inborn songs or copy freely those of other species, being endowed with only a generalized ability and urge to sing (Griffin 1984, 114).

Much is written in the brain by the genes. Tiny salmon need no lessons to attack other young salmon and flee from salmon-eating trout (Alcock 1989, 326). The perceptions of the two slightly different shapes and the appropriate responses are coded in the salmon genome, for which the cooperation of many genes is doubtless necessary. A digger wasp emerges from its pupa without ever having seen its mother, sets out to find a caterpillar, immobilizes it, brings it to an appropriate spot, digs a hole to bury it, lays an egg on it, and carefully closes the hole. Sometimes instinct is blind. An acorn wood-pecker will drop hundreds of acorns through a knothole in a cabin wall that looks like a hole in the side of a pine tree that it would use to store acorns. A robin carried blind instinct further: when a piece of meat stuck in a nestling's throat, the mother bird, programmed to eject extraneous objects from the nest, threw out the meat with baby bird attached (Welty 1982, 190, 409–410).

Instinct ranges from mechanical behavior to facility for learning certain kinds of actions. It can be like the directions for putting together an unassembled desk, it can amount to an ability to guess how parts can be fitted to make a piece of furniture, or it can be an instruction that needs to be sharpened by learning. The ability of sandhoppers (*Talitrus*), which orient by the sun, to head for the beach was found to be partly learned, partly innate. They are born knowing the direction of the ancestral shore, but as they mature, they lose the instinct if it is not exercised, and they are easily trained to take a different heading. *Arctosa* spiders, rather similarly, are born with a direction sense but keep it only if it is practiced and can modify it as necessary (Scapini 1988, 210–220; Scapini 1986, 53–61). This mixture of learning and instinct is well suited for small dwellers on a curving coastline.

Dogs can be taught a great variety of things, but they learn with special ease skills for which they have been bred, such as retrieving and sheep herding, which are modifications of wolf instincts. Sheepdog pups go through the motions of herding before training (Fox 1978, 37). Kittens instinctively go after mice, but they do much better for watching their mother (Barnett 1981, 235). Lion cubs seem to learn hunting by observation, as they trail after their parents (Schaller 1972b, 263). Young beavers know how to build dams by instinct, but they improve by imitation, and every dam is different (Hanney

1975, 50). The animal inherits a capability for a particular activity: the construction of a watertight barrier of sticks and mud in a suitable place to make a pond.

Kingfishers have been observed to beat small fish to a torpor and drop them into the water for the youngsters to practice fishing (Welty 1982, 420). Oystercatchers learn one of two ways of opening mussels: hammering at a weak spot after the mussel has been brought to land or inserting the beak while the mussel is *in situ* to cut its closing muscle. Parents hand one or the other method down to their offspring by example, not by genes; chicks raised by foster parents follow the way of the foster parents. No bird is observed to use both ways (McFarland 1985, 516).

The oriolelike South American caciques, fearful of monkeys and snakes that eat their eggs and young, choose sites near hives of vicious wasps, preferably on islands (Robinson 1986, 42). They also count on the wasps' keeping away botflies, whose maggots often kill the young. (The wasps, inclined to attack anything, do not trouble the caciques.) If they find a properly guarded nesting site, they repel cowbirds. But when they cannot find a wasp-protected location, they tolerate the cowbird's parasitism and raise cowbird young, which eat the fly larvae on cacique chicks and improve their chances of growing up. This is not inherited; cacique chicks from nontolerant colonies transferred to tolerant colonies become tolerant adults (N. Smith 1979, 9–15). Either all the birds in a colony accept cowbird eggs or all reject them (Welty 1982, 388). It would not be suitable for the bird to convert this behavior into instinct because it is to be varied according to circumstances.

The housebuilding ability of the weaverbird has become fixed, although it could hardly have been elaborated without learning. Weavers can be raised artificially for generations without ever seeing nest-building materials, yet their offspring, given materials, weave good nests—but they improve with practice (Welty 1982, 341).

Young birds are born knowing the ancestral migration route, but migration patterns are rather flexible and changeable. In many species, different populations go in different directions, and some may stay put. Moreover, birds often travel much farther than would seem necessary or judicious; migration is dangerous, and a large fraction, especially of small birds, are lost on the way (F. Gill 1990, 244–247). It is difficult to imagine that migration patterns could have been shaped without an element, perhaps a large element, of learning.

Some instincts are a potentiality for a certain behavior that is rarely necessary. Deer and sheep, living on soils extremely deficient in phosphates or calcium, have been observed to eat the bones of bird chicks or small rodents, so far as they can be caught (Furness 1989, 8–12). Such an instinct would be difficult to establish by natural selection. It goes far beyond the adequately mysterious ability of the body to translate shortage of salt into the drive to find a saltlick.

Darwin believed, quite understandably, that learning and habituation led to instincts (Richards 1987, 103). So far as this may be the case, it implies that learning is assimilated genetically through selection, in somewhat the same way as the altered wing venation of *Drosophila*. But the assimilation of an environmentally induced structural modification only means fixing potentialities already present in the genome. Changes induced in *Drosophila* by stress (heat, chemicals) are the same as those frequently occurring by mutation in unstressed animals (Dobzhansky 1970, 41). In the case of learned behavior, the organism must assimilate a new, perhaps quite arbitrary pattern.

The human has about a million neurons for each of the genes that together plan the whole body, and the transmission through DNA of instructions dictating a complex series of actions coordinated with perceptions would seem prima facie impossible. The known means of differentiation in morphogenesis, the diffusion of chemical inducers, is not a conceivable way of wiring instincts. Animal behaviorists record no case of a mutation giving rise to a new behavior, only to pathologies or distortions of behavior. In the much-studied fruit flies, these include paralysis, hyperactivity, defective flight, and inability to mate (Hall 1985, 304–310, 327).

Yet behavioral proclivities are readily accentuated by selection. Eleven generations of selection of *Drosophila* produced one strain about twice as active as the controls and another that hardly moved (Ewing 1963, 369–378). Twenty generations of selection produced mice six times as active as those selected for lethargy (Barnett 1981, 514–515). This may imply that new instincts are built closely on old ones.

Such modification of behavior by selection must be common in the wild. Birds, even strong fliers such a hawks, become reluctant to fly over water, at the price of failing to colonize nearby islands (J. Diamond 1981, 508). Squirrels from areas where there are rattlesnakes distinguish between rattlesnakes and less dangerous gopher snakes, while those raised in areas without rattlesnakes do not (Alcock 1989, 227).

Among the most purposeful of animals is the African honey guide, which has the appropriate scientific name *Indicator*. It is fond of bee larvae and beeswax (which other birds cannot digest) but has difficulty forcing its way into a hive of indignant bees. To recruit help, it locates an animal, such as an African badger (ratel), or a human, flutters about, and calls to it. As the animal follows, the honey guide flies from perch to perch until it reaches a hive it has previously located. Thereupon it changes its tune and waits for the mammal to demolish the hive, leaving plenty for the bird to eat (Isack and Reyes 1989, 1343–1346).

The ancestral honey guide may have profited occasionally from the leavings of animals' assaulting bee colonies, but it is a leap to seek animals and try to lure them. The behavior is thoroughly instinctive. Young honey guides, growing up in nests of other birds, have no opportunity of learning from their elders. Yet the instinct has been extended in recent times to soliciting humans to help. Honey guides have also learned to come to people who whistle for them.

Learning is not simply passed on to offspring. Experiments to demonstrate this have failed. When an animal acquires some habit, it would be pointless for its descendants to inherit the ability unless it were useful, and the only criterion of usefulness is confirmation by selection. But such selection cannot be based on mutations in the ordinary sense. The chance of a random alteration's occurring to make the needed changes in a response chain would be astronomically small. The "mutations" can only be rearrangements in an integrated guidance system, some of which will strengthen the learned pattern.

We have no hint of how genes can dictate the formation of neuronal connections to bring about a complex sequence of actions tied to perceptions. Because of the extreme complexity of instincts, the probability of improvement by random mutation must be minimal. It would seem more difficult to dictate such fine details as interconnections in the brain than to make appropriate modifications in the shape of an organ. The reverse is the case; in higher animals, instinctive behavior is altered much more readily than organs are reshaped.

It is not likely that the formation, imprinting, and transmission of instincts can be understood until much more is known about the functioning of the brain. It is obvious, however, that the brain is plastic. Memory and learning hardly seem to have definite locations in the brain. The ability of rats to run a maze is decreased by the extent of removal of brain tissues more than by the removal of specific parts of the cortex. The motor cortex is organized not around muscles or specific movements but around tasks (Restak 1979, 240), and the genetic directions it receives may be so also.

The ability of the brain to compensate for damage to parts has been mentioned. The development of its neurons is partly dependent on stimulation; indeed, it may be that activation of many genes' programming the brain depends on environmental stimuli "arriving via various sensory modalities" (Chaudhari and Hahn 1983, 927). In laboratory animals, the brain responds physically to a stimulating environment: the cerebral cortex is thickened, and dendrites branch more and develop more spines, making more interconnections of neurons. There are also such changes as an increase of the ratio of nerve cells to supporting cells, the number of synapses, and the amount of neurotransmitters produced. Rats with plenty of toys and playmates had synaptic junctions 50 percent larger than those kept alone in bare cages (Rosenzweig, Bennett, and Diamond 1972, 22–29; M. Diamond 1988, 51–57).

The relation of genes, instinct, learning, and patterns in the brain is complex, a complexity permitted by the very large number of genes concerned with the formation of the brain. In the rat, it is believed that there may be as many genes for the formation and functioning of the brain as for all other organs and tissues together (Chaudhari and Hahn 1983, 924). This is far from sufficient, of course, to dictate the interconnections of hundreds of millions or billions of neurons; these must make their own patterns in conformity to general directions of some kind.

The formation of an instinct seems to require integrative learning. The ability of the archerfish or the heron to compensate for the refraction of light at the water's surface could arise only by guidance to the perceptive faculties of the brain.

A more complex instinct is the frantic reaction of cattle to the presence of warble flies. These relatives of the human botfly lay eggs on the animal's skin; the grubs burrow in, make a tour of the body, and weeks later come to rest in a lump in the hide, to the animal's discomfort. Although the flies cause no pain, the cattle are much more agitated by their presence than by that of biting flies, which mean immediate torment (Klots and Klots 1971, 208). That the cattle have learned to associate the warble fly with a distant woe is hard to believe; that they came to distinguish between the warble flies and not-dissimilar biting flies by an accumulation of mutations in any ordinary sense equally defies belief. The difficulty is increased because the cow's agitation is largely futile; it cannot prevent a determined botfly from leaving an egg on its back.

The instinct of a wasp to find and paralyze a spider on which to lay its egg requires, among other things, that the wasp have in its ganglion-brain a perception of spiders, to recognize a spider over a wide range of size, near or far, from front, back, or side, not only distinguishing spiders from crickets but usually fixing on a particular kind of spider. For the different aspects of spider to be generalized and fitted together to make an image that can somehow be coded is a task for a well-developed nervous system. The wasp has such a capacity for making and storing an image, as it shows by its visual memory of its nesting site. The conclusion that the pattern must be put together in the nervous system is strengthened by the fact that wasps can apparently adjust the instinctive perception to a variety of prey, such as caterpillars, flies, bees, and grubs.

The perception must also be coordinated with a complex behavior pattern. It would be the harder to reach a spider-perceptive pattern by random inputs because it would be useless unless the wasp were prepared to sting in just the right spot with a dose of poison to paralyze without killing. But spiders capture insects and are capable of fighting back with poisonous fangs or entangling a wasp in a web—although the wasp may somehow neutralize its defenses. Many instincts involve a sequence of a half-dozen or so different actions.

So far as behavior may lead to instinct, one might guess that the central nervous system somehow acts on the genetic apparatus, which is, after all, more of a copying machine than a computer. If there is any such feedback, however, it is very subtle. If it is difficult to envisage how the genes transmit instructions to the nervous system, it is even more difficult to guess how the nervous system could give detailed instructions to the genes. The unlikelihood of such feedback is highlighted by the remarkable instincts of social insects: the workers, which explore and may learn, leave no descendants; the queen, sitting in the royal chamber and laying eggs, learns nothing. The instincts that she transmits to her servants are tested only by the success of the colony in breeding new reproductives to start new colonies.

The capacity of social insects for elaborate instincts must have been an important factor, perhaps the most important factor, in their success. This implies an ability of the genome to throw up, probably by chaotic processes, new patterns of behavior and combinations of them. This is partly understandable because the complex existence of social insects offers many possibilities of interactive behavior; hence, mutability of instincts would be especially useful for them. It is also likely that selection for new behaviors leads to greater facility for mutations in the same direction (as suggested by J. Campbell 1985 for somatic mutations).

Groups of animals develop broad faculties for general lines of development. Spiders have different glands for many kinds of silk, which they put to a multitude of uses. Snakes have an aptitude for injecting sundry poisons in various ways. Radiolarians produce a phantasmagoria of lacy skeletons. Something like this doubtless occurs with instincts, such as those of the hunting wasps. Some take a variety of prey, others only a certain species; but it seems that wasps have acquired a facility of hunting instincts, whereby reassortment of modules—perhaps enhanced by hunger—might combine with experimentation to produce new skills.

Behavioral patterns can more readily be reassorted because they are not necessarily of any particular utility, varying more or less freely, like details of leaves. For example, social arrangements of ants are extremely varied without regard to diet, and they are fluid. The recently introduced fire ant, *Solenopsis invicta*, which has become a plague in the southern United States, has gone from having a single queen, the most frequent mode of ants, to having many, even hundreds of queens, in about two decades (Ross and Fletcher 1985, 349), presumably to reduce vulnerability to insecticides. An instinct may be capricious. An ant of the American West, *Conomyrma bicolor*, has the custom of trying to keep a rival species, *Myrmecocystus*, from foraging by dropping pebbles down the entrance to the *Myrmecocystus* nest. The custom is no great rarity because an Australian ant, *Iridomyrmex purpureus*, does the same thing to its competitors. (Hölldobler and Wilson 1990, 387, 424). It would seem more trouble than it is worth, because it is just as easy to remove the pebbles as it is to bring them up to close the entrance.

It is more serious that an opossum (*Didelphis*) usually reacts to being seized by becoming immobile, tail curled and mouth open. This hardly discourages a predator from dinner, and if released the animal does not take the opportunity to scamper away but remains as though paralyzed for some minutes (Nowak and Paradiso 1983, 19). There is no apparent utility in this behavior; called catatonia, it seems to arise from some failure of nerve, like human fainting. There could be no selection for passivity under attack. On the other hand, there would be no strong selection against catatonic immobility; by the time the opossum gives up, there is probably little it could do to save itself.

The American hog-nosed snake (*Heterodon nasicus*) has a similar custom. If cornered, it behaves aggressively, but if this does not deter the attacker, it "pretends" to be gravely injured, opens its mouth, and goes limp on its back. If picked up, it dangles limply; if turned right side up, it flips over to display its belly. An animal does not save itself by surrendering; the normal reaction of a hungry carnivore would be to begin the meal. But something in the genetic makeup must dictate this eccentric and unrewarding behavior; a South American relative, *Lystrophis*, acts similarly (Shaw and Campbell 1974, 69–70). Unlikely instincts, like much else in the makeup of animals, do not necessarily contribute to "fitness."

It is hard to fancy either that a beaver brain conceived the idea of building a dam to facilitate flotation of logs or that the genes hit upon such a scheme. But it is not to be expected that science can approach understanding of the genesis and transmission of instincts when such a basic property of the brain as memory is impenetrable. The animal genome can modify itself only to the extent and in the ways that it has been evolved to do, but how instincts are shaped by the needs of the organism must be infinitely more complex than the ways in which bacteria "learn" to use new nutrients.

Instinct is inseparable from learning and intelligence. That even a very simple animal can learn means that automatic behavior is reinforced or inhibited. If the genome implants a plan of behavior in the brain, such as a hunting instinct, this is accompanied by a capability of refining or strengthening that behavior by practice or observation. For an animal to become more intelligent means to enlarge the open-ended capability at the expense of semi-hardwiring. But if it is incredible that genes should code for instincts, it is a higher wonder that they should produce the generalized faculty of intelligence, which surmounts much of the need for genetically dictated responses.

It was the original miracle of life to use information to propagate itself physically; it has become able to incorporate learning into its being as instinct; its ultimate achievement has been to organize learning to raise its powers indefinitely. We take for granted the condition that the tangled web of neurons forms and manipulates representations of reality and enables us to cope with an infinite variety of problems because we continually experience its workings. But this is the most evolved and most incredible of the abilities of organisms to adapt themselves to their universe.

12

Evolution and Humanity

Becoming Human

Perhaps the most remarkable turn of evolution, certainly the most interesting for us, is the genesis of the human animal, which has become in a split second of geologic time the dominant creature of earth. The humanoid line diverged rather abruptly from the simian stock into a new pattern (or attractor), it continued with something like momentum of development, its evolution has been not biochemical but structural and behavioral, and it has found an amazing adaptive capacity, more than compensating for sundry maladaptive traits. It has become so powerful that its principal problem is self-restraint.

Although monkeys have no obvious need for greater intelligence than squirrels or opossums, primates grew big brained. In arboreal life, the forearms became freely mobile, and the forepaws, turned to grasping branches, picking fruit, and catching small animals, became capable of adeptly handling objects. Their ability to manipulate things made it useful to think about them, perhaps utilize them as tools. The primates became versatile, eating anything from leaves to lizards. They were socially flexible, too, living alone, in biparental families, in single-male harems, or in multimale troops.

Like many other animals, primates tended to increase in size, up to the great apes that laid the basis for the human development. Our biochemically very close relatives, the chimpanzees, have a complex social life, communicate their feelings, use crude tools, occasionally hunt and share meat, and hurl sticks or stones at predators. Because of climatic change, ecological opportunity, or genetic accident, the hominid branch left their way of life behind somewhat over 5 million years ago

(B. Campbell 1988, 35, 162–163). The apes have changed little since then, but the hominid line has changed very much.

It is believed that the protohumans took to more open country in response to a drier climate, and on the grasslands they stood up. The key innovation is believed to have been bipedalism, a stance unique to the human among mammals, which goes back about 4 million years in *Australopithecus afarensis* ("Lucy") (B. Campbell 1985, 374). Upright posture made it easier to look over the grass, to use tools and weapons, to bring food to a home base, and to carry infants during lengthy helplessness—that is, to become human. Although the apelike creature walked like a human, its brain was only about the same size as a chimpanzee's. It may have been more intelligent, however (Jacobs 1985, 281–284), with a brain somewhat larger relative to the body than the chimpanzee's.

The growth of the brain has occurred almost entirely during the last third of the time since humanoids diverged from ape stock, increasing from around 600 cc. to around 1,400 cc. (Simons 1989, 1348). About 2 million years ago, some of the hominids had become sufficiently human in our eyes to be named *Homo habilis*, or "man the toolmaker." With stone tools, they began the ascent that has been going forward ever since. For a long time, it was extremely slow, but it has been accelerating, especially in the last 200,000 years.

Leaving behind the security of the trees, protohumans had to cooperate in defending themselves from hungry beasts twice as fast as they were. At first they were probably primarily scavengers, not hunters. Although they might be unable to bring down a zebra, a small group with the crudest of spears could drive a lion from its kill. And the use of weapons made crude strength less important both for making a living and for dominance.

Behavioral changes—sharing of food and closer relations between sexes and generations—may have been crucial in the transition to humanity. Among most mammals, there is little or no permanent relationship between the sexes (except as they are in contact in a group) and none between males and their offspring. Primates are outstanding among mammals for the extent of paternal care, even among polygynous monkeys. Usually, however, the father plays a small role, if any. Because the female chimpanzee advertises estrus by conspicuous swelling of the genital area and copulates with a large number of males, no one is responsible for any particular baby, and no father transmits learned behavior. Moreover, groups are small and rather fluid. Cultural accumulation is consequently much less than the chimpanzees' considerable intelligence would permit.

The pygmy chimpanzee, or bonobo (*Pan paniscus*), which is rarer than the ordinary chimpanzee (*Pan troglodytes*) and has not been so much studied, has a more humanlike society. Estrus is much less marked; genital swelling of the female is more or less continuous, and she is usually sexually attractive and receptive. Copulation is frequently in human style, face to face, and males and females sleep together (de Waal 1989, 159, 180–181). Male bonobos, unlike male chimpanzees, do not form cliques; most groups are mixed—male and female, old and young. There is little hierarchic order and almost no male aggressiveness. They mutually groom and share food to an extent rare among animals (Goodall 1986, 484–485; Kuroda 1980, 183–185; Badrias and Badrias 1984, 336–341).

The bonobo's skeleton is more like the human than is the chimpanzee's. It has shorter jaws than the chimpanzee, is more slender, and stands quite erect. It walks bipedally more than any other ape, sometimes using its hands to carry things. Unlike chimpanzees, it has a human-sized penis, and it has longer hair on the top of its head. It is at least as intelligent as the chimpanzee. Typically, bonobos make contorted faces and play games covering their eyes (de Waal 1989, 181–198).

Among *Australopithecus,* males were much larger than females, but dimorphism gradually decreased in the hominid lineage; it is a general rule among birds and mammals that the more monogamous they are, the more the sexes resemble each other. Genital swelling and estrus must have disappeared quite early. Pair bonding made competition less severe (McFarland 1985, 154). It also permitted longer immaturity, which made it possible and functional for fathers to help educate their children. It seems that intergenerational coherence allowed compounding of culture. Was familial life the key to human success?

Homo habilis may or may not have been ancestral to *H. erectus* around 1.6 million years ago. A systematic hunter of large animals, a toolmaker, and a regular user of fire, *H. erectus* had a skeleton, except for the head, very like that of modern humans. Its brain, which did not increase much during more than a million years, came up to the modern size range (Stanley 1981, 149).

Hominids in the form of *H. erectus* spread widely to Europe and Asia, the first members of the clan to become notably successful animals. Living in a cooler, seasonal climate presupposed a more planned existence, with storage of food, better shelters, and probably clothing (B. Campbell 1985, 383). However, for over a million years, our ancestors were using practically the same simple stone tools.

Tools began improving more markedly about 100,000 years ago. Somewhat earlier, *H. sapiens,* with a jutting chin, weak brows, and high forehead, appeared in Africa and soon spread to Eurasia. The immediate predecessors of modern humans in Europe, however, were the Neanderthalers, who lived from roughly 100,000 to 35,000 B.C. They were much like *H. erectus* and may well be deemed a separate species from *H. sapiens.* Their skull was longer and lower than that of modern humans, with heavy brows. The Neanderthalers seem in some way advanced; they cared for invalids and gave formal burial to at least some of their dead (B. Campbell 1988, 418–422). They were replaced by the physically less powerful Cro-Magnons about 40,000 years later in the Near East and 5,000 years later in Western Europe (Stanley 1981, 152).

Our Cro-Magnon ancestors were physically weaker than the Neanderthalers, and we would like to believe that they prevailed by superior intellect. Their brains were no bigger than the Neanderthalers' but were more developed in the frontal areas, perhaps making them more imaginative and inventive. Artifacts had been severely utilitarian for 2 million years; in a brief time after the disappearance of the Neanderthalers, a great variety of objects appeared, many of them artistic, ornamental, or symbolic, along with better-finished knives and weapons. Most striking are the many beautiful pictures, mostly of animals, that paleolithic artists laboriously painted deep in the interior of caves (White 1989, 92–93).

Not only did the brain not grow further; it shrank slightly. Either the quality of the brain somehow improved, or human interaction became more effective. Improvement of social organization, like that which seems to have made the human offshoot possible millions of years before, may have encouraged invention and self-expression.

An attractive hypothesis is that the power of speech suddenly expanded, presumably by a mixture of learning and changes in the brain. Language is the most distinctive human trait, an inborn faculty, not merely the result of enlargement of the brain; microcephalic children can speak quite intelligibly. But it is unknown when language in the modern sense appeared. In the ages when the species was being formed, protohumans must have improved communication beyond the vocalizations of their ancestors, such as the distinct calls monkeys use to warn of leopard, snake, or eagle. The expansion and lowering of the larynx made possible a greater variety of speech sounds. But language more or less as we know it may have been the signal invention bringing heightened self-awareness and the formulation of ideas and cultural wants, making possible the new symbolic culture. Language seems to have been a single invention or else to have spread its basic patterns universally, because all the thousands of languages have much in common: the ability to negate, to form questions, and give commands, plus grammatical categories such as noun and verb, and semantic universals such as male-female and human-nonhuman (Fromkin and Rodman 1983, 15–17).

The Evolution of Intelligence

If many things human, such as dreaming or masochism, are inexplicable in terms of survival value, the central human attribute, intelligence, seems at first glance hardly to require explanation. According to the variation-selection theory, "The brain exists in its present form because it permits the survival and multiplication of the genes that direct its assembly" (E. Wilson 1978, 2). Its biological value has been abundantly shown by the huge growth of numbers of our kind. Long before the industrial revolution initiated a population explosion about two centuries ago, even ages before the invention of agriculture, our species had spread from the African homeland to all hospitable regions of the globe. Surely no recondite explanation is needed for the organ that has brought such reproductive success.

Yet the biological value of the brain may not suffice to account for it. It cannot be assumed that something useful for the community or the species necessarily results in biological fitness for the individual, and it has never been shown that higher-than-average intelligence corresponds to reproductive success for humans (Restak 1979, 86). Physical strength and ability to dominate others might be more important than exceptionally good brains in the natural selection of primitive humans. It is not clear why early hominids should have been much more subject to selection for intelligence than chimpanzees, whose mental powers seem to have long been static.

Up to a certain level, intelligence doubtless contributes to reproductive success; above that level, it can interfere. Long ago, hominid society was probably quite structured and had conflicting values; the unusually intelligent would very likely be more concerned with social position, or leading the hunt, or being regarded as the sagest member of the clan or the best caster of spells than with having the maximum number of progeny. Indeed, the outstandingly gifted individual might well have been a social misfit among our forebears, as often occurs today. In modern society, abstract intelligence is not closely correlated with economic or social success, much less with fertility, and there is no reason to suppose that it had much to do with reproductive success 100,000 or 1 million years ago.

Then as now, a lively mind and devotion to intellectual activities would probably distract one from having and caring for children. As Williams pointed out, "There is no reason for believing that a genius has ever been likely to leave more children than a man of somewhat below average intelligence," and ruling classes often fail to reproduce themselves (Williams 1966, 14–15).

The conflict between intellectual-cultural interests and attention to family must be as old as humanity. To be more intelligent is to be more idealistic in a broad sense, that is, more concerned with ideas, things abstract and symbolic, less with matters biological. If the human brain had been forged for its utility in a biological sense, it would surely be much better equipped with practical abilities, especially in relation to reproduction.

Moreover, increase of intelligence seems to have required prolonged immaturity, which greatly reduces reproductive capacity. It would be very difficult for the brighter folk to compete biologically with those who might be a little less sharp mentally but whose children became independent sooner to go on to have children of their own.

The most striking mental powers, such as high ability in literature, music, or mathematics, are not of the kind likely to be rewarded by number of descendants. Strong parental instincts would be much more conducive to reproductive success, but in this area the human endowment is mediocre. Sexual selection could have played a part in the increase of intelligence if females preferred clever suitors or clever women were more attractive to men. This is not impossible, but it seems unlikely unless character has changed sharply. Modern women are not known to be especially drawn to high intelligence, especially not to powers of abstract reasoning, and it is not evident that men usually prefer intellectual women over pretty ones or that the former have more babies.

A variant idea is that intelligence was biologically rewarding because social skills were helpful in winning mates. Male baboons are said to prevail mostly by alliances (Lewin 1989b, 130). But humans, aside from their mastery of language, do not seem exceptionally well endowed with social capacities and frequently have difficulty getting along with their families and fellows. Human intelligence is least impressive in matters most relevant for reproductive fitness: courtship, sex, and parenting. The capacity for seduction hardly entails a capacity for mathematics (and perhaps the contrary).

Intelligence is more likely to be rewarding in the long term than in the short run, and to be more useful for the community than for the individual; such may have been the case since the days of *Australopithecus*. It is plausible to guess that cooperative hunting rewarded the more intelligent group, as did cooperation in the warfare that we may suppose to have been frequent. Stratagems and trickery could give many a victory to our distant ancestors, enabling them to enlarge their hunting grounds. But the individuals who invented the tricks would not necessarily benefit reproductively.

A powerful computer designed to solve mathematical problems can probably handle (with appropriate software) tasks from record keeping to drawing pictures. Nonetheless, there seems to be no good reason for an apelike creature to acquire the capacity to compose a symphony, to reason out abstruse equations, or to speak learnedly for hours on subjects of no immediate biological significance, such as the interpretation of the French Revolution. The human brain has enormous versatility, not only in such areas as music and philosophy but also in ability for figure skating or acrobatic diving.

A leading authority on human evolution, Bernard Campbell, concedes that "it is not easy to account for the capacities of the human brain if it is seen as merely the product of a hunting and gathering life-style" (B. Campbell 1985, 340–341). There is nothing in the life-style of chimpanzees, which we assume not to be extremely different from early hominids, demanding mathematical ability. A bright chimpanzee has trouble counting to more than 4 (King and Fobes 1982, 349), being inferior in arithmetic to crows and other birds.

Even unconditional Darwinians are troubled by this question (Alcock 1989, 520). Alfred Wallace, who supported Darwin on almost all points, differed with him on this (to Darwin's chagrin) and concluded that the human brain could not be a product of adaptation and could not be accounted for in terms of survival value. Wallace was particularly impressed by human powers of abstract thought and the beautiful but useless human capacity for song, which, unlike that of birds, is not related to sexual selection. He saw moral qualities, such as the acceptance of martyrdom, as "the workings within us of a higher nature which has not been developed by means of the struggle for material existence." He was struck by the native capacities of precivilized peoples (whom he knew intimately as Darwin did not), and in Wallace's view, "in his large and well-developed brain [the human] possesses an organ quite disproportionate to his actual requirement—an organ that seems prepared in advance, only to be utilized as he progresses in civilization" (Wallace 1881, 193, 202–203).

There are other reasons to doubt an explanation in terms of random mutation and relatively greater reproductive success of the more intelligent individuals. The rapid growth of the human brain, more than doubling its weight in 2 million years, implies very strong selective pressures, but cultural progress did not correspond to brain enlargement. At the time during which we would have to postulate selection for brains, the glacial progress of toolmaking would indicate relative unimportance of cultural competition. Cultural development began to accelerate slowly about 200,000 years ago, when the long-term expansion of the brain was nearing its end.

Intelligence and Brain

If intelligence is defined as the ability to learn and to modify behavior to suit circumstances, many relatively simple organisms have a good deal of it, although their "brain" may be trivial in size and complexity compared to that of a frog. Birds, which economize weight, seem to have much more intelligence per gram of brain than mammals. Crows and jays, with brains the size of a peanut, can count much better than monkeys (F. Gill 1990, 154). Small birds can tie fibers together to make a dangling nest. We might take less pride in our musical ability if we realize that songbirds can sing a hundred different songs with a brain about 1/2,000 of that of the human. Human memory is not impressive compared with that of Clark's nutcracker (*Nucifraga colombiana*), which stores about 32,000 pine seeds per year in approximately 10,000 caches and months later remembers where most of them are (Tomback 1982, 451–457).

Those who study prehistoric animals, however, ordinarily assume that size of brain (in proportion to body size) roughly indicates intelligence. The component elements, neurons, which are like gates or switches in a computer, are the same in humans and other mammals, and a greater number of switches and larger storage capacity should make a more powerful apparatus. Greater computer power, however, requires not only more elements but more organization. And the larger brain will not necessarily be better organized unless it has capacities of self-organization.

Other things being equal, it apparently does; the larger number of neurons do not usually get in the way of each other (although they sometimes do, in the human case) but manage to cooperate. This is corroborated by the fact that large animals are generally more intelligent than smaller comparable ones—large dogs more than small ones, horses more than donkeys, and so forth. The biggest lizard, the Komodo monitor (*Varanus komodiensis*), is said to be far more intelligent than its small cousins, a crafty hunter that in captivity easily learns to recognize different people (Auffenberg 1981, 121).

We assume that the first mammals, which appeared about the same time as the dinosaurs, must have been more intelligent than their ancestors, the therapsid reptiles, which had brains three to four times smaller in proportion to the body

(Carroll 1988, 412). The supposedly smarter mammals, however, were limited to scurrying in the shadows and were no bigger than a rat, mostly smaller. For 140 million years, neither mammal nor dinosaur brains grew much (Jerison 1973, 257). But upon the extinction of the great reptiles at the end of the Cretaceous, the mammals rapidly radiated into the many modern orders, expanding the brain on the way.

No one has any idea why the mammalian brain swelled or why the primates have developed more than average mammalian intelligence. It is sometimes assumed that their favored habitat is conducive to selection for intelligence, but of nine orders of mammals that have taken to arboreal life, only the primates have become outstandingly intelligent (B. Campbell 1985, 61). Life on the ground is likely to be more varied, perhaps more challenging; it also calls on different senses, hearing and smell, more than does arboreal existence. The tree-dwelling sloth is among the less quick-witted of mammals.

For whatever reason, primates grew steadily more talented. Monkeys are notoriously clever, sometimes capable of manipulating mechanical things as well as a human can. Japanese macaques have been observed acquiring rudiments of culture as they learned to wash potatoes and to separate grain from sand by throwing it into water and picking up the floating grains, making the discoveries more or less by chance but showing the good sense to appreciate them and spread the knowledge (McFarland 1985, 514).

Chimpanzees have brains about twice as large as standard for a mammal of their size, and their abilities are celebrated. More than most other animals, they learn from their elders, and they imitate many actions on first sight. They have numerous meaningful calls, which are different in particular populations. They have beginnings of culture, using simple tools, such as sticks to poke termites out of a nest (Goodall 1986, 536–547). Although they cannot be taught to speak, they can master a hundred or more sign-words, make simple sentences, invent signs of their own, and use the pronouns *you* and *me*. They recognize themselves in a mirror (J. Gould 1982, 483), which seems to indicate self-awareness. They will work for chips with which to buy bananas; in many ways they are like humans without the power of speech.

The daily lives of monkeys and chimpanzees are much more banal than their intelligence would suggest (Lewin 1989b, 129).

Gorillas are even more obviously capable of much more than their simple existence requires. They have rather less need for intelligence than other herbivores because they are hardly endangered by predators. But gorillas seem to be about as well endowed mentally as the chimpanzees, which at least face a challenge of finding all manner of edibles, from figs to small game. A domesticated gorilla, Koko, was credited with understanding 400 words, more than twice as many as any chimpanzee has mastered (Konner 1982, 167). If human intellectual powers cannot be clearly related to pressures of selection, neither can those of other large primates. If it is true, as Roger Lewin suggests (Lewin 1989b, 40), that there has been parallel evolution of the most closely related large primates—gorilla, chimpanzee, and human—other factors than survival of the fittest must have played a part in all three cases.

Other animals also seem to possess intelligence beyond biological needs. The elephant has no natural enemies and lives on simple forage, but its brain is about four times as large as the human, and its capabilities are impressive. Working Indian elephants know some two dozen commands, and this number is limited less by their ability to learn than by the needs of their trainers. They accomplish their tasks, chiefly in logging, with few specific instructions, and they perform complicated circus tricks. Elephants do not learn very rapidly, but they can distinguish many different patterns and remember them at least a year (Rensch 1957, 158–163). They scratch themselves with sticks held in the trunk, use branches to brush off flies, and throw objects at people they dislike (Beck 1980, 32–36).

Marine mammals, except the rather torpid vegetarian sirenians, seem generally to be quite intelligent. Porpoises, sea otters, sea lions, and their relatives are charmingly playful, curious, and teachable. One might expect that dogs should be more intelligent than their pinniped relatives, which live in a much less varied environment and have a less social existence. But a sea lion is capable of following such a partly abstract command as, "Find the larger of two black balls and push it to the smaller ball" (Riedman 1990, 319).

Cetaceans have more intelligence with less apparent need for it. Sharks and tuna, with unimpressive brains, compete successfully with porpoises in catching fish. The huge baleen, or filter-feeding, whales lead an even simpler existence, engulfing huge quantities of water to sift plankton. They need only to find good pastures in the ocean; even their muscular coordination is simple compared with that of a four-legged running and jumping animal, not to speak of a flying bird. Yet some whales have brains four to seven times as large as humans—the sperm whale is in the lead. Dinosaurs of comparable bulk managed with half a pound or less (Jerison 1973, 145). The very large brains of toothed whales and the large brains of baleen whales (which have a simpler existence) must be to some extent independently evolved because the suborders are only rather distantly related.

Since large whales are not kept in oceanariums where they can be psychologically tested, we know of their talents only what can be deduced from distant and sporadic observation. They are frolicsome, migrate long distances, and sometimes engage in something that sounds very much like singing. The humpback whale is the leading vocalist. Its songs are long and highly structured. Groups of notes make "phrases," and "phrases" of one type form a "theme"; repetitions like rhyme and verse in poetry, a series of "themes" make up a "song." All the whales in an area hundreds of miles across sing about the same song, but this gradually changes from year to year (McSweeny et al., 1989, 139–148).

The size of the great whales' brains can be attributed to their overall dimensions. However, small whales, or porpoises, with bodies comparable in size to humans, have brains of roughly human dimensions or a little larger—larger even in proportion to body weight in the case of the best-studied species, the bottlenosed dolphin (*Tursiops truncatus*). Moreover, the porpoise brain is quite as complicated as the human brain, with comparable convolutions of the cortex. The auditory areas, having to do with echolocation, are especially developed (Slijper 1962, 248). The porpoise brain is much more impressive than that of a chimpanzee.

How their intelligences compare is disputed, although porpoises seem capable of understanding about as much as chimpanzees, and they have 20 to 30 distinct signals. John Lilly, an enthusiast, believed they approach genuine language (Lilly 1974, 71–77). They learn tricks with great ease. They can even grasp the idea of inventing new tricks to please the trainer and earn a fish. They have different personalities, some being kind and helpful, others irascible (Pryor 1975, 216–217, 236). What goes on in the porpoise brain we cannot guess. Unless it is much less active than our own, it must generate an interesting mental world.

Cetacean intelligence, like human, is linked to long gestation, slow maturation, and longevity (Würsig 1989, 1550). Brain size grew rather slowly up to about 20 million years ago and then came to a ceiling (Jerison 1973, 348), much as the human brain did more recently, oddly at about the same ratio of brain to body ("encephalization quotient"). It has been conjectured that cetaceans' large brain, like the human, is linked to neoteny: the whale is somewhat like a much overgrown ungulate embryo, with a large tail, poorly developed (or absent) limbs, and simple or (in the baleen whales) nonerupting teeth (Groves 1989, 310–311). But whatever caused its brain to enlarge ceased long ago to have effect.

It may be surmised that the elephant is bright primarily because it is big; as the head of the elephant ancestor grew in pace with the body, there was room for more brain, although the need for intelligence decreased with invulnerability. Baboons are big monkeys, with correspondingly enlarged crania and, according to anecdotal evidence, keen intelligence. The great apes, especially the chimpanzee and gorilla, are like very big monkeys—gorillas may weigh 400 pounds—with heads and brains of corresponding size. They have considerably more intelligence than baboons and other monkeys, although they have no more and probably less need for it.

The mammalian-primate great ape heritage consequently provided the foundation for human intelligence. But the hominid tribe, starting out on a solid basis at about the level of the chimpanzee in *Australopithecus,* broke away from the pattern and tripled the size of the brain without much growth of the body, and emerged with the faculty of which we are proud.

For reasons stated above, natural selection does not seem to deserve much credit, perhaps very little credit, for this achievement. A further reason to doubt the role of natural selection is that intelligence does not represent a single trait that might be subject to mutation and selection, but a set of more or less independent faculties, perception, interpretation, learning, memory, motivation, and coordination of expression, made effective through a special capability, language (Gardner 1985). Selection for particular intellectual traits would require a whole set of suitable and probably unlikely mutations.

It seems more probable that heightened intelligence was the beneficiary of an integrated evolutionary process, the crucial element of which was the evolutionary mode of neoteny, or the retention of immature characteristics into adulthood.

A baby chimpanzee is much more like a human baby than an adult ape is like an adult human. Its profile is almost that of a human infant. Among its traits suggestive of the human are reduced body hair, tender skin, nonopposable big toe, flat face, short jaws, small teeth, thin skull, and high relative brain weight (S. Gould 1977, 357; Groves 1989, 311–314). The maturing chimpanzee changes more than the human, developing protruding jaws with powerful canines and skull ridges, and the ratio of brain to body weight declines. The human skull differs in that the sutures close much later, and the opening for the spinal cord (foramen magnum) remains at the bottom instead of shifting to the back. Most important, the brain keeps on growing much longer in humans than in apes. Brain growth in the chimp slows down at birth and is nearly completed in a year; in the human, it continues rapidly for two years and requires another four or five to come near completion.

Juvenilization is obvious in puppies and kittens, which have less snouted faces and more rounded heads than those of adults. Monkeys appear neotenic, at least in regard to the head, in comparison with more primitive primates, such as lemurs. The bonobo is somewhat neotenic in comparison with chimpanzees, with a rounder head and shorter jaws. It is also behaviorally neotenic that the young and adults of the bonobo mingle more congenially than young and older chimpanzees, and the bonobo males are less aggressive.

Like the suppression of estrus and genital swelling, the delayed maturation of humans—we take twice as long as chimpanzees to grow up—is an essential part of neotenic development. It is difficult to account for in terms of natural selection. On the one hand, individuals that mature more rapidly have a headstart in reproduction. On the other hand, the need to care for children for many years makes it impossible to raise many of them. But long childhood gave the brain time to mature and learn, and it was probably crucial for the development of culture and the progress of the species. The human also remains behaviorally immature into adulthood in the sense of retaining the ability to learn and more or less flexibility of behavior, which chimpanzees largely lose after adolescence (P. Wilson 1983, 37).

Human neoteny refers primarily to the head and to behavior (Montague 1962, 326–327). But hairlessness and soft skin are also conspicuous neotenic traits. It is unlikely that the human loss of the coat of hair that protects apes, carnivores, and almost all mammals is adaptive. Hair may provide a hiding place for lice and fleas, but animals with short, dense pelage are not usually overrun with parasites, and the loss of useful hair is the more eccentric because a mop is retained on top where it is not especially needed. Hairlessness, moreover, should be compensated by a reasonably tough hide, not necessarily like that of the rhinoceros but at least comparable to that of the pig. Hairlessness, tender skin, and exceptional intelligence seem all to be parts of an evolutionary package, elements of which are evidently unadaptive.

The tendency to neoteny with enlargement of the brain may have been related to the new social order of the hominids. It would not be surprising if our ancestors continued the bonobo trend to male nonaggressiveness and increased intergenerational sociality. The development of effective and easily available weapons reduced the importance of crude strength in the competition of males for mates; at the same time, so far as food was derived from hunting large animals, it became important for the female to secure the loyalty of her mate and his support for herself and her children.

It may be surmised that for this reason the female became more attractive. She developed permanent breasts and remained sexually accessible nearly all the time, even in pregnancy, instead of inviting sex only at estrus. The face-to-face position made sex more personal, and the female orgasm encouraged her to seek it regularly. To have some assurance of paternity, the male had to have frequent intercourse and keep the female as much to himself as possible.

The female fortified bonding by taking on a more juvenile-feminine appearance. Much of the charm of the human female lies in her neotenic traits, with less and finer body hair, smaller feet and hands, more delicate nose and ears, and higher pitched voice. Like a child, she has less muscle, more fat, and pleasing contours. An understandable by-product of the neotenic pattern would be the prolongation of immaturity, which per se could not come about by natural selection.

Feedback made this development circular. As neoteny progressed, it became still more necessary for the male to support his mate and the young; as the father became more important for the long-helpless young, there was more competition among females to secure and hold helpful mates, that is, to make herself attractive, which meant more neotenic. It also made the reproductive success of the male dependent not only on insemination but on seeing his children to maturity. If he became fonder of his children and his somewhat childlike mate, the family was complete.

Natural selection, in this view, deserves credit not for driving this development but for making it possible. The rising intelligence of the expanding brain had to be sufficiently useful to offset the detriments of the neotenic tendency, including not only prolonged helplessness of the young but also loss of body hair and the weakening of the jaws and teeth. This could occur only when cultural acquisitions had attained sufficient importance. This may be the reason that chimpanzees do not go in the same direction: their immediate gain from a somewhat enlarged brain would probably be less than the costs of the neotenic changes. It was only well after the *Australopithecus* acquired a hominoid posture that brain development began in earnest, presumably because of gradually increased use of tools and weapons.

The idea that a confluence of changes, central to which was neoteny, brought about the flowering of humanity is unattractive. It fits, however, with the fact that human intelligence does not seem to be especially tailored to the practical needs of a hunter-gatherer or to be conducive to maximization of reproduction. On the contrary, the human mental endowment is suggestive of an information processor that expanded its capacities generally through growth of components and consequent complexity. Growing with the neotenic head, the brain became capable of far more than it had to do for biological fitness, capable of the imagination of a Shakespeare, the inventiveness of a Da Vinci, the martyrdom of saints, and the coordination of gymnasts. If human intelligence had been fashioned for its utilities for an apelike creature, it could not have become capable of making and enduring modern civilization.

We would prefer to think that our talent is the prize of victory in the struggle for existence, the reward of its naturally selected utility. But reality does not seem to have been so simple. Intelligence could come into its own only when it became worth the cost, but the basic factors in its genesis are probably the big-headedness of vertebrate embryoes, coupled with the need for love.

The Mind

Intelligence is our distinction and means of livelihood and power, but we think of the mind, consciousness or self-awareness, the spirit if one will, as the preeminent mark of our kind. A young child may know little and be capable of less, but we regard him or her as a sentient, priceless person like ourselves, with the indefinable quality of selfhood that we feel.

But if human intelligence is an enhancement of animal intelligence, other animals also must possess something of mind. In our personal experience, there is a continuum from the fullest conscious awareness to drowsiness, half-sleep, and deep sleep. It is rational to assume that the mind had no beginning point but grew gradually as the complexity of the nervous system increased, eventually (probably around the level of the chimpanzee) adding self-perception to perception of the world. It contradicts the sense of evolution to suppose that what we sum up as mind appeared suddenly in the human. It is not an invention of our species but the culmination of the autonomy of life. The development of mind must have been subject to the same causation that has led to the evolution of the weaverbird's building powers and the marvel of the eye. Yet physicists (as Penrose 1989) are more disposed to theorize about it than are biologists.

Without direct knowledge of any sensations except our own, we rightly credit other people with feelings like ours. It is equally reasonable to attribute feelings not wholly unlike ours to animals behaving in ways akin to our own. (For an extensive argument, see Griffin 1981). Animals act very much as though they feel fear, pain, curiosity, and anger. When animals learn in nonautomatic fashion, one must suppose that this involves something analogous to the pain and pleasure that mediate learning in humans. A dog's growl, yelp, or whine clearly conveys anger, joy, or pain.

Language gives coherence to mind, helping to develop and fix mental experiences, but animals have much symbolic communication. Ground squirrels have different calls to warn of hawks, coyotes, or snakes (Voelker 1986, 113). Lions have half a dozen distinct vocalizations, (Schaller 1972, 104–105), and monkeys and chimpanzees have as many as a dozen different screams, hoots, or grunts. After the baboons in an African park had quite lost their fear of cars, a visitor shot two. The word was spread; for nearly a year, all the baboons in the vicinity stayed well away from cars (Grant 1985, 419).

Monkeys solve puzzles for no reward, and they make a great effort to watch interesting scenes. Birds' singing or less melodious cawing, screeching, hooting, or squalling certainly exceeds the bounds of the utilitarian. Birds spend up to four-fifths of their waking time singing (Welty 1982, 238); it would seem that they must enjoy it. Small birds gather around and mob dangerous owls and hawks; it may be supposed, if one is sympathetic, that they are taking pleasure in harassing the enemy, albeit at some risk. One may guess that whales like their choruses and that birds take pleasure in their courtship rituals. Bowerbirds seem to have an aesthetic sense not totally unlike that of unsophisticated humans, as they collect shiny objects and replace withered petals (Griffin 1981, 136). Chimpanzees must have fun in their festivals of shouting, chasing around, and shaking branches (E. Wilson 1980, 109). They may dance ecstatically on seeing a waterfall (Konner 1982, 431). Dogs seem to be in high spirits when they race to return a stick thrown by the master. A panda, instead of walking down a snowy slope, lay down and slid on its belly; having done so, it walked back up and repeated the stunt (S. Gould 1987b, 21).

The howling of a tormented young elephant sounds like that of a child in pain. Elephants cluster around a wounded member of the herd, try to help it to its feet, and are slow to leave the corpse when it dies. Elephants are fascinated by their dead, sometimes laying branches over them, and examining elephant bones (Moss 1975, 34). Wounded African buffalo, instead of withdrawing to rest and recuperate, lie waiting to ambush the hunter, as though thirsting for vengeance. They may be so infuriated as to persist in their assault until killed (Miloszewski 1983, 140–141). Mother chimpanzees, monkeys, and baboons will carry a dead baby around for days (J. Gould 1982, 483), surely grieving; less intelligent animals quickly forget a dead baby.

The fact that animals sleep, that is (in our experience) suppress consciousness, implies that they must have consciousness

(Penrose 1989, 449). Sleeping dogs often sound as though they were dreaming of a chase, and the brain waves of cats respond to stimuli in much the same way as humans (Griffin 1984, 150–151, 203). The pattern of socialization of puppies is like that of human infants, with basically the same nervous structures capable of generating emotional responses, and both have similar nervous-behavioral disorders (Fox 1978, 258). Animals can be made neurotic by causing them anxieties with which they cannot cope. When animals act as though they were curious or afraid, it is reasonable to assume that their internal sensations are not wholly unlike ours.

Monkeys may have rather complicated minds. A male vervet gives an alarm call to frighten an approaching rival. A baboon's raising its tail reflects anxiety and subordination, so a female, apparently wishing to seem nonchalant, reached back to push her tail down as she approached the dominant male. A chimpanzee similarly tried to conceal nervousness by covering its grin with its hand (Cheny and Seywarth 1990, 42–44). Chimpanzees are almost as varied in personality as humans (B. Campbell 1985, 363). Darwin explored animal emotion (in *The Expression of Emotion in Man and Animals*) as a subject important in its own right without reference to natural selection.

The intangible mind, an adjunct of a plastic intelligence, even more than intelligence defies explanation in terms of evolutionary theory. Many things entering the mind have clear biological utility: we feel thirsty when the body needs water, and pain tells of injury or damage to the body and calls for attention to it. But the mind is much more than a bundle of utilitarian traits fixed by selective evolution.

For many mental traits it is hard to imagine a contribution to reproductive success. Not only are there abstract facilities, far beyond the needs of our ancestors' (or our) biological needs, but many quirks of the mind are puzzling. Hypnotic suggestion can induce enlargement of breasts or cause or prevent skin rash (Ornstein and Sobel 1987, 102–103). Pain seems simple and functional, but the mind-brain can register it or not, for reasons beyond present understanding. Through biofeedback, one may learn to modify one's brain waves (Restak 1979, 330–331).

Physicians know well that emotions have much to do with the functioning of the body, for both good and ill. Loneliness is bad for you, and depression weakens the immune system, while positive thinking improves longevity. Such a linkage is incomprehensible within the reductionist scheme, but it indicates that the mind is enigmatic not only in its origins but in its reality, and it seems to be impossible to explain in purely material terms.

The mind is no more independent of body than living creatures are independent of their physiology; but psychology is more complex than physiology just as physiology is more complex than inorganic chemistry. The mind in its autonomy represents a higher stage in order building, something to which the brain gives rise in the same way that elementary particles produce new properties when assembled in elements, which make the complexities of molecules, which are the basis of life, which has evolved animals with large central nervous systems.

Although subject to physical causation, the mind has its own causation. Conditioned reflexes, beloved of many psychologists because of their definiteness and seeming simplicity, tell nothing of interest about the mind. Whatever correspondence there may be between mental states and states of the brain or interconnections of neurons, it is largely unknowable and complexly chaotic, and any attempt to monitor it must disturb the relation. So far as is known, the mind does nothing contrary to physical or physiological laws, but it is free within the indeterminacy of the turbulence of the brain. The mind, even more than the weather, is unsimulatable. It is a controlled and order-creating essence over a chaotic substrate, the indefinitely complex interaction of elements in the brain.

Self-awareness is a special quality of the mind. A computer may be able to analyze difficult problems, but we do not suppose that it is self-aware, that is, has a mind. Self-awareness is different from information processing; even when confused and unable to think clearly, one may be vividly aware of one's self and one's confusion. The essence of mind is less data processing than will, intention, imagination, discovery, and feeling. If some kinds of thinking can be imitated by a computer, others cannot (Penrose 1987, 117).

The mind is fluid, flexible, and inventive, as is life. It not only understands but creates. It has indefinite capacities for change, growth, and complexification, as it adapts to needs and generates conditions for its own further expansion. It is the acme of the plasticity and restorative-integrative powers of the organism. Different persons have different abilities, sometimes to a phenomenal degree. Idiot savants, although in most ways dull, may be able to carry on many-digit calculations in their heads. Such abilities testify to the plasticity of the mind; it becomes powerful as it is intensely focused.

The mind wants to be in command, and it is in charge within its material limits. The personality would be "master of my fate and captain of my soul," and it finds health and strength in this exaltation. The mind is essentially integrated, not merely a set of specific functions such as might be built into a computer. It is epitomized in the phrase "my mind": being a mind, I possess a mind.

Life has a dual nature: its material basis and the essence of functionality and responsiveness that distinguishes living things and flourishes at higher levels of evolution. The material and the mental are both real, just as are causation and will. The mind derives richness from these two sides, like feeling and bodily function, love and sex, the spiritual and the carnal, the joy of creation and the satisfaction of bodily wants. The ideal and the material are not opposed but inseparable although deeply different. Religious feeling can be treated as an epiphenomenon of the activity of the brain or as a partial insight into deeper realities of the universe; both are doubtless partly correct or at least defensible.

In the opinion of physicist Freeman Dyson, "The mind, I believe, exists in some very real sense in the universe. But is it primary or an accidental consequence of something else? The prevailing view among biologists seems to be that the mind arose accidentally out of molecules of DNA or something. I find that very unlikely. It seems more reasonable to think that mind was a primary part of nature from the beginning and we are simply manifestations of it at the present stage of history. It's not so much that mind has a life of its own as that the mind is inherent in the way the universe is built, and life is nature's way to give mind opportunities it wouldn't otherwise have" (Dyson 1988, 72). Many biologists find this view distasteful, but it must be taken into account if we are seeking a realistic comprehension of life.

Sociobiology

Evolutionary theory cannot be deemed complete unless it can give an accounting of the mind and the behaviors it has created.

This is difficult; the mind does not fit in the framework of biology. A number of biologists, however, led by Edward O. Wilson, have grappled with the problem and tried to wrap up the human reality within the logic of natural selection.

They would deny that the mind has any existence subject to scientific investigation. According to the reductionist approach, the human is exceptional only for the hypertrophied brain. *Homo sapiens* is characterized as "the naked ape," as by Desmond Morris in his book of that title, and human mental processes, like those of the ape, are in principle as understandable as the functioning of a computer.

This means that sensations and emotions, pleasures and sorrows, are purely instrumental. According to E. O. Wilson, "Human behavior—like the deepest capacities for emotional response which drive and guide it—is the circuitous technique by which human genetic material has been and will be kept intact" (E. Wilson 1978, 167). Similarly, Michael Ghiselin says, "We have evolved a nervous system that acts in the interests of our gonads, and one attuned to the demands of reproductive competition" (Ghiselin 1974, 263). The mind is only an adjunct of the body, which is only a means of perpetuating and replicating the genes, their survival machine. The purpose of culture can and by implication should be simply to maximize reproductive success, and it is to be understood in these terms (Alexander 1979, 131, 144).

In this spirit, Wilson, a specialist in the study of ants, promoted a new discipline, "sociobiology," to analyze human behavior in terms applicable to social animals. Sociobiologists would dissect infinitely complex human behavior into specific traits and find selective value for them.

They are encouraged in this effort by the failure of anthropologists, sociologists, and psychologists to make much sense of the dizzying phantasmagoria of human existence, and they have produced, especially in the 1970s and early 1980s, an ample literature on the fascinating topic of why we are as we are. Sociobologists can doubtless offer insights. Humans, like monkeys and dogs, have inborn behavioral tendencies, and for almost any human trait one can find a more or less convincing animal counterpart. For example, people are generally fascinated with sports that are objectively childish; they make considerable sacrifices to watch grown persons hit, throw, or carry a ball around according to wholly artificial rules. It is reasonable to relate this predilection to a conflictive hominid past in which the "team" was a group fighting for its existence. If human societies are disposed to internal cohesion and hostility to outsiders, group loyalty has been inherited from forefathers who fought together in defense or offense, for territory and perhaps for women—as in Homer's *Iliad*—a heritage that may be held responsible for the otherwise unaccountable frequency and bloodiness of wars in history. Pride and lust for power are equivalent to the animal drive for hierarchic dominance.

In this vein we learn that natural selection favored religion because of a need for social cohesion, or, as Wilson has it, indoctrinability arose because of its utility for group survival (E. Wilson 1980, 286). Heterosexual love is easily interpreted: "We have evolved to search for mates who will help us foster our inclusive fitness" (M. Smith 1987, 234), a formulation that not many lovers and no homosexuals would accept.

The institution most cited as demonstrating the utility of the sociobiological approach is the incest taboo. The prohibition is general in human societies; it has been formally set aside only in a few cases of hereditary rulers. The evolutionary reason given for the taboo is that matings of close relatives bring together recessive harmful or lethal genes and lead to frequent birth defects (E. Wilson 1978, 37).

Among our nearest relatives, the chimpanzees, however, sibling mating is fairly common, and father-daughter incest must be frequent in the small band, paternity being unknown (Goodall 1986, 466–470). If there was much inbreeding among our early ancestors, there would be little selective pressure against incest. On the other hand, if an instinctive repugnance for incest came about because our hominid ancestors observed that unions of close relatives often had defective offspring, it would demonstrate not Darwinian but Lamarckian evolution.

Whatever instinct humans may have against incest, it is ambiguous. The temptation to violate the taboo is widespread. Psychologists commonly believe that morals prohibit those actions to which one may be strongly tempted; if there were an instinctive aversion, the taboo would be superfluous. Freudians believe that incestuous lusts, far from being instinctively negated, lurk powerfully in the unconscious (Godelier 1989, 69).

If humans were molded entirely by the reproductive imperative, sexual and reproductive behavior should be easiest to analyze in biological terms. Sociobiologists try to do so. For example, breasts advertise that women are well nourished and hence probably fertile (Barash 1982, 267). Women are more desirous of romance, men of sex, because men have more to gain reproductively from philandering, while women have more interest in the fullness and permanence of bonding. Sexual behavior is considerably leveled, however, by the fluidity of modern society and by easily available contraception. As modern women lose fear of pregnancy (which sociobiologically should hardly be feared), the frequency with which they stray from monogamy approaches that of their husbands.

In practice, people seem to be more confused about matters related to sex than anything else. They are natively clumsy at mating. It is a skill to be learned (chimpanzees also lack inborn proficiency). Falling in love is very human, but the romantic approach is antithetical to the biological. Parental affection is strongly mandated by natural selection, but humans are not well endowed with instincts for parenting.

The prime achievement of sociobiology is the reduction of altruism to inclusive fitness, that is, to genetic selfishness. According to Alexander (Alexander 1979, 143), altruism in human society is solely directed toward relatives. In the opinion of Ghiselin, " 'altruism' is a metaphysical delusion," and "what passes for cooperation turns out to be a mixture of opportunism and exploitation" (Ghiselin 1974, 25, 247). But if humans can willfully renounce reproduction, they can surely be altruistic beyond the dictates of reproductive success. It is more realistic to relate altruism not to inclusive fitness but to training and personality traits, such as empathy and self-confidence (Mussen and Eisenberg 1977, 159). It is not remarkable that people tend to favor their kin; it would be more interesting if sociobiology could tell us why people often act on behalf of nonkin, even other races and humanity as a whole.

If our ancestors had really dedicated themselves to competitive breeding, they would never have had the leisure or surplus energy and resources to develop the cultural foundations of civilization. We are programmed to seek and enjoy sex (much more than reproductive success). We are also programmable to turn away from sex, to vary it from biologically mandated expressions, to redirect and spiritualize it, sometimes to be revolted by it. Celibacy is esteemed in many societies of both East and West.

Even if we have a duty to reproduce genes, the question arises, which genes? There can be no special value in genes for green eyes or long noses. But if one would promote genes for creativity, intelligence, or good character (if genes for such things exist), one resorts to human values only indirectly related to inclusive fitness. It may be questioned why it should be more important to promote the information contained in the reproductive cells than information in the mind. Are ideas that can marry other ideas and have an indefinite progeny of idea offspring less important than children of the flesh?

Hardly any proposition is universally valid for all societies. The diversity of family structure among precivilized peoples is complete. People share in causes they find useful, with little regard to relatedness of beneficiaries. Siblings may love and help siblings or detest and fight them. There are hedonists and ascetics, monks and libertines. People fall in love with their likes or sometimes choose racially different mates, genes notwithstanding. (For a critique, Kitcher 1985).

Humans often behave in ways contrary to biological imperatives. Some fast; others, in socially approved ceremonies or spiritual ecstasy, torment their bodies in self- or mutual flagellation. Worst is the practice of female circumcision, or clitoridectomy. Maturing girls in much of Africa are subjected to this very painful, dangerous, and purposeless mutilation: the clitoris is more or less excised and the labia are cut and sewn together, leaving only a small orifice, all without anesthesia, precautions against infection, or even good instruments. This is not even understandable as imposition of brutal men; it is done by women, perhaps compensating for the sufferings of their youth by inflicting it or forcing their daughters to undergo it. Its purpose is to undo something of what natural selection has brought about, reducing sexual pleasure and making intercourse difficult. It has not only this result but frequently causes infection, sometimes death. The height of the barbarity is that women are sometimes resewn after childbirth to simulate virginity (Lightfoot-Klein 1990).

It is probably true that "no behavior can ever be explained by natural selection alone" (Baldwin and Baldwin 1981, 64). Sociobiology fundamentally fails because its basic ideas—"the gene's the thing," relatedness, and inclusive fitness—are flawed and incomplete. Traits may be adaptive or unadaptive in different conditions, depending largely, in the human case, on the cultural heritage. Psychologists would do well to take evolution more into account when they try to understand us, but the fact that understanding human nature requires awareness of biological foundations does not mean that it can be analyzed in biological terms. The more complex the animal is and the larger is the learned component of behavior, the less applicable is the variation-selection theory of evolution (Oxnard 1984, 333). The more powerful are the cultural forces and the more complex is the society, the less do humans conform to socio-biological expectations. Physical traits lose importance; the more intelligent and educated people are, the less likely they are to be concerned about their genes and the more inclined they are to seek symbolic successes.

Sociobiology tells more about the sociology of biology than about the biology of society. *Homo sapiens* is a unique creature, no more a "naked ape" than apes are "furry people."

Cultural Evolution

Life receives, creates, and propagates information, and certain mammals have attained extraordinary capabilities of learning about and dealing with their world. One of these, the human animal, has hands with which to make things and by combining hands and brain has reshaped its life and has made cultural dominant over genetic information. This evolutionary innovation ranks with such monumental transformations as life's beginning, photosynthesis, the nucleated cell, and multicellularity. Creating its new world and thereby revolutionizing the history of life, this creature has become more than an animal. The possessor and product of a huge and ever growing mass of knowledge is a strange type of being with a new universe of potentialities.

However novel in its dramatic manifestations, this metamorphosis is the culmination of a long trend. Choices and behavior play an increasing part in the development of animals as their data-processing capacities increase, and in some of them one sees beginnings of culture. Bears, for example, teach the cubs how to catch fish and otherwise get along in the world. The art of the beaver is passed from generation to generation partly by nucleic acid, partly by example. It is possible that cultural attainments have given certain groups of animals a competitive advantage long ago, as they have shared knowledge of food sources by imitation. Some bands of chimpanzees have skills, such as the use of simple tools to extract termites, that other bands lack.

It may be that intelligent animals can perform better by learning than by instinct. Sea lion pups have to learn to swim; perhaps reliance on learning makes for a higher level of skill. If orphaned sea otters are raised by hand, they cannot survive in the wild unless a diver acts as a surrogate mother, showing them such things as how to detach abalone and crack the shell. Learning makes greater flexibility in a complex world, and it became an important part of evolution long before bipedal primates made stone tools.

The human has gone far in renouncing instinct. In its first months, the newborn baby does not improve instincts but loses them. The neonate can make walking and swimming movements and will reach for an object or turn toward a sound; as the brain grows rapidly, such abilities disappear, only to be recovered by learning after a number of months or a year (Restak 1979, 109). This abdication of instinct has given unique flexibility. *H. sapiens,* like the cockroach and the pigeon, is an extraordinarily adaptable species, able not only to cope effectively with the natural environment—some would say outrageously—but to make and adapt to a radically new environment called civilization. The species does not fit into an environmental niche but makes all manner of niches for itself.

Genetic evolution is left behind, although we do not know what changes there may have been in the genetic composition of the human species during the last 10,000 years of ever more artificial existence. As soon as humans began living in villages of more than a few hundred, they were dependent for their livelihood on a highly ordered society, and the change of environment from the life of the hunter-gatherer to the farmer and town or city dweller has been profound. This new mode of existence has lasted long enough for important genetic change to occur, if there were clear-cut pressures. New species have arisen in much less time. It is not known, however, what selective pressures may have existed, and any genetic changes have been swallowed up by cultural change and buried under learned capacities and self-inflicted ailments.

Biological limitations have largely been transcended. At least in the modern countries, reproduction is no longer limited by the food supply; even most of the diseases that normally prune populations have been controlled, and many genetic shortcomings (such as diabetes) have lost importance. There are probably technological solutions for all material needs of humanity; the problems are social and political.

The differences between biological and cultural evolution are so great that it may be improper to use the same word for them. The tempo of cultural evolution is tremendously rapid, limited mostly by the ability of people to readjust their thinking. Cultural evolution is detached from physical bodies; its achievements may be transmitted to normal humans anywhere in accordance with their intellectual capacities. If there are units of cultural evolution, they are undefinable to a much higher degree than the ill-defined units of biological evolution. Some writers have discussed the rise and fall of civilizations as more or less integrated competitive entities; others think in terms of local cultures; yet others in terms of particular ideas, customs, and institutions.

Yet the principles of cultural evolution, however complicated and poorly understood, are not totally different from those of genetic evolution. The same principles of entropy (that is, decadence) and growth of complexity (that is, creativity) prevail. Ideas, although they are not fixed or stable, are a little like organisms in having something of a life of their own in the environment constituted by an information-soaked civilization. Ideas are born of ideas, often with cross-fertilization. They have phylogenies, like animal species, and their development is constrained by both their origins and external conditions. As in biological evolution, there is a struggle for survival and propagation among ideas. Competition is perhaps clearest among scientific theories (Popper 1984, 239). Some of them are rigid like fossil species; others make leaps to new integrations or give rise to many related ideas, like species swarms.

Ideas and other elements of culture may be likened to species; they are also analogous to genes. Dawkins and others call self-replicating bits of culture "memes," finding them subject to some of the same dynamics as the genes to which they are successors (Dawkins 1989, 189–201). Ideas are subject to cultural selection. Although the reasons for the prevalence of certain inventions, mores, institutions, or methods are often obscure, they relate to something like utility, which can be material, psychological, or social. Even in paleolithic times a better tool or weapon would spread and replace inferior ones. Not only have practical things, such as crops, methods of mining and manufacturing, and so forth imposed themselves. So also have countless customs, forms, styles, and beliefs, such as, in prehistoric times, the odd habit of marking off days by groups of seven. People learn not only how to do things but how to regard themselves and to relate themselves to their fellow humans. Nowadays the electronic diffusion of fashions, intellectual modes, attitudes, and even philosophies of government is converting most of the world into a single society with increasingly homogenized values.

On balance, the sieve of survival in the cultural universe has been sufficiently efficient to give humanity growing understanding and a large degree of control over the physical conditions of life. To a great extent, ideas prosper because those who hold them prosper, that is, show themselves most "fit." This selection, natural or not, has brought biological triumph. *Homo sapiens* has swollen in numbers as probably no large species ever before, explosively in the last century and recently dangerously.

But no one would contend that the ideas that succeed in propagating themselves are necessarily the fittest for the future of the biological species. Cultural evolution, like genetic evolution, is largely a trial-and-error process, with many dead ends, with times of stagnation as well as of burgeoning. The life of a culture is a ceaseless contest between creative and destructive forces, activity and will to improvement against decadence and social, moral, and even technological degeneration. Chance or chaos plays a major role in cultural as well as genetic change, as historically given conditions, human psychology, and natural and economic environment make for utterly unpredictable outcomes. Much cultural variety and creativity has no readily discernible causation, perhaps no discoverable concrete cause at all. There is a tremendous chaotic medley of feedbacks. These may be positive or negative according to the human response: careless farming leads to erosion and abandonment of fields or to measures of soil conservation; careful farming raises fertility. Violence generates violence or raises demands for order. Scientific discovery encourages more research and more discovery or leads to disillusionment with technology.

Social factors have played a large role in biological evolution; similarly, the development of culture-civilization has always gone hand in hand with the evolution of social patterns and values. There must have been many steps from the loose ape society to the family bonding that made humanity possible. Some innovation must have brought an upsurge of creativity about 35,000 years ago. New institutions probably made possible the neolithic beginnings of agriculture and villages or towns about 10,000 years ago, the rise of systematic thought in ancient Greece, the development of a critical mentality in the sixteenth century, and the new science and rationalism of the seventeenth century, which led directly to the contemporary outburst of information.

In the growth of culture, as in genetic evolution, forms attained serve as the basis for new growth. Intelligence provides new means of bettering intelligence. It especially adds to the means of transmitting and coordinating information, which multiply the power of intelligence. Language, writing, printing, and recently electronic information processing have each successively raised exponentially the ability to generate, combine, and use information. The mind makes for itself a new and immensely rich environment, which enlarges the mind by the wealth of data at its disposal, expanding with no apparent limit. Humans become products of their own products, as well as of their genes. Intensity of positive feedback has brought about an acceleration of change, rising to the volcanic upheaval of today, in which the world of the children is strange to the parents.

The evolution of culture may be compared with biological evolution in that there is invention (mutation) and selection on all levels. There is more or less random or directionless movement (genetic drift), stasis (as cultures have vegetated for centuries), and extinction (as cultural traits, and sometimes whole cultural complexes or civilizations, disappear). Cultural creativity flourishes in partly independent centers, like speciation in an archipelago. For long ages cultural progress was glacially slow, in somewhat the same fashion that bacteria were the acme of existence for 2 billion years. Moreover, prior to the recent rise of all-engulfing Western civilization, there have been numerous times of more rapid change, comparable to evolutionary radiations, and of stagnation.

Civilization building has been mostly the work of small groups of competing states, a little like competing populations. Such open systems have seen a ferment of ideas and innovation, new syntheses from which a new civilization emerged. They have worn themselves out, however, and have been followed by imperial unification and uninventive, static societies. The chief creators of civilization were the Sumerians in southern Mesopotamia, who invented such basics as writing, the wheel, sailing ships, and (as far as we know) the organized state. But after 2300 B.C., their land was ruled by a series of uncreative empires; great centralized states, like large species, resist change. India, China, and Peru similarly underwent a succession from a loose order of states with high creativity—chaotic, if one will—to big, static empires (Wesson 1967). The best example is the intellectual ferment of classic Greece, followed by the dullness of the Roman Empire. One is reminded of the early radiation of reptiles in the Triassic followed by the dominion of the dinosaurs.

Somewhat as the Cretaceous-Tertiary extinctions and the passing of the dinosaurs permitted the radiations of birds and mammals, the fall of the Roman Empire gave way to disorders that ultimately permitted the growth of a new civilization. This new civilization of the West has enjoyed almost continual and usually accelerating expansion for a thousand years, most strongly since the fifteenth century. From the seventeenth century it has been unchallenged in the world; as the twentieth century wanes, it overwhelms independent cultural life around the globe.

Why Western civilization has reaped this cultural victory, comparable to the biological victory of *H. sapiens*, is not easy to explain. The principal factor, however, has probably been that since the disintegration of the Roman Empire in the fifth century, Europe, especially in its western parts, has escaped the suffocation of an imperial unification. Not a few leaders have sought to regain the glory of the Caesars; however, from Charles V or Philip II of Spain in the sixteenth century, the Hapsburgs in the Thirty Years War, Napoleon, and most recently Hitler, all failed. The independence of European nations was saved only by exceptional geography, making major units, such as France, Spain, and England fairly defensible entities. Without such a fortuitous detail as the English Channel, the freedom of Europe could not have been sustained. In the dividedness of the Continent, especially its western extensions, no political unification proved possible, and the interaction of sovereignties maintained an openness permitting continual experimentation and change (Wesson 1978, 110–111).

The intellectual and political openness of Europe led to the industrial revolution, which greatly increased the practical value of knowledge, and to the electronic revolution. It also favored the responsibility of the state to its people, an old idea that could be applied on a large scale only when intensity of international communication reached an adequate level. The difficulties of finding a viable balance of freedom and authority and maintaining it long enough to permit the emergence of an electronic civilization are great. *Homo sapiens* has won a lottery (Wesson 1979, 22).

In the contemporary world, independent sovereignty has lost much of its meaning, and the role of the rivalries of nations in invigorating societies has much diminished. Modern civilization is like the genome of a single enormous superorganism, all parts of which affect one another. Or it may be regarded as a superconsciousness or collective mind developing from the interactions of individuals, corresponding to the neurons that together make the mind of a single brain. Significantly, the greatest strengths and most rapid advances of contemporary civilization are in information handling. There has even arisen a new brand of human being, marked by no genetic change but by symbiosis with an artificial near intelligence. People "marry" computers.

New vistas open, with new paraphernalia of living, new materials, new amusements, and especially new means of supplementing the human brain. Accessories to human mental powers increase not only knowledge but insights and discoveries, relieving the brain of drudgery but making new demands on it for guidance, raising the level of both problems and opportunities.

Yet human civilization may exhaust its resources or commit a sort of suicide. Technological progress is marred by great unevenness between the more developed and less developed worlds. The advanced society may be too much bent on satisfactions and have too little self-control. It may be unable to improve the social order sufficiently to adapt to technological change and make possible further improvement. In most of the world, the social-political-economic order is a failure; people are far poorer than they need be with the means of production at hand.

Our species has been caught up in a chaotic turmoil of powerful, seemingly uncontrollable feedback, from which we cannot escape without biological disaster and which we must, for survival, learn to manage. If there is to be a next stage for humanity, it has to rest on new, broader, more humane values, in dedication to common needs of this possibly imperiled species. It must build a society less domineering and more favorable to productive interaction and the expansion of the mind. Amid the ever-growing complexity of problems and interactions, it becomes more necessary to apply intelligence to the structure of society, which is the determinant of the fate of intelligence.

13

Conclusions and Perspectives

The Autonomy of the Genome

Propelled by the energies of the universe, life has brought about its miracles through natural selection plus self-ordering. But nature is too profound to be neat and easily accessible to finite minds, and if many things in the being and ways of life are more or less explainable, hardly anything is really understood. The universe is incomprehensibly complex in its foundations, and organisms are the acme of complexity. Evolution involves many kinds of systems, from macromolecules to species, each interacting with innumerable systems at its own and other levels. If electrons are unfathomable, can we pretend to a full understanding of how animals came to be?

It is doubtful that there can be any theory both logically coherent and broad enough to cover the manifold aspects of evolution. It is much if biology, like other sciences, can find illuminating descriptions. Evolution is the result of at least four major factors—environment, selection, random-chaotic development, and inner direction—and one might no more expect any law to govern it all than to find a law of the mind.

Conventional evolutionary theory, indulging the old daydream of a universe like a great clockwork, is incomplete, despite or because of its neatness and the logical charm of building on a few axioms. The difficulties that it confronts, many of them touched upon in previous chapters, are various. But most of them seem to indicate that the genome is less the passive subject of environment and random variations sifted by natural selection than an active maker of a future that is chaotically open, however constrained by the material environment and its past. The evidence given in previous chapters that the genome has much more autonomy than usually been credited to it may be summarized.

First, adaptation often fails to correspond to habit, as shown by the (1) success of widespread species, such as eucalyptus, cockroaches, or sparrows, over a wide range of conditions; (2) incompleteness of modification to fit a changed environment, as of marine iguanas and semiaquatic wasps; (3) incompleteness of adjustment to a changed way of life, as seen in pandas and gorillas; (4) failure of species to track closely the environment, remaining static until replaced by a new species; and (5) near invariance of "living fossils," such as dragonflies and ginkgo.

Second, many traits are seemingly independent of adaptive needs, as one sees in (1) the endless variety of traits of no evident utility, such as shapes of leaves within a single environment; (2) the high degree of protein polymorphism in animals, change of structural genes being effectively decoupled from organismic evolution; (3) rapid diversification in some families (species swarms), typified by African cichlids and Hawaiian drosophilids; (4) extreme diversity of sexual organs and behaviors, especially among arthropods; and (5) antithetic behaviors in similar conditions—some male monkeys kill young not their own, others tolerate or help them.

Third, evolution often has apparently maladaptive outcomes, as in (1) negative traits, such as the eccentric mating of bedbugs; (2) failure of organisms to exercise maximum powers of reproduction, as in the low fecundity of chimpanzees, condors, and many orchids; (3) seemingly needless early death, as of salmon, opossums, and tarantulas; (4) the persistence of sex, despite its high costs and the ability of a number of fish and reptiles to develop parthenogenetic species.

Fourth, patterns impose themselves, as one sees in a (1) coherence of pattern in taxa, contrary to the implication of the theory of natural selection that forms are conglomerates of useful traits; (2) lack of fossil evidence of gradual transitions between patterns; (3) integrity of pattern shown by capacity for self-repair; (4) maintenance of very complex organs and instincts in the face of inevitable deleterious mutations, especially in animals of low fecundity; (5) the ability of genes governing an organ to evolve harmoniously together; (6) parallel evolution, involving traits not dictated by natural selection, such as the blindness of army ants; and (7) development in certain directions beyond requirements of adaptation, as in exaggerated number of legs of millipedes and the total loss of hind limbs of cetaceans.

Finally, evolution achieves many adaptations—things apparently beyond the powers of natural selection, such as (1) organs and instincts for which it is difficult to envision viable intermediate stages or which require different unlikely changes to come together for utility, such as the electric apparatus of fish or the botfly's habit of laying its egg on a mosquito; (2) complex instinctual patterns unattainable by random input; and (3) the unaccountable achievement of high intelligence, as in porpoises and humans.

To stress the autonomy of the genome does not answer specific questions about evolution, and it suggests that many matters may be unfathomable because they arise from a largely unknowable history. But many problems become less mysterious if one accepts what is evident to intuition: that living beings are very highly organized in supremely complex ways, capable of interacting with the world around but always driven by their own dynamics, basically stable but subject to chaotic redirection. This conception applies to many broad phenomena, such as the categories of organisms of different levels of generality, the fixity of some species, the diversification of others under certain circumstances, the possibility of taking new directions (macroevolution), lag of structural adaptation behind behavioral change, and the tendency to persist in certain directions.

The autonomy of the genome indicates that evolutionary innovation is to be understood less as channeled by natural selection, more as a chaos-mediated realization of inherent potentialities within environmental limits. Until much more is known how genes shape organs and transmit instincts, there can be little real understanding of their evolution, but patterns clearly have great importance in ways difficult to fathom.

Natural selection can be envisaged as the dominant force in the stability of species, but its powers are limited. Living nature as we observe it could not have come about if natural selection did not permit chaotic innovation channeled by internal directives sometimes to prevail. Unviable organisms leave no descendants, but natural selection cannot dictate just how the organism is to evolve.

To some degree, organisms are the makers of their own evolution by their behavior, the choices that lead to new adaptations. This raises the question of the role of will and direction

in evolution—how will itself has evolved and how it is related
to adaptation. Will is an element of mind, which we must
assume to exist in higher animals. Mind, with its nonadaptive
qualities, does not fit well in the biological picture, and it has
been generally neglected in biology (Eccles 1989, 176), understandably in view of the difficulty of investigating it. But it must
have had a role in shaping highly intelligent creatures, especially ourselves.

The difference between the active, self-driven and the passive, mechanistic sides of evolution is not sharp, but it is real.
The genome is not only subject to environmental forces but is
creative. How this comes about is largely mysterious. But perhaps life is inherently mysterious, as the most highly ordered
aspect of a universe that can only incompletely comprehend
itself.

The Direction of Evolution

Evolution can be conceived as a goal-directed process insofar
as it is part of a goal-directed universe, an unfolding of potentialities somehow inherent in this cosmos. There has been
direction in the history of life on earth just as there has been
in the history of the universe, from fireball to solar system.

In fulfillment of its primary condition, self-perpetuation and
multiplication, life has explored the roads open to it. It has
expanded through the eons, moving into new environments
or niches, increasing numbers of both individuals and species.
As Darwin wrote, "Every step in the natural selection of each
species implies improvement in that species in relation to its
conditions of life" (F. Darwin 1897, 2:177). The vision of inevitable progress was doubtless a major reason, in the optimistic
Victorian age, for the rapid popularization of his theory. Life
has taken more and more of the earth to itself, and it remakes
its world, becoming its own environment—a trend culminated
today in the exorbitant florescence of humanity.

The corollary of reproduction and variation is increase of
diversity and complexity. It is more likely that a structure
remains viable when elements are added than when they are
taken away; a subtraction is likely to remove a necessary link,
but an addition does not necessarily interfere with the effec-
tiveness of the whole. Organisms become simpler only when
functions are no longer needed, as when an animal becomes a

parasite that sucks its nourishment from another or when a burrower loses eyesight. Although the shapes of modern bacteria are like those of billions of years ago, it is reasonable to guess that they too have found more effective ways of utilizing materials in their environment.

Whether or not the broad base of the pyramid of life has risen greatly, more and more is built atop it. Progress in evolution is a slippery concept, and we might logically define progress differently for every species. It cannot be regarded as inevitable, because improvement is always contested by decay, the entropic effect of random mutations not necessarily being overcome by excess reproduction and selection. But procaryotic cells have given rise to more highly structured protists; these have become differentiated to form larger and more complex aggregates; and multicellular creatures have created more and more layers of interacting complexities. This may be called progress in the minimal sense that one stage is a preparatory precondition for another (Maynard Smith 1988, 219). Creatures at each level can do things those before or below them could not, perhaps making a breakthrough to a new pattern, leading to greater flexibility and the possibility of more breakthroughs.

There are a multitude of progressions (Simpson 1974, 35): termites become more effective consumers of cellulose, and anteaters become better raiders of termite nests. There are widespread trends, however. Some are very obvious, such as a growing ability to use resources of the environment. Other trends are less so. For example, animals, becoming less bound to or conditioned by the environment, become more dependent on others. Whereas reptiles generally deposit their eggs and abandon them, birds and mammals give much care to their young. An essential characteristic of humans is the long immaturity during which the new generation can assimilate the cultural heritage of the community. In modern civilization, dependence, or interdependence, has become total; if families had to subsist on their own, nearly all humans would perish.

A broader trend in animals is toward an increased ability to respond to external conditions. Complexification, learning, and responsiveness lead to awareness and responsiveness. Intelligence has come in high degree to large primates. It has also come in good measure to the largest mammals on land

and in the ocean, not because of need but because it somehow became possible to actualize a potentiality.

The intelligence of whales, which have no means of making anything, is intrinsically limited; what might their musical genius be if they had music schools, instruments, and scores? But big-brained primates had hands to make things for the brain to work with and tools to supplement their hands and make more tools. In the latest flicker of geological time, information has taken on a life of its own, thanks to language, writing, and ever-improving means of gathering and handling data. The thrust of evolution has become the cultural development of humanity, and information stored in nucleic acid has been supplemented or swamped by information humanly produced and symbolically stored.

We have thus largely freed ourselves from the givens of the biological past. We have even partly freed ourselves from our own cultural past as we fashion a changing present. To be modern and civilized, or to be cultivated, is almost equivalent to liberation from ways and ideas inherited from cruder times, to an ability to choose what befits the new needs of present and future. The new freedom means not being a slave of the reproductive urge but managing it; treating our conspecifics according to their moral, intellectual, or cultural worth, not their bodily characteristics; and valuing others more for their spiritual than their genetic relatedness. It means esteeming progress in social, ethical, and intellectual terms.

The Evolution of Evolution

Despite the amazing commonality of life, it is not to be assumed that generalizations about evolution must be applicable to all living things, to animals as to bacteria, protozoa, and plants, not to speak of fungi, which seem rather eccentric and have been neglected in this book. The genetics of these kingdoms differ. Bacteria are haploid, and they sometimes receive injections of DNA. Protozoa have various odd practices, such as self-renewal of the nucleus. Plants multiply chromosomes (polyploidy), widely dispense with sex, and have double fertilization, somatic mutations, and other complications. Most fungi are haploid, and some have plural nuclei or different kinds of nuclei or tissues not divided into cells. In relatively evolved animals, the germ line is segregated, as Weismann insisted.

There may be different nuances of evolution between vertebrates and insects, perhaps in even subtler ways between amphibians and mammals. Natural selection also operates differently if only one in hundreds or thousands of offspring can be expected to survive to reproduce (as in the case of most invertebrates) or if the female produces only a few young in her lifetime (as in the case of many large birds and mammals).

Evolution is not a single process but a conglomerate of interactions, changing as the organism becomes more complex in its structures, its needs, and its relations with other organisms and the physical environment. At the beginnings of life and for over 2 billion years, it was largely biochemical, consisting mostly of simple changes of genes making different proteins; bacteria have only elementary structures. The crucial inventions were enzymes; bacteria have discovered most of those used in living organisms, and it has become ever more difficult to make significant changes in physiology (J. Campbell 1987, 299–300). For those interested in biochemistry, the story was almost finished a billion years ago; for the student of morphology, it was only beginning several hundred million years later.

Eucaryotic cells have much more complex plans than bacteria, and they have hundreds or thousands of times as many genes, doubtless engaged in much more complex cooperation. Change accelerated with improved ability to make larger and more intricate shapes than bacteria. In the small bodies of protists, life attains great variety and complexity. Yet what could be achieved within a single compartment was limited. Change accelerated from the bacterial level, but it was still very slow.

With the advent of multicellularity, the basic structural forms were rapidly invented. In a few tens of millions of years, a new world of much larger and more highly organized organisms appeared. Since then change has continued irregularly but sometimes rapidly. Overall, the span from the earliest metazoans to our present complexity is roughly the same as that from the earliest eucaryotes to the first metazoans. However, no new phyla have appeared (so far as is known) since the early Cambrian. Apparently no new classes have arisen for at least 150 million years and no new orders since the postdinosaurian radiations. Evolution has left behind the stages when it was possible to bring forth new phyla and new classes. The meaning of phylum has changed. The Burgess Shales creatures, however different in basic plans, must have been quite similar in proteins. If one could have observed them in life, many of them may well have looked rather alike in their weirdness, less diverse in appearance than modern vertebrates, ranging from lampreys to bats. In more than half a billion years, the framework has been filled in with countless add-ons.

In metazoans, change became structural at a higher level, presumably the work of gene families and regulators, not ordinarily making new proteins but selectively expressing or repressing large numbers of genes. The formation of organs must be coded differently from the synthesis of proteins. It is characteristic that many mutations are known to modify the location, length, and shape of *Drosophila* bristles, but the genes that actually produce bristles are only hypothetical (Manning 1975, 80).

As structures become more evolved, that is, more complex, innovations appear later in ontogeny and are necessarily less profound. The oceans, where life began, have long been barren of important innovation. The striking additions to marine fauna have been invasions of land animals, reptiles and mammals, which were enabled by terrestrial conditions to make adjustments that gave them advantages in the ocean—the ways of evolution are not straightforward. On land, mammals and birds are the most creative groups, but their basic radiations lie about 60 million years in the past.

Organisms have become structurally more complicated and, we may assume, perfected, approaching physiological limits. The better use of organs consequently takes priority over modifying them, and evolution has become increasingly behavioral. Higher animals respond to the environment much less by modifying organs than by better means of reacting to information, more directedness, control, and functionality of actions. Change is less guided or compelled by the physical environment; the higher organism may be defined as the one less bound to or controlled by the environment (Grant 1985, 375). Learning, becoming ever more important, comes to lead evolutionary change. The genetics of behavioral evolution are obscure, but it must be far more complex than that needed for the differentiation of organs. In the last instant of geological time, cultural evolution has become dominant, and the focus of evolution has shifted to the symbolic storehouse of information outside the individual body.

Life has thus advanced in levels of reaction, which are also levels of integration: the bacterium forms a simple self-organizing and self-maintaining system; the protist is a much more complex unity with the capacity of self-renewal; multicellular animals form a whole on a much higher level of complexity, with faculties of self-repair. On a yet higher level, the mind-brain forms a whole, integrating information, motivation, and awareness into a personality. Beyond this, genetic evolution gives way to cultural evolution, the entirety of a vast agglomeration of knowledge and behaviors making possible a new kind of existence.

The Anthropic Principle

A simple idea gives a better perspective on life and the universe. This is the modern equivalent of Descartes's dictum, "I think; therefore I exist." It is reformulated: "Thinking beings exist; therefore the cosmos must be such as to bring about the existence of thinking beings." The conditions and natural laws of the universe had to be precisely correct to generate creatures capable of studying it. Our rise demanded countless adaptations to the needs of the environment; it also required that the environment in countless ways be fitted for our rise. This illuminating and profound truism is called the weak anthropic principle. The more controversial "strong anthropic principle" goes beyond it to affirm that consciousness is inherent in the universe (Barrow and Tipler 1986).

However truistic the weak anthropic principle, it gives substance for reflection. It turns the question of evolution upside down. Instead of asking how living forms were produced by the inanimate universe, leading to thinking beings, it poses the question—potentially a scientific project—what the existence of creatures able to ponder their ancestry implies: what was necessary for *Homo sapiens* to come into existence and become capable of philosophy?

The requisites have been considered mostly in terms of physical laws of nature. Physicists conclude that the cosmos had to be perfectly adjusted to an incredible degree of precision, a trillionth of many trillions, to make extremely complex structures possible: there had to be just the right excess of matter over antimatter—one part in a billion—in the fireball in which it all began about 15 billion years ago; the relation of gravitational and other forces had to be precisely right in order to make stable stars; and nuclear forces had to be exactly adjusted to cause the fusion of hydrogen into helium in the interior of stars, providing a steady source of energy. An indispensable detail is that carbon had to have a resonance (at 7.6 million electron volts) in order to permit the formation inside stars of heavier nuclei from helium by way of beryllium; otherwise there would be a gaseous universe with no way to make complex structures. It was also necessary that oxygen not have a comparable resonance; if it did, it would destroy carbon too rapidly and prevent the accumulation of heavier elements.

Once potentially useful elements were cooked up in stellar interiors, they could provide the basis for the generation of life on a solid planet only if shot out into the cool of space. But ejection of material out of the innards of a huge star with intense gravity can be accomplished only by an incredible explosion, a supernova. For a star spontaneously to explode and blow away a large proportion of itself, many conditions have to exist, one being that the interaction of neutrinos with matter has to be precisely correct (Gribbin and Rees 1989, 245–247, 253).

For most of these and many other adjustments making possible the development of a habitable universe, there is no way to state exactly the probability of their occurring by chance, if parameters could vary freely. For one condition, however, it can be calculated: the probability that stable galaxies and stars could be formed in the universe coming out of the Big Bang. If the expansion were a little too rapid, gas clouds could not condense but would become ever thinner; if it were a little too slow, condensing galaxies would draw each other together and cause the universe to collapse. The universe had to balance, so to speak, on the edge of a knife, and a very sharp edge indeed. During the 15 billion years of its existence, the tiniest imbalance would rapidly compound itself. It turns out that, going back to the earliest time that has physical meaning, the balancing had to be correct to one part in 10^{60}, which is a million repeated ten times (Gribbin and Rees 1989, 17). It seems that one must either accept the strong anthropic principle of a purposefully designed universe or postulate an infinity of universes to produce one suitable for complex beings like ourselves able to study it.

Various more specific conditions were also necessary for the production of intelligent organisms. The sun had to be far enough from the crowded parts of the galaxy that its planets were not pulled out of orbit by other stars during these billions of years. The radiant output of the sun had to be reliably steady for 4 billion years. Most remarkable, while solar radiation has increased about 25 percent, the surface temperature of the earth has remained livable. By means not well understood, in which the reduction of the amount of carbon dioxide has played a part, the atmosphere has closely compensated for the rising input of energy. It is also phenomenal that for hundreds of millions of years, the proportion of oxygen, a very active gas, has remained around 21 percent.

Many more general realities have also been indispensable for the evolution of observers of the universe. For example, there had to be chaos. This is a fact of nature for which there is no necessity in the more specific laws of physics. It is part of the potential for complication that appears everywhere, from mathematics to the diversity of galaxies. Chaos must have played a large role in the diversification of life.

It was equally necessary that there exist an opposite tendency to simplification or degradation of order. The second law of thermodynamics decrees the increase of entropy, or disorder in any closed system. Life can defy the second law locally because it gets energy from without, that is, from the sun, the radiation of which is made possible ultimately by the expansion of the universe. Energized matter tends toward complexity and structure through self-organization; life is a process of taking in energy to dissipate entropy (Weber et al. 1989, 375). It is a detour that a tiny fraction of the sun's radiation follows on its way to entropic degradation.

If there were only complexification brought about by an influx of radiation energizing chaos—making dissipative structures, as organisms are sometimes called—the energy-processing systems would soon reach a plateau; it is necessary to degrade some or most of the old in order to make new beginnings. Through these opposing tendencies toward higher order (stabilized by selection) and disorder, or toward building up and tearing down, generation and extinction of forms, life raised itself. By corollary, there is as much randomness in life as is consistent with its progress.

This was possible because there were means of structures' coming into existence and reproducing themselves with occasional variations, which could be sieved in ways improving their capacities of self-reproduction. For this purpose, a minute part of the organism—its genetic equipment—had to be capable of directing the construction of an entire new organism much more complicated than the reproductive apparatus. There is no way of judging the likelihood of this, but one might imagine it to be impossible.

When life had achieved this self-replicative capability, it was on its way, but many conditions were indispensable for its progress. There had to be ample water, which happens not only to be the most abundant fluid at moderate temperatures but the best of solvents. Water had to have the rare characteristic of expanding on solidification, so that ice floats; if ice sank, the oceans would have been liquid only in an upper layer and inhospitable for life. Supplies of carbon-rich compounds were necessary, for only carbon has the combinatorial powers to make life possible. For our kind of life, it was essential that photosynthesis soon began, producing a surplus of oxygen, making ozone and a blanket absorbing ultraviolet light of the wavelengths most damaging to DNA.

At best, evolutionary progress has no straight path. New integrations seem to have been necessary to rise to new levels. But departures from established and successful ways have probably always been very improbable, requiring special circumstances or special means to make difficult transitions. It may be that sex—something disadvantageous for individuals and not necessary for particular species—played an essential role. That sex makes possible new combinations does not account for its prevalence but may be a by-product of a deeper necessity. At least, if there were no sex, evolution would have followed very different paths and could not have had the results it has achieved. The question, Why sex? can be answered partially by saying that without it, we would not be here to ask the question.

In a different dimension, massive periodic extinctions may have been necessary to enable evolution to make fresh starts. Classes dominant for many million years, from trilobites to dinosaurs, seem to be able to perpetuate themselves more or less indefinitely, but they seem to lose vitality and perish when conditions somehow become adverse. Conceivably if external events or conditions had not brought the dinosaurs to an end, rat-sized mammals would still be sniffing around dinosaur nests.

It also had to be possible for a network of protoplasmic cables not only to carry messages, coordinate actions, and learn appropriate responses to stimuli but also to achieve a high degree of information-integrative and pattern-inventive capacity, thereby producing high intelligence. Further, it had to be possible for the apparatus of heredity to encode complex patterns of perception and response, that is, instinct, difficult as this appears to be.

It is not intuitively apparent that this near-miracle should be the case. A computer, with its software, would seem to be beyond the capabilities of genes, no matter how long they might experiment. That a set of instructions for the manufacture of proteins, which is all the DNA can directly carry, can produce a first-rate computer with its software would not be credible if we did not have the evidence at hand. This superb apparatus had to be the gift of an animal of approximately the right size with means of facile manipulation. Neither a rat-sized nor an elephant-sized creature, nor an animal lacking adept hands could master fire, smelt metals, and produce basic requisites of technological civilization.

It was necessary not only that evolution should create intelligence but that it should to have labile and generalized capacities not likely to contribute to reproductive success. Not only was the neural supercomputer necessary; it had to rise to something that may be called mind, a somewhat different, mostly nonutilitarian entity that not only processes information but seeks it and plays with it. Cultural creativity requires self-awareness, indefinite complexity of motivation, and an urge for coherence of ideas—that is, for truth and abstract thinking.

Not least, it was necessary that many intelligences could come together in a social existence hospitable for innovation and the accumulation of culture infinitely above the capacity of any single individual. From such social existence there had to come about an inherently improbable cultural-historical evolution, permitting a hunting-gathering species to become a producer of foodstuffs and crafts and to live crowded in huge agglomerations called cities. It was also necessary for authority and freedom to be blended, however irregularly, in such fashion as to enable humans eventually to discover the usefulness of organized pursuit of learning and understanding, science and technology. As beneficiaries of this self-compounding culture building, we take it for granted, but there were many aborted starts toward higher civilization. The breakthrough to electronic capacities became possible only in the unique conditions of Europe. Possibly it is hardly less improbable than the correct ratio of gravitational and nuclear forces.

The anthropic principle tells us nothing, of course, about the likelihood of life, intelligence, or technical-electronic civilization or supercivilization on distant planets. However, because there are many potential cutoffs, the probability of a manifestation of life is inversely proportional to its advancement. The number of planets with microscopic life is certainly smaller than that of planets with conditions favorable to the formation of large molecules; those with bacterialike life must be far more numerous than those with amoebalike life, and these than worlds graced with larger, multicellular creatures or the equivalent. Planets with animals capable of limited learning must far outnumber those with animals capable of cumulative culture. How many may have an electronic civilization we can hardly guess. We know only that no lifelike form has arisen capable and desirous of colonizing the whole galaxy.

Our existence was made possible by the joint occurrence of an enormous set of improbable conditions. One may regard it as destined, as though inherent in the nature of things. Or one may suppose that life has by quite extraordinary luck found itself in the precisely necessary conditions and has taken the only possible road to higher intelligence, however difficult, through bacteria, protists, multicellular organisms, and responsive animals to builders of culture. Life, evolution, intelligence, and modern civilization all seem unlikely to the edge of impossibility, but perhaps they are part of the improbability of the genesis of the universe itself. Only a truly marvelous universe could create the marvel of the mind, contemplating its universe and itself.

The Essence of Evolution

Evolutionary innovation is a little like the development of new branches of mathematics. What mathematicians can do is endless, but they can advance only in ways permitted by the nature of things and by building on prior achievements. Most advances are only slight additions and amplifications; sometimes new insights open up new fields. Like undiscovered theorems and unsynthesized compounds, the potentialities of new living forms await discovery or realization. The process has been slow and the failures infinitely more numerous than the successes, but life has, step by step, found combinations of nucleic acids and their accessories, making it possible for proteins to guide the formation of something so complex and creative as the human mind.

Are we merely eccentrically successful animals, made aware of our meaninglessness through chance-built intelligence? Was it tragic or just shortsighted of Jacques Monod to lament that "man at last knows that he is alone in the unfeeling immensity out of which he emerged by chance" (Monod 1972, 167)? If we had, in the name of truth, to believe that humanity is the insignificant by-product of random change, selected by chance and material conditions, we should accept this valiantly and intelligently, although the truth could be dangerous for civilization and our well-being. But we have to believe no such thing. The simile of the blind watchmaker is false not in the sense that the watchmaker is sighted but that living organisms are not watches, made by external forces or design, not artifacts, but self-organizing and partly self-directed users of energy (Weber et al. 1989, 400).

The way we have come about does not prove the emptiness of humanity. It is less likely that we are an accidental product of an indifferent universe, as mechanistic philosophy would have it, or insignificant survival machines for genes, as sociobiology asserts, than that the richness of nature and the human achievement show the fallacy of the philosophy. If there is what may be called a spiritual component in humans, evolution cannot have been a wholly mechanistic process (Bowler 1989, 8). If reproduction does not constitute the whole meaning of life, the ways in which our ancestors raised themselves should inspire rather than demoralize.

It is honorable to form part of the miraculous community of life. Being the outcome of a very long and labored process is no reason to feel worthless in the grandeur of the universe, a trivial excrescence on an insignificant planet. We are a fantastic culmination of an engimatic process. This does not mean that the process was in any way designed for our benefit, only that we are, so far as we know, the first and only results of that process able to study its own origins and ponder their meaning.

Life is not an epiphenomenon, a minor incidental of the material universe, but a manifestation of the profoundest reality, part of the history from the fireball of 15 billion years ago to the computer-assisted technology of today. Accidental in detail, always within physical laws, the broad tendency toward the greater complexity and efficacy of life conforms to the way of the universe, which by its expansion enables the powers of self-organization to defy the dominion of entropy.

The creation of the cosmos itself was only the beginning of an indefinitely continuing process. At the beginning of space-time, the fireball is believed to have carried no information except mass and energy, which were equivalent. But differentiation, heterogeneity, and complexification began, it seems, immediately. The differentiation and divergence of different kinds of particles subject to different forces was followed by the emergence of atoms, the condensation of gas clouds into galaxies and stars, and the formation of heavier elements in stars. Eventually there came to be a solar system with cool planets, one of which had suitable temperatures for liquid water, in which complex compounds could be formed, chains of molecules growing into clumps or globules, from which eventually self-reproducing entities, or life, emerged. In the view of the theorists of nonequilibrium thermodynamics, life is part of a continual generation of complexity through competitive dissipation of energy into entropy and had to come when physical conditions permitted its degree of complication.

Evolution is a one-way journey, like the formation of galaxies. If a tendency to complexity permeates existence, one may legitimately view life as a realization of the drive or force that generated the cosmos in the first place—the matrix of order and direction behind the material universe. Biology should be quite as revealing of the depths of reality as physics or cosmology, perhaps more so. So far as we know, life is the most highly structured part of the order building of the universe. Every living creature is a little rebellion against the rule of increasing entropy, a brief victory for the generative principle of order—a victory that is only temporary in each individual but can be extended by propagation of its design. The complexification and increasing capacities of life may be taken as movement toward the fulfillment of potentialities for order-creation, as life has become stronger and better able to make use of its environment, better able to choose and act on its choices, up to the present height of learning, transmitted culture, and modern science.

Moral Meaning of Evolution

How we have come about cannot be understood in isolation from the character of the universe, and this should be the beginning of wisdom about ourselves. The Darwinists appreciated this as they set out to vanquish superstition and mysticism, but they succeeded too well and came to moral negativism.

The concept of survival of the fittest leads toward crude individual or group selfishness and a narrow notion of success as self-imposition, with implications of amorality, if not immorality. Morality is the subordination of individual drives or benefits to rules or needs of a higher order—higher, that is, than the unit of natural selection. If one consistently adhered to the Darwinist canon, the logical social ethic (beyond being merely egocentric) would be to join with genetically kindred persons to get the better, reproductively, of all others, ultimately to replace them by whatever means available. Although Dawkins judiciously refrains from advocating universal selfishness, his doctrine would excuse it. If the gene's the thing, might rules, victory justifies the victor, and there is no right for the weak.

But the study of life is edifying and deepening for the spirit, and biologists are usually quite humane, more concerned with the preservation of endangered species than with maximum propagation of the biologists' genes. They find various ways of getting over the contradiction between survival of the fittest and social morality. They may take pleasure in deflating the human hubris by pointing out that we are only one species among many, seeing in the evolutionary scheme cause for sober humility, as does Stephen J. Gould. Other writers would neutralize the materialistic implications of Darwinism by bringing back the notion of "higher." G. G. Simpson conceded in his major work that "These concepts [of survival of the fittest] had ethical, ideological, and political repercussions which were and continue to be, in some cases, unfortunate, to say the least." But he sought to get away from the unfortunate aspect by distinguishing "differential reproduction," usually nonviolent, from the violent contest implied by "struggle for existence," although it is not much better to be starved by the competition than to be eaten. In his epilogue, after having described the mechanistic process and having demolished the fallacy of teleology, Simpson tried to save dignity by calling the human uniqueness "the highest form of organization of matter and energy that has ever appeared" (Simpson 1949, 221, 344).

But it is difficult to salvage what has been thrown into the abyss of mechanistic materialism, to harmonize the ruthless competition of Darwinism with the need for cooperation (Ho 1988, 107). Violent or not, in a world of competition for limited resources, natural selection means that the gain of one is the loss of others. And what is the importance of our being "higher?" If there has been nothing in the genesis of mind beyond impersonal forces and the laws of physical nature, what intrinsic value can anything have, even though somehow "higher?" Simpson goes on to state that "his [man's] essential nature is defined by qualities found nowhere else," thereby locating our species outside the framework of the processes that gave rise to it.

There is reason in the creationists' linking ideas of human origins with religion. More broadly, philosophical idealists doubt that living creatures, including ourselves, are merely the product of environmental and accidental sifting of random variations. To insist on strict Darwinism is to be a philosophic materialist; a mechanistic or reductionist idea of our origins leads straight to a mechanistic or reductionist view of ourselves.

There is something of self-hate in the materialist approach. It depreciates the life of the mind and works of imagination and character. It demeans the richness and wonder of nature. It seems to make unnecessary further thinking about the mysteries of existence, of life and the universe. If one is gripped by the idea that we were made by chance (an unlovable deity) and are not intrinsically superior to amoebas (which by the same logic are not superior to bacteria or grains of sand), one is not prepared to cope with the responsibility of intelligence and power.

As our life becomes more complicated and confusing and as traditional guides to conduct become less persuasive, it is the more urgent to have a sound comprehension of our origins and the road to the present in order to orient ourselves on the road to the future. The understanding of evolution should help to heal the breach between the natural and the moral order that is indispensable for civilization. Placing humanity in its context as part of the biological scheme should give a better appreciation, not a poorer one, of our role. The evolutionary view should be a moral guide, enabling us to surmount limitations of our heritage to meet the new needs we have created for ourselves and to adapt more intelligently to our ever-changing self-made environment.

We should not resist our community with nature or find it demoralizing but rather appreciate our phenomenal role, not as the goal of evolution—there surely can be no final goal—but as the winner of a stellar role in the present and perhaps into the future that is ours to make. To treat life mechanistically is to deny its essence, and humanity, as an achievement of living nature, is part of something the greatness of which we cannot estimate.

We are one with other living things; love of life is one of our noblest attributes. If we permit ourselves a little pride, we may accept the honor of our past and present, with all its meaning in sensation, passion, and creativity, using the best we have for a better future.

Though subject to the limitations of our materiality and origins, we have indefinite capacities of improvement; it is the special quality of life that it can always grow and better itself. Trial and error bring more failures than successes, but the future has been forever open for inventive, exploring, changing living things.

As evolution advances, it derives less from chance or random variation, more from choice. Life ever more makes its own future. We have not the faintest notion what our long-term role in the drama of evolution may be, how it may appear a million or even a thousand or a hundred years from today. But it can hardly be unimportant; we seem to be the makers of a turning point. Ours is the moment when biological evolution gives way to cultural-informational evolution, with all its explosive potential. Human civilization is not an end but a vaulting into the unknown. It is a supreme glory that humans can decide what destiny they desire and, if wise enough, can make their own evolution.

References

Abbot, Larry. 1988. "The Mystery of the Cosmological Constant." *Scientific American* 258 (May):106–113.

Alcock, John. 1988. *The Kookaburra's Song: Exploring Animal Behavior in Australia.* Tucson: University of Arizona Press.

Alcock, John. 1989. *Animal Behavior.* 4th ed. Sunderland, Mass.: Sinauer Associates.

Alcock, John. 1990. *Sonoran Desert Summer.* Tucson: University of Arizona Press.

Alexander, Richard D. 1964. "The Evolution of Mating Behavior in Arthropods." In *Insect Reproduction,* ed. K. C. Highnam, 78–94. London: Royal Entomological Society.

Alexander, Richard D. 1979. *Darwinism and Human Affairs.* Seattle: University of Washington Press.

Alexander, Richard D. 1982. "Biology and the Moral Paradoxes." *Journal of Social and Biological Structures* 5 (October):389–395.

Alford, D. V. 1975. *Bumblebees.* London: Davis Poynter.

Alloway, Thomas A. 1988. "Behavioral Development and Colony Life Cycles in Ants." In *Evolution of Social Behavior and Integrative Levels, Evolution of Social Behavior and Integrative Levels,* ed. Gary Greenberg and Ethel Tobach, 165–176. Hillsdale, N.J.: Laurence Erlbaum Associates.

Alvarez, Fernando, Luis A. de Reyna, and Myriam Segura. 1976. "Experimental Brood Parasitism of the Magpie. *Pica pica.*" *Animal Behaviour* 24 (November):907–916.

Alvarez, L. W. 1983. "Experimental Evidence That an Asteroid Impact Led to the Extinction of Many Species 65 Million Years Ago." *Proceedings of the National Academy of Sciences USA* 80 (January):627–642.

Anderson, D. T. 1987. "Development and Evolution." In K. S. W. Campbell and M. F. Day, eds., *Rates of Evolution,* 143–155. London: Allen and Unwin.

Anderson, David J. 1990. "Evolution of Obligate Siblicide in Boobies." *American Naturalist* 135 (3) (March):334–350.

Anderson, Malte. 1983. "Brood Parasitism within Species." In C. J. Bernard, ed., *Animal Behavior, Ecology, and Evolution.* New York: John Wiley.

Anderson, O. Roger. 1983. *Radiolaria.* New York: Springer Verlag.

Aneshansley, D. J., and T. Eisner, J. M. Widom, and B. Widom. 1969. "Biochemistry at 100 C: Explosive Secretory Discharge of Bombardier Beetles (*Brachinus*)." *Science* 165, 61–63.

Arnold, S. J. 1987. "Quantitative Genetic Models of Sexual Selection: A Review." In S. C. Stearns, ed., *The Evolution of Sex and Its Consequences,* 283–315. Boston: Birkhäuser Verlag.

Askew, R. R. 1971. *Parasitic Insects.* New York: Elsevier.

Auffenberg, Walter. 1981. *The Behavioral Ecology of the Komodo Monitor.* Gainesville: University of Florida Press.

Austad, Steven N. 1988. "The Adaptable Opossum." *Scientific American* 258 (February):98–105.

Avers, Charlotte L. 1989. *Process and Pattern in Evolution.* New York: Oxford University Press.

Ayala, Francisco J. 1985. "Reduction in Biology." In David J. Depew and Bruce H. Weber, eds., *Evolution at a Crossroads: The New Biology and the New Philosophy of Science,* 65–80. Cambridge, Mass.: The MIT Press.

Badrias, Alison, and Noel Badrias. 1984. "Social Organization of *Pan paniscus*." In Randall L. Susman ed., *The Pygmy Chimpanzee: Evolutionary Biology and Behavior,* 325–346. New York: Plenum Press.

Bakker, R. 1986. *Dinosaur Heresies.* England: Longman Harlow.

Baldwin, John D., and Janice I. Baldwin. 1981. *Beyond Sociobiology.* New York: Elsevier.

Barash, David. 1982. *Sociobiology and Behavior.* New York: Elsevier.

Barinaga, Marcia. 1990. "Where Have All the Froggies Gone?" *Science,* 247 (March 2):1033–1034.

Barnaby, J. Feder. 1990. "Toxic Waste: Bacteria to the Rescue." *New York Times,* June 27, C1, C5.

Barnard, C. J. 1983. *Animal Behavior and Evolution: Ecology and Behavior.* New York. John Wiley.

Barnett, S. A. 1975. *The Rat: A Study in Behavior.* Chicago: University of Chicago Press.

Barnett, S. A. 1981. *Ethology: The Science of Animal Behavior.* New York: Oxford University Press.

Barrett, Spencer C. H. 1989. "Waterweed Invasion." *Scientific American* (October):90–97.

Barrow, John D., and Frank J. Tipler. 1986. *The Anthropic Cosmological Principle.* New York: Oxford University Press.

Barryman, A. A., and J. A. Millstein. 1989. "Are Ecological Systems Chaotic—and If Not, Why Not?" *Trends in Ecology and Evolution* 4(1):26.

Barzun, Jacques. 1941. *Darwin, Marx, Wagner.* Boston: Little, Brown.

Bass, Andrew H. 1986. "Evolution of a Vertebrate Communication and Orientation System." In Theodore H. Bullock and Walter Heiligenberg, eds., *Electroreception*. New York: Wiley.

Bates, N. H. 1989. "Founder Effect Speciation." In Daniel Otte and John Endler, eds., *Speciation and Its Consequences*. Sunderland, Mass.: Sinauer.

Batra, Suzanne W. T. 1987. "Deceit and Corruption in the Blueberry Patch." *Natural History* 96 (August):56–59.

Beardsley, Tim. 1986. "Fossil Bird Shakes Evolutionary Hypothesis." *Nature* 322:677.

Bechants, Jonathan R., and David Zipses, eds. 1970. *The Lactose Operon*. Cold Spring Harbor, N.Y.: Cold Spring Harbor Laboratory, 1970.

Beck, Benjamin B. 1980. *Animal Tool Behavior: The Use and Manufacture of Tools by Animals*. New York: Garland STPM Press.

Beehler, Bruce M. 1989. "The Bird of Paradise." *Scientific American* 261 (December):117–123.

Begon, Michael, John L. Harper, and Colin R. Townsend. 1986. *Ecology: Individual, Populations, and Communities*. Sunderland, Mass.: Sinauer Associates.

Bell, Peter, and Christopher Woodcock. 1983. *The Diversity of Green Plants*. 3d ed. London: Edward Arnold.

Bellairs, Angus. 1970. *The Life of Reptiles*. New York: Universe Books. 2 vols.

Bennett, M. V. L. 1971. "Electric Organs." In W. S. Hoar and D. J. Randall, eds., *Fish Physiology*, 5:347–492. New York: Academic Press.

Benson, Spencer A. 1988. "Is Bacterial Evolution Random or Selective?" *Nature* 330 (November 3):21–22.

Benton, M. J. 1990. "Evolution of Large Size." In *Palaeobiology: A Synthesis*, ed. Derek E. G. Briggs and Peter R. Crowther, 147–152. Oxford: Blackwell Scientific.

Bernhardt, Peter. 1989. *Wily Violets and Underground Orchids*. New York: William Morris.

Bildstein, Kurt L. 1983. "Why White-Tailed Deer Flag the Tails." *American Naturalist* 121 (May):709–715.

Bock, Walter J. 1986. "The Arboreal Origin of Flight." In *The Origin of Birds and the Evolution of Flight*, ed. Kevin Padian. San Francisco: California Academy of Sciences.

Bohm, David. 1987. "Hidden Variables and the Implicate Order." In *Quantum Implications: Essays in Honour of David Bohm*, ed. B. J. Hiley and F. David Peat, 33–45. London: Routledge and Kegan Paul.

Bonner, John T. 1980. *The Evolution of Culture in Animals*. Princeton: Princeton University Press.

Bonner, John T. 1982. "Comparative Biology of Cellular Slime Molds." In *The Development of Dictyostelium discoideum*, ed. W. Loomis, 1–33. San Diego: Academic Press.

Bonner, John T. 1988. *The Evolution of Complexity by Natural Selection.* Princeton: Princeton University Press.

Bornman, Chris H. 1978. *Welwitschia.* Cape Town: C. Struik.

Borror, Donald J., and Dwight M. DeLong, 1964. *An Introduction to the Study of Insects.* Rev. ed. New York: Holt Rinehart and Winston.

Boucher, Douglas H. 1986. *The Biology of Mutualism: Ecology and Evolution.* New York: Oxford University Press.

Bouliere, F. 1975. "Mammals Small and Large: The Ecological Implications of Size." In *Small Mammals: Their Productivity and Population Dynamics,* ed. Frank B. Golley, K. Petrusewiez, and L. Ryszkowski, 1–8. New York: Cambridge University Press.

Bowler, Peter J. 1989. *Evolution: The History of an Idea.* Rev. ed. Berkeley: University of California Press.

Braam, J., and R. W. Davis. 1990. "Rain-, Wind-, and Touch-Induced Expression of Calmodulin and Calmodulin-Related Genes." *Cell,* 60 (3) (February 9):357–364.

Brian, M. V. 1981. "Caste Differentiation and the Division of Labor in Hymenoptera." In *Social Insects,* ed. Henry R. Hermann, 1:122–222. New York: Academic Press.

Brian, M. V. 1985. "Comparative Aspects of Caste Differentiation in Social Insects." In *Caste Differentiation in Social Insects,* ed. J. A. L. Watson, B. M. Okoto-Kotber, and C. Noirot, 385–398. New York: Pergamon Press.

Briggs, John, and F. David Peat, 1989. *Turbulent Mirror.* New York: Harper & Row.

Brooks, Danile R. 1988. *Evolution as Entropy: Toward a Unified Theory of Biology.* 2d ed. Chicago: University of Chicago Press.

Brown, Charles R., and Mary B. Brown. 1990. "The Great Egg Scramble." *Natural History* (February):34–41.

Browne, Malcolm W. 1989. "Toxic Feast for Microbes," *New York Times* May 23, III, 13.

Budker, P. 1971. *The Life of the Sharks.* New York: Columbia University Press.

Bull, James J. 1983. *Evolution of Sex Determining Mechanisms.* Menlo Park, Calif.: Benjamin/Cummings.

Bull, James J. 1987. "Sex Determination Mechanisms. In S. C. Stearns, ed., *The Evolution of Sex and Its Consequences,* 93–115. Boston: Birkhäuser.

Bunge, Mario. 1989. "From Neuron to Mind." *News in Physiological Sciences* 4 (October).

Bunnell F. L., and D. E. Tart. 1981. "Population Dynamics of Bears." In Charles Fowler and Tim D. Young, eds., *Dynamics of Large Mammal Populations,* 75–98. New York: Wiley.

Bunnell, Sterling. 1974. "The Evolution of Cetacean Intelligence." In Joan McIntyre, ed., *Mind in the Water,* 52–67. New York: Charles Scribner's Sons.

Bürgin, Toni, and Olivier Riebbel, P. Martin Sanders, and Karl Tschanz. 1989. "The Fossils of Monte San Giorgio." *Scientific American* 260 (6) (June):74–81.

Burnham, Charles R. 1988. "The Restoration of the American Chestnut." *American Scientist* 76 (September–October) 478–487.

Burton, Maurice. 1969. *Animal Partnerships.* London: Friedrich Warner.

Burton, P. 1973. "What Makes an Owl." In *Owls of the World: Evolution, Structure, and Ecology,* ed. J. A. Burton, 37–40. Italy: A&W Visual Library.

Bush, G. L., S. M. Case, A. C. Wilson, and J. L. Patton, 1977. "Rapid Speciation and Chromosomal Evolution in Mammals." *Proceedings of the National Academy of Sciences USA* 74 (9) (September):3942–3946.

Cairns, John, Julie Overbaugh, and Stephen Miller. 1988. "The Origin of Mutants." *Nature* 335 (September 8):142–145.

Campbell, Bernard G. 1985. *Human Evolution: An Introduction to Man's Adaptations.* New York: Aldine.

Campbell, Bernard G. 1988. *Humankind Emerging.* 5th ed. Glenview, Ill.: Scott Foresman.

Campbell, John H. 1985. "An Organizational Interpretation of Evolution." In *Evolution at a Crossroads,* ed. David J. Depew and Bruce H. Weber, 133–167. Cambridge, Mass.: The MIT Press.

Campbell, John H. 1987. "The New Gene and Its Evolution." In K. S. W. Campbell and M. F. Day, eds., *Rates of Evolution,* 283–309. London: Allen and Unwin.

Carl, S. M., G. L. Bush, A. C. Wilson, and J. L. Patton. 1977. "Rapid Speciation and Chromosomal Evolution in Mammals." *Proc. National Academy of Science USA* 74 (9):3945.

Carroll, Robert L. 1988. *Paleontology and Vertebrate Evolution.* New York: W. H. Freeman.

Carson, Hampton L., et al. 1970. "The Evolutionary Biology of Hawaiian Drosophilidae." In Max K. Hecht and William G. Steere, eds., *Essays in Evolution and Genetics in Honor of Theodosius Ddobzhansky.* New York: Appleton-Century-Crofts.

Carson, Hampton L., and Kenneth Y. Kaneshiro. 1976. "*Drosophila* of Hawaii." *Annual Review of Ecology and Systematics* 7:311–345.

Cava, Robert J. 1990. "Superconductors beyond 1-2-3." *Scientific American* 263 (2):42–49.

Charlesworth, Deborah. 1979. "The Evolution and Breakdown of Tristyly." *Evolution* 33 (March):486–498.

Chaudhari, Nipura, and William E. Hahn. 1983. "Genetic Expression in the Developing Brain." *Science* 220 (May 27):924–928.

Cheney, Dorothy, and Robert Seyfarth. 1990. "In the minds of monkeys." *Natural History* (September):38–46.

Cheng, Thomas C. 1970. *Symbiosis: Organisms Living Together.* New York: Pegasus.

Clarke, C. A., G. S. Mani, and G. Wynne. 1985. "Evolution in Reverse: Clean Air and the Peppered Moth." *Biological Journal of the Linnean Society* 26 (October):189–199.

Clarkson, E. N. K. 1986. *Invertebrate Paleontology and Evolution*. London: Allen and Unwin.

Cloudsley-Thompson, John L. 1988. *Evolution and Adaptation in Terrestrial Arthropods*. New York: Springer.

Clutton-Brock, Juliet. 1989. *The Natural History of Domesticated Animals*. Austin: University of Texas Press.

Cockburn, Andrew, and Anthony K. Lee. 1988. "Marsupial Femmes Fatales." *Natural History* 97 (March):40–47.

Colbert, E. H. 1980. *Evolution of the Vertebrates: A History of the Backboned Animals through Time*. 3d ed. New York: John Wiley.

Cole, B. J. 1981. "Dominance Hierarchies in Leptothorax Ants." *Science* 212:83–84.

Cole, Charles J. 1978. "The Value of Virgin Birth." *Natural History* 87 (January):56–63.

Colinvaux, Paul. 1986. *Ecology*. New York: John Wiley.

Collier, John. 1986. "Entropy in Evolution." *Biology and Philosophy* 1(1):5–24.

Conrad, M. 1986. "What Is the Use of Chaos?" In *Chaos*, ed. A. V. Holden, 3–14. Princeton: Princeton University Press.

Coope, G. R. 1979. "Late Cenozoic Fossil Coleoptera: Evolution, Biogeography, and Ecology." *Annual Review of Ecology and Systematics* 10:247–267.

Corner, E. J. H. 1966. *The Natural History of Palms*. Berkeley: University of California Press.

Coulter, G. W. et al. 1986. "Unique Qualities and Special Problems of Africa's Great Lakes." *Environmental Biology of Fishes* 17 (3) (November):161–183.

Cowen, R. 1990. "Jumping Gender: Frogs Change from She to He." *Science News* 137 (March 3):134.

Craig, Catherine L., and Gary D. Bernard. 1990. "Insect Attraction and Ultraviolet Reflecting Spider Webs." *Ecology* 71(2) (April):616–624.

Crew, David. 1987. "Courtship in Unisexual Lizards." *Scientific American* 257 (December):116–121.

Crick, Francis. 1966. *Of Molecules and Men*. Seattle: University of Washington Press.

Crick, Francis. 1981. *Life Itself: Its Origin and Nature*. New York: Simon and Schuster.

Crick, Francis. 1988a. "Lessons from Biology." *Natural History* 97 (November):32–39.

Crick, Francis. 1988b. *What Mad Pursuit: A Personal View of Scientific Discovery*. New York: Basic Books.

Crompton, John. 1950. *The Spider*. London: Collins.

Cronquist, Arthur. 1988. *The Evolution and Classification of Flowering Plants.* New York: New York Botanical Garden.

Crook, John M. 1972. "The Rites of Spring." In Thomas B. Allen, ed., *The Marvels of Animal Behavior,* 287–308. Washington, D.C.: National Geographic Society.

Crow, James F., and Motoo Kimura. 1970. *An Introduction to Population Genetics.* New York: Harper & Row.

Crutchfield, James P., J. Doyne Farmer, and Norman H. Packard. 1986. "Chaos." *Scientific American* 255(6) (December):46–57.

Cullis, C. A. 1987. "The Generation of Somatic and Heritable Variation in Response to Stress." *American Naturalist* 130 (July 1987):Supplement, 562–573.

Cullis, C. A. 1988. "Control of Variations in Higher Plants." In *Evolutionary Processes and Metaphors,* ed. Mae-wan Ho and Sidney W. Fox, 49–62. New York: John Wiley.

Culver, David C. 1982. *Cave Life: Evolution and Ecology.* Cambridge: Harvard University Press.

Darlington, C. D. 1961. *Darwin's Place in History.* New York: Macmillan.

Darwin, Charles. 1871. *The Descent of Man and Selection in Relation to Sex.* London: John Murray.

Darwin, Charles. 1902. *The Formation of Vegetable Mould through the Action of Worms.* New York: Appleton.

Darwin, Charles. 1960. *The Origin of Species by Means of Natural Selection.* 6th ed. Garden City N.Y.: Doubleday.

Darwin, Charles. 1964. *On the Origin of Species by Natural Selection.* 1st ed. London: John Morrow, 1859; facsimile.

Darwin, Charles. 1982. *Different Forms of Flowers on Plants of the Same Species.* Reprint: Chicago, University of Chicago Press.

Darwin, Francis. 1897. *Life and Letters of Charles Darwin.* London: D. Appleton.

Davies, Paul. 1988. *The Cosmic Blueprint: New Dimensions in Nature's Creative Ability to Organize the Universe.* New York: Simon and Schuster.

Dawkins, Richard. 1976. *The Selfish Gene.* New York: Oxford University Press.

Dawkins, Richard. 1986. *The Blind Watchmaker.* New York: Norton.

Dawkins, Richard. 1989. *The Selfish Gene.* Rev. ed. New York: Oxford University Press.

Dean, Jeffrey, Daniel J. Aneshansley, Harold E. Edgerton, and Thomas Eisner. 1990. "Defensive Spray of the Bombardier Beetle." *Science* 248 (June 8):1219–1221.

Deeming, Charles, and Mark Ferguson. 1989. "In the Heat of the Nest." *New Scientist* 121 (March 25):33–38.

del Pino, Eugenio M. 1989. "Marsupial Frogs." *Scientific American* 220 (May):110–118.

Depew, David J., and Bruce H. Weber. 1988. "Consequences of Nonequilibrium Thermodynamics for Darwinism." In *Entropy, Information, and Evolution*, ed. Bruce H. Weber, D. J. Depew, and J. D. Smith, 317–354. Cambridge, Mass.: The MIT Press.

Depew, David J., and Bruce H. Weber. 1989. "The Evolution of the Darwinian Research Tradition." *Systems Research* 6(3):255–263.

De Robertis, Eddy M., Guillermo Oliver, and Christopher V. Wright. 1990. "Homeotic Genes and the Vertebrate Body Plan." *Scientific American* 263(1) (July):46–52.

de Waal, Frans B. M. 1989. *Peacemaking among Primates*. Cambridge: Harvard University Press.

Diamond, Jared M. 1981. "Flightlessness and Fear of Flying in Island Species." *Nature* 293 (October):507–508.

Diamond, Jared M. 1990a. "Nature's Infinite Book." *Natural History* (July):20–26.

Diamond, Jared M. 1990b. "Biological Effects of Ghosts." *Nature* 345 (June 28):769–770.

Diamond, Marian C. 1988. *Enriching Heredity: The Impact of the Environment on the Anatomy of the Brain*. New York: Free Press.

Dickemann, Mildred. 1981. "Paternal Conidence and Dowry Competition." In Richard D. Alexander and Donald W. Tinkle, eds., *Natural Selection and Social Behavior*, 405–416. Oxford: Blackwell Scientific.

Ditmars, Raymond L. 1933. *Reptiles of the World*. New York: Macmillan.

Dixon, A. F. G. 1985. *Aphid Ecology*. Glasgow. Blackie and Son.

Dixon, D. 1988. *The New Dinosaurs: An Alternative Evolution*. Topsfield, Mass.: Salem House.

Dobzhansky, Theodosius. 1956. "What Is an Adaptive Trait?" *American Naturalist* 90 (November–December):337–347.

Dobzhansky, Theodosius. 1970. *Genetics of the Evolutionary Process*. New York: Columbia University Press.

Dodson, Stanley I. 1989. "Predator-Induced Reaction Norms." *Bioscience* 39 (July):447–452.

Dominey, Wallace J. 1984. "Effects of Sexual Selection and Life History on Speciation." In Anthony A. Echelle and Irv Kornfield, eds., *Evolution of Fish Species Flocks*, 231–250. Orono: University of Maine Press.

Donovan, S. K. 1989. "Introduction to *Mass Extinctions: Processes and Evidence*, ed. S. K. Donovan, xi–xiv. New York: Columbia University Press.

Dover, Gabriel A. 1986. "Molecular drive in multigene families: how biological novelties arise, spread and are assimilated." *Trends in Genetics* 2(6):159–165.

Duellman, William E., and Lindo Trueb. 1986. *Biology of Amphibians*. New York: McGraw-Hill.

Dykhuizen, Daniel E. 1990. "Mountaineering with Microbes." *Nature* 346(5) (July):15–16.

Dyson, Freeman. 1988a. *Infinite in All Directions.* New York: Harper & Row.

Dyson, Freeman, 1988b. "Mankind's Place in the Cosmos." *U.S. News and World Report,* April 18, 72.

Eberhard, William G. 1985. *Sexual Selection and Animal Genitalia.* Cambridge: Harvard University Press.

Eccles, John C. 1989. *Evolution of the Brain: Creation of the Self.* London: Routledge.

Economist. 1990. "Money and Mayhem." 315, April 21, 93–94.

Edwards, C. A., and J. R. Lofty. 1972. *Biology of Earthworms.* London: Chapman and Hall.

Ehrlich, Paul R. 1986. *The Machinery of Nature: The Living World around Us and How It Works.* New York: Simon and Schuster.

Ehrlich, Paul, David S. Dobkin, and Darryl Wheye. 1988. *The Birder's Handbook.* New York: Simon and Schuster.

Eigen, Manfred, et al. 1989. "How Old Is the Genetic code?" *Science* 244 (May 12):673–679.

Eisenberg, John F. 1972. "The Elephant: Life at the Top." In *Marvels of Animal Behavior,* ed. Thomas B. Allen, 181–208. Washington, D.C.: National Geographic Society.

Eldredge, Niles. 1985. *Time Frames: The Rethinking of Darwinian Evolution and the Theory of Punctuated Equilibrium.* New York: Simon and Schuster.

Eldredge, Niles. 1987. *Life's Pulse: Episodes from the Story of the Fossil Record.* New York: Facts on File.

Eldredge, Niles. 1989. *Macroevolutionary Dynamics: Species, Niches, and Adaptive Peaks.* New York: McGraw-Hill.

Emlen, Stephen T. 1972. "Exploring the Mysteries of Migration. In *The Marvels of Animal Behavior,* ed. Thomas B. Allen, 270–286. Washington, D.C.: National Geographic Soceity.

Emlen, Stephen T. 1990. "White-Faced Bee-Eaters: Helping in a Colonial Nesting Species." In *Cooperative Breeding in Birds,* ed. Peter B. Stacey and Walter D. Koenig, 487–526. Cambridge: Cambridge University Press.

Endler, John A. 1986. *Natural Selection in the Wild.* Princeton: Princeton University Press.

Endler, John A., and Tracy McLellan. 1989. "The Process of Evolution: Toward a Newer Synthesis." *Annual Review of Ecology and Systematics* 19:395–421.

Erwin, D. H. 1990. "Mass Extinction Events: End Permian." In *Palaeobiology: A Synthesis,* ed. Derek E. G. Briggs and Peter R. Crowther, 187–193. Oxford: Blackwell Scientific.

Estes, Richard et al. 1988. "Phylogenetic Relations within the Squamata." In Richard Estes and Gregory Pregill, eds., *Phylogenetic Relationships of the Lizard Families,* 119–281. Stanford: Stanford University Press.

Etkin, William. 1981. "A Biological Critique of Sociobiological Theory." In Elliott White, ed., *Sociobiology and Human Politics,* 45–97. Lexington, Mass.: Lexington Books.

Evans, Howard E. 1984. *Insect Biology: A Textbook of Entomology.* Reading, Mass.: Addison-Wesley.

Evans, Howard E., and Mary Jane Eberhard. 1970. *The Wasps.* Ann Arbor: University of Michigan Press.

Ewing, Arthur W. 1963. "Attempts to Select for Spontaneous Activity in *Drosophila melanogaster.*" *Animal Behavior* 11:369–378.

Faegri, K., and L. van der Pijl. 1979. *The Principles of Pollination Ecology.* 2d ed. New York: Pergamon Books.

Feldman, Marcus W., ed. 1988. *Mathematical Evolutionary Theory.* Princeton: Princeton University Press.

Fenchel, Tom. 1987. *Ecology of Protozoa: The Biology of Free-Living Phagotrophic Protists.* Madison, Wis.: Science Technology.

Fenton, Carroll, and Mildred A. Fenton. 1989. *The Fossil Book: A Record of Prehistoric Life.* Garden City, N.Y.: Doubleday.

Fisher, R. 1930. *The Genetical Theory of Natural Selection.* Oxford: Clarendon Press.

Fitch, Walter M. 1988. "The Evolution of Life." In Mae-wan Ho and Sidney W. Fox, eds., *Evolutionary Process and Metaphor,* 35–48. New York: John Wiley.

Fleck, Ludwik. 1979. *Genesis and Development of a Scientific Fact.* Chicago: University of Chicago Press.

Foelix, Rainer. 1982. *The Biology of Spiders.* Cambridge: Harvard University Press.

Forsyth, Adrian. 1986. *A Natural History of Sex: The Evolution and Ecology of Sex.* New York: Scribners.

Forsyth, Adrian, and Kenneth Miyata. 1984. *Tropical Nature.* New York: Scribners.

Fox, Michael W. 1978. *The Dog: Its Domestication and Behavior.* New York: Garland STPM Press.

Fox, Sidney W. 1988. "Evolution Outwards and Forwards." In Mae-wan Ho and Sidney W. Fox, eds., *Evolutionary Process and Metaphors,* 17–34. New York: John Wiley.

Frankel, Joseph. 1983. "What Are the Developmental Underpinnings of Evolutionary Changes in Protozoan Morphology?" In B. C. Goodwin, N. T. Holder, and C. C. Wylie, eds., *Development and Evolution,* 279–314. New York: Cambridge University Press.

Franks, Nigel, R. 1987. "The Parasitic Strategies of a Cuckoo Bee." *Trends in Ecology and Evolution* 2 (November):324–326.

Franks, Nigel R. 1989. "Army Ants: A Collective Intelligence." *American Scientist* 77 (March—April):138–145.

Frisch, Karl von. 1974. *Animal Architecture.* New York: Harcourt Brace Jovanovich.

Frisch, O. V. 1973. *Animal Camouflage.* New York: Collins Publ.

Fromkin, V., and R. Rodman. 1983. *An Introduction to Language.* 3d ed. New York: Holt, Rinehart and Winston.

Fryer, Geoffrey, and T. D. Iles. 1972. *The Cichlid Fishes of the Great Lakes of Africa: Their Biology and Evolution.* Edinburgh: Oliver and Boyd.

Fuller, John L., and William R. Thompson, 1978. *Foundations of Behavior Genetics.* St. Louis: C. V. Mosby Co.

Furness, Robert W. 1989. "Not by Grass Alone." *Natural History* (December):8–12.

Futuyma, Douglas J. 1986. *Evolutionary Biology.* 2d ed. Sunderland, Mass.: Sinauer Associates.

Gadgil, Madhav, and W. H. Bossert. 1970. "Life Historical Consequences of Natural Selection." *American Naturalist* 104:1–29.

Ganders, Fred R. 1979. "The Biology of Heterostyly." *New Zealand Journal of Botany* 17:607–635.

Gardner, Howard. 1985. *The Mind's New Science: A History of the Cognitive Revolution.* New York: Basic Books.

Gazzaniga, Michael. 1988. *Mind Matters: How the Mind and Brain Interact to Create Our Conscious Lives.* Boston: Houghton Mifflin.

Geest, V. 1986. "The Paradox of the Great Irish Stags." *Natural History* 95 (March):54–65.

Georghiu, George H. 1986. "The Magnitude of the Resistance Problem." In *Pesticide Resistance: Strategies and Tactics for Management,* 14–43. Washington, D.C.: National Academy Press.

Ghiselin, Michael. 1974. *The Economy of Nature and the Evolution of Sex.* Berkeley: University of California Press.

Gilbert, Scott F. 1985. *Developmental Biology.* Sunderland, Mass.: Sinauer Associates.

Gill, Donald E. 1989. "Fruiting Failure, Pollinator Insufficiency, and Speciation in Orchids." In Daniel Otte and John Endler, eds., *Speciation and Its Consequences,* 458–481. Sunderland, Mass.: Sinauer Associates.

Gill, Frank B. 1990. *Ornithology.* 3d ed. New York: Freeman.

Gingerich, Philip D., B. Holly Smith, and Elwyn L. Simons. 1990. "Hind Limbs of Eocene *Basilosaurus:* Evidence of Feet in Whales." *Science* 249 (July 15):154–157.

Gittleman, J. L. 1985. "Communal Care in Mammals." In *Evolution: Essays in Honor of John Maynard Smith,* ed. P. J. Greenwood, P. H. Harvey, and M. Slotkin, 187–208. Cambridge: Cambridge University Press.

Glaessner, Martin F. 1984. *The Dawn of Animal Life*. New York: Cambridge University Press.

Glagolev, A. N. 1984. *Motility and Taxis in Procaryotes*. New York: Harwood Academic Publ.

Gleick, James. 1988. *Chaos: Making a New Science*. New York: Viking.

Glen, William. 1990. "What killed the dinosaurs?" *American Scientist* 78 (4):354–370.

Godelier, Maurice. 1989. "Incest Taboo and the Evolution of Society." In Alan Grafen, ed., *Evolution and Its Influence*, 63–92. Oxford: Clarendon Press.

Godfrey, Laurie R. 1985. "Perspectives on Progress." In Laurie Godfrey, ed., *What Darwin Began: Modern Darwinism and Non-Darwinist Perspectives in Evolution*, 40–60. Boston: Allyn and Bacon.

Gold, A. R. 1988. "Forests of Vermont Severely Damaged by Flea-sized Insect." *New York Times*, January 22, XXIII, 9:1.

Goldberger, Ary L., and Bruce J. West. 1987. "Chaos in Physiology: Health or Disease?" In H. Degn, A. V. Holden, and L. F. Olsen, eds., *Chaos in Biological Systems*, 1–4. New York, Plenum Press.

Goldberger, Ary L., David R. Rigney, and Bruce J. West. 1990. "Chaos and Fractals in Human Physiology." *Scientific American* 262 (February):42–49.

Goldsmith, Edward. 1990. "Evolution, Neo-Darwinism, and the Paradigm of Science." *Ecologist* 20 (2) (March–April):67–73.

Goodall, Jane. 1986. *The Chimpanzees of Gombe*. Cambridge: Harvard University Press.

Gorczynski, R. M., and E. J. Steele. 1980. "Immunological Inheritance." *Proceedings National Academy of Sciences USA* 77:2371–2375.

Gotwald, William. 1979. "Phylogenetic Implications of Army Ant Zoogeography." *Annals of the Entomological Society of America* 72:462–467.

Gould, James L. 1982. *Ethology: The Mechanisms and Evolution of Behavior*. New York: W. W. Norton.

Gould, James L., and Carol G. Gould. 1988. *The Honey Bee*. New York: Scientific American Library.

Gould, Stephen J. 1977. *Ontogeny and Philogeny*. Cambridge: Harvard University Press.

Gould, Stephen J. 1980a. "Is a New and General Theory of Evolution Emerging? *Paleobiology* 6:115–136.

Gould, Stephen J. 1980b. *The Panda's Thumb: More Reflections in Natural History*. New York: Norton.

Gould, Stephen J. 1983. *Hens' Teeth and Horses' Toes*. New York: Norton.

Gould, Stephen J. 1987a. "Life's Little Joke." *Natural History* 96 (April):16–25.

Gould, Stephen J. 1987b. *An Urchin in the Storm*. New York: Norton.

Gould, Stephen J. 1989a. "Full of Hot Air." *Natural History* (October):28–38.

Gould, Stephen J. 1989b. *Wonderful Life: The Burgess Shale and the Nature of History.* New York: Norton.

Govind, C. K. 1989. "Asymmetry in Lobster Claws." *American Scientist* 77 (September–October):468–474.

Grant, Verne. 1985. *The Evolutionary Process: A Critical Review of Evolutionary Theory.* New York: Columbia University Press.

Green, Gregory A. 1988. "Living on Borrowed Turf." *Natural History* 97 (September):58–65.

Greene, Eric. 1989. "A Diet-Induced Polymorphism in a Caterpillar." *Science* 243 (February 3):643–646.

Greene, John C. 1961. *Darwin and the Modern World View.* Baton Rouge: Louisiana State University Press.

Greenwood, Paul G., and Richard N. Mariscal. 1984. "The Utilization of Cnidarian Nematocysts by Aeolid Nudibranchs." *Tissue and Cell* 16 (5):719–730.

Greenwood, Peter H. 1984. "African Cichlids and Evolutionary Theories." In A. A. Echelle and I. Kornfield, eds., *Evolution of Fish Species Flocks,* 141–154. Orono: University of Maine Press.

Greenwood, Peter H. 1974. *The Cichlid Fishes of Lake Victoria: Biology and Evolution of a Species Flock.* London: British Museum, Natural History.

Gribbin, John, and Martin Rees. 1989. *Cosmic Coincidences: Dark Matter, Mankind, and Anthropic Cosmology.* New York: Bantam Books.

Griffin, Donald R. 1981. *The Question of Animal Awareness: Evolutionary Continuity of Mental Experiences.* New York: Rockefeller University Press.

Griffin, Donald R. 1984. *Animal Thinking.* Cambridge: Harvard University Press.

Griffin, Donald R. 1986. *Listening in the Dark: The Acoustic Orientation of Bats and Men.* Ithaca: Cornell University Press.

Griswold, Charles E., and Teresa C. Meikle. 1990. "Social life in a Web." *Natural History* (March):6–10.

Gross, Richard J. 1983. *Deer Antlers: Regeneration, Function, and Evolution.* New York: Academic Press.

Groves, Colin P. 1989. *A Theory of Human and Primate Evolution.* New York: Oxford University Press.

Grundfest, Harry. 1967. "Comparative Physiology of Electric Organs of Elasmobranch Fishes." In *Sharks, Skates, and Rays,* ed. Perry W. Gilbert, Robert F. Mathewson, and David P. Rall, 399–432. Baltimore: Johns Hopkins University Press.

Grzimek, Bernhard, ed. 1974. *Grzimek's Animal Life Encyclopedia.* New York: Van Nostrand Reinhold.

Guyton, Arthur C. 1986. *Textbook of Medical Physiology.* 7th ed. Philadelphia: Saunders.

Haldane, J. B. S. 1955. "Population Genetics." *New Biology* 18:34–51.

Hall, Barry G. 1988. "Adaptive Evolution That Requires Spontaneous Mutations." *Genetics* 120:887–897.

Hall, Barry G. 1989. "Selection, Adaptation, and Bacterial Operons." *Génome* 31:265–271.

Hall, J. C. 1985. "Genetic Analysis of Behavior in Insects." In *Comprehensive Insect Physiology, Biochemistry, and Pharmacology,* ed. G. A. Kerkut and L. I. Gilbert, 9:287–384. Oxford: Pergamon Press.

Hall, John E., and James D. Stuart. 1984. *Bats: A Natural History.* Austin: University of Texas Press.

Halvorson, Herlyn O. 1985. "The Beginnings of Sexuality in Procaryotes." In Herlyn Halvorson and Albert Monroy, eds., *The Origin and Evolution of Sex,* 3–6. New York: A. R. Liss.

Hamilton, William D. 1964. "The Genetic Evolution of Social Behavior." *Journal of Theoretical Biology* 7:1–52.

Hamilton, William D. 1972. "Altruism and Related Phenomena, Mainly in Social Insects." *Annual Review of Ecology and Systematics* 3:193–232.

Handel, Steven N., and Andrew J. Beattie. 1990. "Seed Dispersal by Ants." *Scientific American* 263 (2) (August):76–83.

Hanney, Peter W. 1975. *Rodents: Their Lives and Habits.* New York: Taplinger.

Hapgood, Fred. 1979. *Why Males Exist: An Inquiry into the Evolution of Sex.* New York: William Morrow.

Hart, Stern. 1989. "Mystery Amoebae." *Science News* 136 (September 30):216–219.

Hartley, B. S. 1984. "Experimental Evolution of Rabitol Dehydrogenase." In Robert P. Mortlock, ed., *Microorganisms as Model Systems for Studying Evolution,* 23–54. New York: Plenum Press.

Hartman, Daniel S. 1978. *Ecology and Behavior of the Manatee in Florida.* Pittsburgh: American Society of Mammalogists.

Hawking, Stephen W. 1988. *A Brief History of Time.* New York: Bantam Books.

Hedrick, Philip W. 1983. *Genetics of Populations.* Boston: Science Books.

Hegstrom, Roger A., and Dilip K. Kondespudi. 1990. "The Handedness of the Universe." *Scientific American* 262 (January):108–115.

Herbert, Nick. 1985. *Quantum Reality: Beyond the New Physics.* Garden City N.Y.: Doubleday.

Herrnstein, R. J. 1989. "Darwinism and Behaviorism." In Alan Grafen, ed., *Evolution and Its Influence,* 35–61. Oxford: Clarendon Press.

Hinton, H. E. 1964. "Sperm Transfer in Insects and the Evolution of Haemocoelic Insemination." In *Insect Reproduction,* ed. K. C. Highnam, 95–107. London: Royal Entomological Society.

Ho, Mae-wan. 1986. "Heredity as Process." *Rivista de Biologia—Biology Forum* 79 (4):407–427.

Ho, Mae-wan. 1988. "Genetic Fitness and Natural Selections: Myth or Metaphor." In *Evolution of Social Behavior and Integrative Levels,* ed. Gary Greenberg and Ethel Tobach, 87–112. Hillsdale N.J.: Erlbaum.

Ho, Mae-wan, Peter Saunders, and Sidney W. Fox. 1986. "Evolution beyond Neo-Darwinism. *New Scientist* 109 (February 27):41–43.

Hoare, Cecil A. 1972. *The Trypanosomes of Mammals.* Oxford: Blackwell.

Hogue, Charles L. 1972. *The Armies of the Ant.* New York: World Publishing Co.

Holden, A. V., ed. 1988. *Chaos.* Princeton: Princeton University Press.

Holden, Constance. 1980. "Identical Twins Reared Apart." *Science* 207 (March 21):1323–1328.

Holder, Nigel. 1983. "The Vertebrate Limb Patterns and Constraints." In B. C. Goodwin, N. Holder, and C. C. Wylie, eds., *Development and Evolution,* 399–425. Cambridge: Cambridge University Press.

Hölldobler, Bert, and Edward O. Wilson. 1990. *The Ants.* Cambridge: Harvard University Press.

Holliday, Robin. 1988. "A Possible Role for Meiotic Recombinations in Germ Line Reprogramming and Maintenance." In Richard E. Michod and Bruce R. Levin, eds., *The Evolution of Sex: An Examination of Current Ideas,* 45–55. Sunderland, Mass.: Sinauer Associates.

Holliday, Robin. 1989. "A Different Kind of Inheritance." *Scientific American* 260 (June):60–73.

Holm, Eigil. 1979. *The Biology of Flowers.* Penguin Books.

Holmes, John C. 1983. "Evolutionary Relations between Parasitic Helminths and Their Hosts." In *Coevolution,* ed. Douglas J. Futuyma and Montgomery Slatkin, 161–185. Sunderland, Mass.: Sinauer Associates.

Hoser, Raymond T. 1989. *Australian Reptiles and Frogs.* Sydney: Pierson and Co.

Howlett, Rory. 1990. "A Chaotic Synthesis." *Nature* 346 (July 12):104–105.

Huxley, Anthony. 1974. *Plant and Planet.* New York: Viking.

Huxley, Julian S. 1942. *Evolution in Action.* London: Allen and Unwin.

Inglis, W. G. 1965. "Evolution in Parasitic Nematodes." In Angel E. R. Taylor, *Evolution of Parasites,* 79–124. Oxford: Blackwell.

Isack, H. A., and H-U. Reyes. 1989. "Honey Guides and Honey Gatherers." *Science,* 243 (March 10):1343–1346.

Isom, Billy G., and Robert G. Hudson. 1984. "Freshwater Mussels and Their Fish Hosts." *Journal of Parasitology* 70 (2):318–319.

Jablonski, D. 1990. "Extraterrestrial Causes." In *Palaeobiology: A Synthesis,* ed. Derek E. G. Briggs and Peter R. Crowther, 164–170. Oxford: Blackwell Scientific.

Jacobs, Kenneth H. 1985. "Human Origins." In Laurie R. Godfrey, ed., *What Darwin Began: Modern Darwinism and Non-Darwinist Perspectives,* 274–292. Boston: Allyn and Bacon.

Janson, Charles H. 1986. "Capuchin Counterpoint." *Natural History* 95 (February):44–53.

Janvis, Christine. 1982. "Evolution of Horns in Ungulates: Ecology and Paleoecology." *Biological Reviews* 57:261–318.

Janzen, Daniel H. 1976. "Why Bamboos Wait So Long to Flower." *Annual Review of Ecology and Systematics* 7:347–391.

Jarvis, J. U. M. 1981. "Eusociality in a Mammal: Cooperative Breeding in Native Mole-rat Colonies." *Science* 212:571–573.

Jerison, Harry J. 1973. *The Evolution of the Brain and Intelligence.* New York: Academic Press.

John, Bernard, and George L. G. Miklos. 1988. *The Eukaryote Genome in Development and Evolution.* London: Allen and Unwin.

Jones, William E. 1971. *Genetics of the Horse.* Ann Arbor: Caballus Publ.

Kaston, B. J. 1978. *How to Know the Spiders.* Dubuque, Iowa: William C. Brown.

Katz, Michael J. 1987. "Is Evolution Random." In Rudolf M. Raff and Elizabeth C. Raff, eds., *Development as an Evolutionary Process*, 235–315. New York: Alan R. Liss.

Kauffman, Stuart A. 1985. "Self-Organization, Selective Adaptation, and Its Limits." In David J. Depew and Bruce H. Weber, eds., *Evolution at a Crossroads: The New Biology and the New Philosophy of Science*, 169–208. Cambridge, Mass.: The MIT Press.

Keddy, Paul A. 1989. *Competition.* London: Chapman and Hall.

Kemp, T. S. 1982. *Mammal-like Reptiles and the Origin of Mammals.* New York: Academic Press.

Kenyon, Karl W. 1975. *The Sea Otter in the Eastern Pacific Ocean.* New York: Dover Publications.

Kerr, Richard A. 1989. "Does Chaos Permeate the Solar System?" *Science* 244 (April 14):144–145.

Kevles, Bettyann. 1986. *Females of the Species: Sex and Survival in the Animal Kingdom.* Cambridge: Harvard University Press.

Kimura, Motoo. 1985. "Natural Selection and Neutral Evolution." In Laurie R. Godfrey, ed., *What Darwin Began: Modern Darwinism and Non-Darwinist Perspectives*, 73–93. Boston: Allyn and Bacon.

King, James E., and James L. Fobes. 1982. "Complex Learning by Primates." In *Primate Behavior*, ed. J. E. Fobes and J. L. King, 327–360. New York: Academic Press.

King, Mary-Claire, and Allan C. Wilson. 1985. "Evolution at Two Levels in Humans and Chimpanzees." *Science* 188 (April 11):107–116.

Kirkpatrick, Mark. 1987. "Sexual Selection by Female Choice in Polygynous Animals." *Annual Review of Ecology and Systematics* 18:43–70.

Kitcher, Philip. 1985. *Vaulting Ambition: Sociobiology and the Quest for Human Nature.* Cambridge, Mass.: The MIT Press.

Klauber, L. M. 1972. *Rattlesnakes: Their Habits, Life Histories, and Influence on Mankind.* Berkeley: University of California Press.

Klots, Alexander B., and Elsie B. Klots. 1971. *Insects of North America.* Garden City, N.Y.: Doubleday.

Koenig, Walter D., Ronald L. Mumme, and Frank A. Pitelka. 1983. "Female Roles in Cooperatively Breeding Acorn Woodpeckers." In Samuel K. Wasser, ed., *Social Behavior of Female Vertebrates,* 235–261. New York: Academic Press.

Koestler, Arthur. 1967. *The Ghost in the Machine.* New York: Macmillan,.

Koestler, Arthur. 1971. *The Case of the Midwife Toad.* New York: Random House.

Konishi, Masakazu. 1983. "Night Owls Are Good Listeners." *Natural History* 92 (September):56–59.

Konner, Melvin. 1982. *The Tangled Web: Biological Constraints on the Human Spirit.* New York: Harper & Row.

Kroodsma, Donald E. 1983. "Marsh Wrenditions." *Natural History* 92 (September):42–47.

Kroodsma, Donald E. 1989. "What, When, Where, and Why Warblers Warble." *Natural History* (May):50–59.

Kruuk, Hans. 1972. *The Spotted Hyena: A Study of Predation and Social Behavior.* Chicago: University of Chicago Press.

Kuhn, Thomas S. 1962. *The Structure of Scientific Revolutions.* Chicago: University of Chicago Press.

Kikalova-Peck, J. 1987. "New Carboniferous Diplura and the Role of Thoracic Side Lobes in the Origin of Wings Insecta." *Canadian Journal of Zoology* 65 (10):2327–2345.

Kuroda, Suehisa. 1980. "Social Behavior of the Pygmy Chimpanzee." *Primate* 21:181–197.

Lack, David. 1947. *Darwin's Finches.* Cambridge: Cambridge University Press.

Lack, David. 1966. *Ecological Adaptations for Breeding in Birds.* London: Methuen.

Lancaster, Jane B., and Chet S. Lancaster. 1983. "The Parental Investment: The Hominid Adaptation." In Donald J. Ortner, ed., *How Humans Adapt: A Biocultural Odyssey,* 33–65. Washington, D.C.: Smithsonian Institution Press.

Langridge, J. 1987. "Old and New Theories of Evolution." In K. S. W. Campbell and M. F. Day, eds., *Rates of Evolution,* 248–262. London: Allen and Unwin.

Lee, Anthony, and Roger Martin. 1990. "Life in the Slow Lane." *Natural History* (August):34–42.

Leggett, A. J. 1987. *Problems of Physics.* New York: Oxford University Press.

Levin, Bruce R. 1984. "Science as a Way of Knowing—Molecular Evolution." *American Zoologist* 24:451–464.

Lewin, Roger. 1981. "Lamarck Will Not Lie Down." *Science* 213 (July 17): 316–321.

Lewin, Roger. 1983. "How Mammalian RNA Returns to Its Genome." *Science* 219:1052–1054.

Lewin, Roger. 1989a. "New Look at Turtle Migration Mystery." *Science* 243 (February 24):1009.

Lewin, Roger. 1989b. *Human Evolution: An Illustrated Introduction.* 2d ed. Boston: Blackwell.

Lewis, W. M. Jr. 1987. "The Costs of Sex." In *The Evolution of Sex and Its Consequences,* ed. S. C. Stearns, 33–57. Boston: Birkhäuser Verlag.

Lightfoot-Klein, Hanny. 1990. *Prisoners of Ritual: An Odyssey into Female Genital Circumcision in Africa.* Binghamton, N.Y.: Haworth Press.

Lilly, John. 1974. "A Feeling of Weirdness." In Joan McIntyre, ed., *Mind in the Water,* 71–77. New York: Charles Scribner's Sons.

Lindow, S. E., N. J. Panopolous, and B. L. McFarland. 1989. "Genetic Engineering of Bacteria from Managed and Natural Habitats." *Science* 244 (June 16):1300–1307.

Lloyd, J. W. 1979. "Mating Behavior and Natural Selection." *Florida Entomologist* 62:17–34.

Loomis, William F. 1988. *Four Billion Years: An Essay on the Evolution of Genes and Organisms.* Sunderlund, Mass.: Sinauer Associates.

Lovelock, James. 1988. *The Ages of Gaia: A Biography of Our Living Earth.* New York: Norton.

Loyn, Richard H. 1987. "The Bird That Farms the Dell." *Natural History* 96 (June):54–60.

Lutz, Paul E. 1986. *Invertebrate Zoology.* Reading, Mass.: Addison-Wesley.

McClintock, James B., and John Janssen. 1990. "Pteropod Abduction as a Chemic Defence in a Pelagic Antarctic Amphipod." *Nature* 346 (August 2):462–464.

McClure, G. A. 1966. *The Bamboos.* Cambridge: Harvard University Press.

McDonald, John F., et al. 1987. "The Responsive Genome: Evidence and Evolutionary Implications." In Rudolf A. Raff and Elizabeth C. Raff., eds., *Development as an Evolutionary Process,* 239–263. New York: Alan Liss.

McFarland, David. 1985. *Animal Behavior: Psychology, Ethology, and Evolution.* Menlo Park, Calif.: Benjamin Cumming.

McFarland, William N., F. Harvey Pough, Tom J. Cade, and John B. Heiser. 1985. *Vertebrate Life.* 2d ed. New York: Macmillan.

McMenamin, Mark A. 1990. "Mass Extinction Events: Vendian." In *Palaeobiology: A Synthesis,* ed. Derek E. G. Briggs and Peter R. Crowther, 179–180. Oxford: Blackwell Scientific.

McMenamin, Mark A., and Dianna Shulte McMenamin. 1990. *The Emergence of Animals: The Cambrian Breakthrough.* New York: Columbia University Press.

McNulty, F. 1974. *The Great Whales.* Garden City N.Y.: Doubleday.

McQueen, D. J., and C. L. McLay. 1983. "How Does the Intertidal Spider *Desis marina* Remain under Water for Such a Long Time?" *New Zealand Journal of Zoology* 10:382–392.

McSweeny, D. J. et al. 1989. "North Pacific Humpback Whale Songs." *Marine Mammal Science* 5 (2) (April):139–148.

Malgren, Björn, W. A. Berggen, and G. P. Lohmann. 1984. "Species formation through Punctuated Gradualism in Planktonic Foraminifera." *Science* 225 (July 20):317–319.

Manning, A. 1975. "Behaviour Genetics and the Study of Behavioural Evolution." In Gerard Baerends, Colin Beer, and Aubrey Manning, eds., *Function and Evolution in Behaviour*, 71–91. Oxford: Clarendon Press.

Manton, S. M. 1977. *The Arthropoda: Habits, Functional Morphology, and Evolution*. Oxford: Clarendon Press.

Margulis, Lynn, Dorion Sagan, and Lorraine Olendzenski. 1985. "What Is Sex?" In Harlyn O. Halvorson and Albert O Monroy, eds., *The Origin and Evolution of Sex*, 69–85. New York: Alan R. Liss.

Margulis, Lynn, and Dorion Sagan. 1986. *Origin of Sex*. New Haven: Yale University Press.

Marler, Peter R. 1972. "The Drive to Survive." In *Marvels of Animal Behavior*, ed. Thomas B. Allen, 19–47. Washington, D.C.: National Geographic Society.

Matsuura, M., and S. Yamane. 1984. *Biology of the Vespine Wasps*. Berlin: Springer Verlag.

Mattison, Chris. 1987. *Snakes of the World*. New York: Facts on File.

May, Robert M. 1974. "Biological Populations with Non-overlapping Generations," *Science* 186 (November 15):645–647.

May, Robert M. 1976. "Simple Mathematical Models with Very Complicated Dynamics." *Nature*, 261 (June 10):459–467.

May, Robert M. 1988. "How Many Species Are There on Earth?" *Science* 241 (September 16):1441–1449.

Maynard Smith, John. 1978. "Optimization Theory in Evolution." *Annual Review of Ecology and Systematics* 9:31–56.

Maynard Smith, John. 1986. *The Problem of Biology*. New York: Oxford University Press.

Maynard Smith, John. 1988. "Evolutionary progress and Level of Selection." In Matthew H. Nitecki, ed., *Evolutionary Progress*, 219–230. Chicago: University of Chicago Press.

Mayr, Ernst. 1960. "The Emergence of Evolutionary Novelties." In Sol Tax, ed., *Evolution after Darwin*, 1:349–380. Chicago: University of Chicago Press.

Mayr, Ernst. 1963. *Animal Species and Evolution*. Cambridge: Harvard University Press.

Mayr, Ernst. 1982. *The Growth of Biological Thought: Diversity, Evolution, and Inheritance*. Cambridge: Harvard University Press.

Mayr, Ernst F. 1983. "How to Carry Out the Adaptationist Program." *American Naturalist* 121 (March):324–334.

Mayr, Ernst. 1984. "The Unity of the Genotype." In Robert N. Brandon and Richard N. Burian, eds., *Genes, Organisms, and Populations*, 69–84. Cambridge, Mass.: The MIT Press.

Mayr, Ernst. 1985. "How Biology Differs from Physical Science." In *Evolution at a Crossroads: The New Biology and the New Philosophy of Science*, ed. David J. Depew and Bruce H. Weber. Cambridge, Mass.: The MIT Press.

Mayr, Ernst. 1988. *Toward a New Philosophy of Biology*. Cambridge: Harvard University Press.

Michener, Charles D. 1974. *The Social Behavior of Bees*. Cambridge: Harvard University Press.

Miklos, George L. G., and Bernard John. 1987. "From Genome to Phenotype." In K. S. W. Campbell and M. F. Day, eds., *Rates of Evolution*, 263–282. London: Allen and Unwin.

Miller, Benjamin F., and Ruth Goode. 1960. *Man and His Body*. New York: Simon and Schuster.

Miloszewski, Mark J. 1983. *The Behavior and Ecology of the African Buffalo*. New York: Cambridge University Press.

Moffat, Anne S. 1989. "A Challenge to Evolutionary Biology." *American Scientist* 77 (May–June):224–226.

Moffett, M. W. 1989. "Samurai Aphids: Survival under Siege." *National Geographic* 176 (3) (September):406–422.

Monastersky, Richard. 1989. "Deep-See Shrimp." *Science News* 135 (February 11):90–91.

Monod, Jacques. 1972. *Chance and Necessity*. London: Collins.

Montagu, M. F. Ashley. 1962. "Time, Morphology, and Neoteny in the Evolution of Man." In M. F. Ashley Montagu, ed., *Culture and the Evolution of Man*, 324–342. New York: Oxford University Press.

Moriarty, Christopher. 1978. *Eels*. New York: Universe Books.

Morowitz, Harold J. 1985. "The Origin of Life." In Laurie J. Godfrey, ed., *What Darwin Began: Modern Darwinism and Non-Darwinist Perspective*, 243–257. Boston: Allyn and Bacon.

Morris, S. Conway. 1990. "Early Cambrian Diversification." In *Palaeobiology: A Synthesis*, ed. Derek E. G. Briggs and Peter R. Crowther, 30–36. Oxford: Blackwell Scientific.

Mortlock, Robert P. 1982. "Regulatory Mutations and the Development of New Metabolic Pathways by Bacteria." In *Evolutionary Biology*, ed. Max K. Hecht, Bruce Wallace, and Chillean T. Prance, 14:205–267. New York: Plenum.

Mortlock, Robert P. 1990. Letter, July 11.

Moss, Cynthia. 1975. *Portraits in the Wild: Behavior Studies of Eight East African Mammals*. Boston: Houghton Mifflin.

Mosse, Douglas H. 1981. "Behavior and Ecology of Bumble Bees." In *Social Insects,* ed. Henry R. Hermann, 3:243–258. New York: Academic Press.

Müller-Schwarze, Dietland. 1984. *The Behavior of Penguins.* Albany: State University of New York Press.

Munro, P. M., F. Launod, and M. J. Gauthier. 1987. *Letters in Applied Microbiology* 4 (1987):121–124.

Murray, J. D. 1981. "On Pattern Formation Mechanisms for Lepidopteran Wing Patterns and Mammalian Coat Markings." *Philosophical Transactions Royal Society of London* 295:473–496.

Mussen, P., and N. Eisenberg. 1977. *Roots of Caring, Sharing, and Helping.* San Francisco: W. H. Freeman.

Neel, J. V. 1989. "Human Evolution and the Founder-Flush Principle." In L. V. Giddings, K. Y. Kaneshiro, and W. W. Anderson, eds., *Genetics, Speciation, and the Founder Principle,* 299–313. New York: Oxford University Press.

Nelson, Margaret C. 1976. "Classical Conditioning in the Blowfly *Phormia regina)." Journal of Comparative Physiology and Psychology* 7:353–368.

Nicolis, Grégoire, and Ilya Prigogine. 1989. *Exploring Complexity.* New York: Freeman.

Nicholls, Robert D., et al. 1989. "Genetic Imprinting Suggested by Maternal Heterodisomy in Prader-Willi Syndrome." *Nature* 342 (November 16):281–284.

Noble, Elmer R., and Glenn A. Noble. 1982. *Parasitology: The Biology of Animal Parasites.* 5th ed. Philadelphia: Lea and Febliger.

Nolan, Sheila M. 1983. "Environmental Factors." In Paul D. Holprich, ed., *Infectious Diseases.* 3d ed. New York: Harper & Row.

Novacek, Michael J. 1988. "Navigators of the Night." *Natural History* 97 (October):66–71.

Nowak, Ronald M., and J. L. Paradiso. 1983. *Walker's Mammals of the World.* 4th ed. Baltimore: Johns Hopkins University Press.

Numata, Makoto. 1979. "The Structure and Succession of Bamboo Vegetation." In *Ecology of Grasslands and Bamboolands in the World,* ed. M. Numata, 237–257. The Hague: W. Junk.

O'Brien, Stephen J., and David E. Wildt. 1986. "The Cheetah in Genetic Peril." *Scientific American* 254(5) (May):84–90.

O'Brien, Stephen J. 1987. "The Ancestry of the Giant Panda" *Scientific American* 257 (November):102–107.

Oettinger, Marjorie A., David G. Schatz, Carolyn Gorka, and David Baltimore. 1990. "RAG-1 and RAG-2, Adjacent Genes That Synergistically Activate V(D)J Recombination." *Science* 248 (June 22):1517–1523.

Ohno, S. 1988. "Codon Preference Is But an Illusion Created by the Construction Principle of Codon Sequence." *Proceedings of the National Academy of Science USA* 85:4378–4386.

Opadia-Kadima, G. Z. 1987. "How the Slot Machine Led Biologists Astray." *Journal of Theoretical Biology,* January 21, 127–135.

Orgel, L. E., and F. H. C. Crick. 1980. "Selfish DNA: The Ultimate Parasite." *Nature* 284 (April 17):604.

Ornstein, Robert, and David Sobel. 1987. *The Healing Brain.* New York: Simon and Schuster.

Oster, G. F., and E. O. Wilson. 1978. *Caste and Ecology in Social Insects.* Princeton: Princeton University Press.

Ostrom, John H. 1986. "The Cursorial Origin of Avian Flight." In *The Origin of Birds and the Evolution of Flight,* ed. Kevin Padian. San Francisco: California Academy of Sciences.

Owen, D. F. 1966. *Animal Ecology in Tropical Africa.* San Francisco: W. H. Freeman.

Oxnard, Charles E. 1984. *The Order of Man: A Biomathematical Anatomy of the Primates.* New Haven: Yale University Press.

Pagels, Heinz R. 1989. *The Deams of Reason: The Computer and the Rise of the Sciences of Complexity.* New York: Bantam.

Patterson, Colin. 1990. "Metazoan Phylogeny: Reassessing Relationships." *Nature* 344 (March 15):199–200.

Paul, Gregory S. 1988. *Predatory Dinosaurs of the World.* New York: Simon and Schuster.

Peacocke, Arthur. 1986. *God and the New Biology.* San Francisco: Harper & Row.

Pearse, John S., and Vicki B. Pearse. 1978. "Vision in Cubomedusan Jellyfishes." *Science,* January 27, 258.

Pearse, Vicki, John Pearse, Mildred Buchsbaum, and Ralph Buchsbaum. 1987. *Living Invertebrates.* Palo Alto: Blackwell.

Pedigo, Larry P. 1989. *Entomology and Pest Management.* New York: Macmillan.

Penrose, Roger. 1987. "Quantum Physics and Conscious Thought." In B. J. Hiley and F. David Peat, *Quantum Implications: Essays in Honor of David Bohm,* 105–125. London: Routledge and Kegan Paul.

Penrose, Roger. 1989. *The Emperor's New Mind: Concerning Computation, Minds, and the Laws of Physics.* New York: Oxford University Press.

Perry, Nicolette. 1983. *Symbiosis: Close Encounters of the Natural Kind.* Poole, Dorset: Blandford Press.

Pettigrew, J. S. 1986. "Flying Primates? Megabats Have Advanced Pathway from Eye to Midbrain." *Science* 231 (March 14):1304–1306.

Plapp, Frederick W., Jr. 1986. "Genetics and Biochemistry of Insecticide Resistance in Arthropods." In *Pesticide Resistance: Strategies and Tactics for Management,* ed. National Research Council, 74–86. Washington, D.C.: National Academy Press.

Pollard, Jeffrey N. 1988. "New Genetic Mechanisms and Their Implications." In Mae-wan Ho and Sidney W. Fox, eds., *Evolutionary Processes and Metaphors,* 63–84. New York: Wiley.

Pool, Robert. 1989a. "Ecologists Flirt with Chaos." *Science* 243 (January 20):310–311.

Pool, Robert. 1989b. "Chaos Theory: How Big an Advance?" *Science* 245:26–28.

Pool, Robert. 1990. "Closing the Gap between Proteins and DNA." *Science* 248 (June 29):1609.

Popper, Arthur N. 1980. "Sound Emission and Detection by Delphinids." In *Cetacean Behavior, Mechanisms and Functions,* ed. Louis M. Herman, 1–52. New York: John Wiley.

Popper, Karl. 1984. "Evolutionary Epistemology." In J. W. Pollard, ed., *Evolutionary Theory: Paths into the Future,* 239–255. New York: Wiley.

Popper, Karl. 1988. "Darwinism as a Metaphysical Research Program." In Michael Ruse, ed., *But Is It Science?* 144–155. New York: Prometheus Books.

Pough, F. Harvey, John B. Heiser, and William McFarland. 1989. *Vertebrate Life.* 3d ed. New York: Macmillan.

Procter-Gray, Elizabeth. 1990. "Kangaroos up a Tree." *Natural History* (January):60–66.

Pruett-Jones, Melinda, and Stephen Pruett-Jones. 1983. "The Bowerbirds' Labor of Love." *Natural History* 92 (September):48–55.

Pryor, Karen. 1975. *Lads before the Wind: Adventures in Porpoise Training.* New York: Harper & Row.

Ptashne, Mark. 1989. "How Gene Activators Work," *Scientific American* 260 (January):40–47.

Radinsky, Leonard B. 1987. *The Evolution of Vertebrate Design.* Chicago: University of Chicago Press.

Radman, Miroslav, and Robert Wagner. 1988. "The High Fidelity of DNA Duplication." *Scientific American* 259 (August):40–47.

Raff, Rudolf A., et al. 1987. "Molecular and Developmental Correlates of Macroevolution." In Rudolf A. Raff and Elizabeth E. Raff, eds., *Development as an Evolutionary Process.* New York: Alan Liss.

Rensch, Bernard. 1957. "The Intelligence of Elephants." *Scientific American* 196 (February).

Restak, Richard M. 1979. *The Brain: The Last Frontier.* New York: Warner Books.

Restak, Richard M. 1988. *The Mind.* New York: Bantam Books.

Richards, A. J. 1986. *Plant Breeding Systems.* London: Allen and Unwin.

Richards, Robert J. 1987. *Darwin and the Emergence of Evolutionary Theories of Mind and Behavior.* Chicago: University of Chicago Press.

Ridley, Mark. 1985. *The Problems of Evolution.* New York: Oxford University Press.

Riedman, Marianne. 1990. *The Pinnipeds: Seals, Sea Lions, and Walruses.* Berkeley: Univesity of California Press.

Riopelle, Arthur J. 1972. "Learning How Animals Learn." In *The Marvels of Animal Behavior*, ed. Thomas B. Allen, 358–384. Washington, D.C.: National Geographic Society.

Robinson, Scott K. 1986. "Social Security for Birds." *Natural History* 96 (March):38–47.

Rood, Jon P. 1988. "Whatever Happened to the C Pack Sisters?" *Natural History* 97 (February):40–47.

Rosenzweig, Mark R., Edward Bennett, and Marian C. Diamond. 1972. "Brain Change in Response to Experience." *Scientific American* 226 (February).

Ross, K. G., and D. J. C. Fletcher. 1985. "Comparative Study of Genetic and Social Structure in Two Forms of the Fire Ant." *Behavioral Ecology and Sociobiology* 17. 349–356.

Rothwell, Norman V. 1983. *Understanding Genetics*. 3d ed. New York: Oxford University Press.

Ruse, Michael. 1979. *Sociobiology: Sense or Nonsense*. London: D. Reidel.

Ruse, Michael. 1982. *Darwinism Defended*. Reading, Mass.: Addison-Wesley.

Ruse, Michael. 1988. *Philosophy of Biology Today*. Albany: State University of New York Press.

Sage, Richard D., Paul V. Loiselle, and A. C. Wilson. 1984. "Molecular versus Morphological Change in Cichlid Fish of Lake Victoria." In A. A. Echelle and I. Kornfield, eds., *Evolution of Fish Species Flocks*, 185–201. Orono: University of Maine Press.

Salisbury, Frank B., and Cleon Ross. 1985. *Plant Physiology*. Belmont, Calif.: Wadsworth.

Sancho, E. 1988. "Dermatobia, the Neotropical Warble Fly." *Parasitology Today* 4 (9):242–246.

Sander, Klaus. 1983. "The Evolution of Patterning Mechanisms." In B. C. Goodwin, N. Holder, and C. C. Wylie, eds., *Development and Evolution*, 137–159. London: Cambridge Univesity Press.

Sanderson, Ivan T. 1961. *Living Mammals of the World*. New York: Hanover House.

Sapienza, Carmen. 1990. "Parental Imprinting of Genes." *Scientific American* 263 (4) (October):52–60.

Saunders, Peter T. 1988. "Sociobiology, a House Built on Sand." In Maewon Ho and Sidney W. Fox, eds., *Evolutionary Processes and Metaphors*, 275–294. New York: Wiley.

Saunders, Shelly R. 1985. "The Inheritance of Acquired Characteristics, a Concept That Will Not Die." In Laurie H. Godfrey, ed., *What Darwin Began: Modern Darwinism and Non-Darwinist Perspectives in Evolution*, 148–161. Boston: Allyn and Bacon.

Scapini. Felicita. 1986. "Inheritance of Solar Direction Finding in Sandhoppers." *Monitore zoologico italiano* 20:53–61.

Scapini, Felicita. 1988. "Heredity and Learning in Animal Orientation." *Monitore zoologico italiano* 22 (2):203–234.

Schaffer, W. M., and M. Kot. 1986. "Chaos in Ecological Systems: The Coals That Newcastle Forgot." *Trends in Ecology and Evolution* 3(1):58–63.

Schaller, George B. 1972a. "The Sociable Kingdom." In *The Marvels of Animal Behavior*, ed. Thomas b. Allen, 67–87. Washington, D.C.: National Geographic Society.

Schaller, George B. 1972b. *The Serengeti Lion, A Study of Predator Prey Relations.* Chicago: University of Chicago Press.

Schaller, George G. et al. 1989. "Feeding Ecology of Giant Pandas and Asiatic Black Bears." In John Gittleman, *Carnivore Behavior, Ecology, and Evolution*, 212–241. Ithaca: Cornell University Press.

Schimke, Robert T. 1982. "Studies on Gene Duplication and Amplification: Historical Perspective." In Robert Schimke, ed., *Gene Amplification*, 1–6. Cold Spring Harbor, N.Y.: Cold Spring Harbor Laboratory.

Schlipp, P. A. 1974. ed., *The Philosophy of Karl Popper.* LaSalle, Ill.: Open Court.

Schram, Frederick R. 1986. *Crustacea.* New York: Oxford University Press.

Schwenk, Kurt. 1988. "Comparative Morphology of the Lepidosaurus Tongue and Its Relevance to Squamate Phylogeny." In Richard Estes and Gregory Pregill, eds., *Phylogenetic Relations of the Lizard Families*, 569–597. Stanford: Stanford University Press.

Selander, Richard D., and Robert K. Selander. 1975. "Trophic Radiation through polymorphism in Cichlid Fish." *Proceedings of the National Academy of Sciences USA* 72(11) (November):4669–4673.

Serpell, James A. 1988. "The Domestication and History of the Cat." In Dennis C. Turner and Patrick Bateson, eds., *The Domestic Cat*, 151–158. New York: Cambridge University Press.

Shapiro, James A. 1988. "Bacteria as Multicellular Organisms." *Scientific American* 258 (June):82–89.

Shaw, Charles E., and Sheldon Campbell. 1974. *Snakes of the American West.* New York: Alfred Knopf.

Shear, William. 1986. "The Evolution of Web-building Behavior in Spiders." In William Shear, ed., *Spiders: Webs, Behavior, and Evolution*, 364–400. Stanford: Stanford University Press.

Sheldon, P. R. 1990. "Microevolution and the Fossil Record." In *Palaeobiology: A Synthesis*, ed. Derek E. G. Briggs and Peter R. Crowther, 106–110. Oxford: Blackwell Scientific.

Shields, William M. 1988. "Sex and Adaptation." In *The Evolution of Sex*, ed. R. E. Michod and B. R. Levin, 253–269. Sunderland, Mass.: Sinauer Associates.

Sibatini, A. 1989. "How to Structuralize Biology." In Brian Goodwin, A. Sibatini, and Gerry Webster, eds., *Dynamic Structures in Biology*, 16–30. Edinburgh: Edinburgh University Press.

Simons, E. L. 1989. "Human Origins." *Science* 245 (September 22):1343–1350.

Simpson, George G. 1949. *The Meaning of Evolution.* New Haven: Yale University Press.

Simpson, George G. 1974. "The Concept of Progress in Organic Evolution." *Social Research* 41.

Simpson, George G. 1980. *Splendid Isolation: The Curious History of South American Mammals.* New Haven: Yale University Press.

Skutch, Alexander F. 1973. *The Life of the Hummingbird.* New York: Crown Publishers.

Skutch, Alexander F. 1987. *Helpers at Birds' Nests.* Iowa City: University of Iowa Press.

Sleigh, Michael. 1989. *Protozoa and Other Protists.* London: Edward Arnold.

Slipjper, E. J. 1962. *Whales.* Ithaca: Cornell University Press.

Sloan, R. E., J. K. Ribgy, L. M. Van Valen, and D. Gabriel. 1986. "Gradual Dinosaur Extinction and Simultaneous Ungulate Radiation in the Hell Creek Formation." *Science* 232:629–633.

Smith, D. C., and A. E. Douglas. 1987. *The Biology of Symbiosis.* London: Edward Arnold.

Smith, Joseph W. 1984. *Reductionism and Cultural Being.* The Hague: Martinus Nijhoff.

Smith, Martin S. 1987. "Sociobiology of Human Development." In Charles C. Crawford, Martin Smith, and Dennis Krebs, eds., *Sociobiology and Psychology: Ideas, Issues, and Applications,* Hillsdale, N.J.: Erlbaum.

Smith, Neal G. 1979. "Coevolution of Oropendola and Cowbird Nest Parasites. Alternate Responses by Hosts to Parasites Which May Be Helpful or Harmful." In Brent B. Nikol, ed., *Host-Parasite Interfaces.* New York: Academic Press.

Smith-Keary, P. F. 1988. *Genetic Elements in Escherischia coli.* London: Macmillan.

Smithson, T. R. 1989. "The Earliest Known Reptile." *Nature* 42 (December 7):676–678.

Stanley, Steven M. 1979. *Macroevolution: Pattern and Process.* San Francisco: W. H. Freeman.

Stanley, Steven M. 1981. *The New Evolutionary Timetable.* New York: Basic Books.

Stanley, Steven M. 1988. *Extinctions.* New York: Scientific American Books.

Starr, C. K. 1979. "Origin and Evolution of Insect Sociality." In H. R. Hermann ed., *Social Insects.* New York: Academic Press.

Stearns, S. C. 1987. "Why Sex Evolved." In *The Evolution of Sex and Its Consequences,* ed. S. C. Stearns, 15–31. Boston: Birkhäuser Verlag.

Stebbins, G. Ledyard. 1974. *Flowering Plants: Evolution above the Species Level.* Cambridge: Harvard University Press.

Stebbins, G. Ledyard, and Francisco J. Ayala. 1985. "The Evolution of Darwinism." *Scientific American* 253 (July):72–62.

Stein, Daniel L. 1989. "Spin Glasses." *Scientific American* 261 (July):52–59.

Stolzenburg, W. 1990. "Hypermutation: Evolutionary Fast Track?" *Science News* 137 (June 23):391.

Strassman, Joan A. et al. 1989. "Genetic Relatedness in Primitively Eusocial Wasps." *Nature* 342 (November 16):268–269.

Strickberger, Monroe W. 1990. *Evolution.* Boston: Jones and Bartlett.

Stryer, Lubert. 1987. "The Molecules of Visual Excitation." *Scientific American* 257 (July):42–51.

Suga, Nobuo, 1990. "Biosonar and Neural Computation in Bats." *Scientific American* 262 (June):60–68.

Thomas, Barry A., and Robert A. Spicer. 1987. *The Evolution and Paleobiology of Land Plants.* London: Croom and Helm.

Thompson, D'Arcy W. 1961. *On Growth and Form.* Cambridge: Cambridge University Press.

Thomson, Keith S. 1988. *Morphogenesis and Evolution.* New York: Oxford University Press.

Thornhill, R. 1979. "Adaptive Female Mimicking in a Scorpion Fly." *Science* 205:412–414.

Thornhill, R., and J. Alcock 1983. *The Evolution of Insect Mating Systems.* Cambridge: Harvard University Press.

Tomback, D. F. 1982. "Dispersal of Whitebark Pine Seeds by Clark's Nutcracker." *Journal of Animal Ecology* 51 (2):451–467.

Tomita, K. 1988. "Periodically Forced Nonlinear Oscillators." In *Chaos,* ed. A. V. Holden, 211–236. Princeton: Princeton University Press.

Trewavar, A. J. 1988. "The Evolution Controversy: A Network View." *Evolutionary Trends in Plants* 2 (June):1–5.

Trivers, Richard L. 1971. "The Evolution of Reciprocal Altruism." *Quarterly Review of Biology* 46:35–57.

Trivers, Richard L. 1976. Foreword to Richard Dawkins, *The Selfish Gene.* Oxford: Oxford University Press.

Trivers, Richard L. 1981. "Sociobiology and politics." In Elliott White, ed., *Sociobiology and Human Politics,* 1–43. Lexington, Mass.: Lexington Books.

Tsonis, Panagiotis A., and Anastasios A. Tsonis. 1989. "Chaos: Principles and Implications in Biology." *Computer Applications in the Biosciences* 5(1):27–32.

Tucker, John M. 1990. "Hybridization in California Oaks." *Fremontia* 18 (3):13–19.

Tyler, Michael J. 1983. "Evolution of Gastric Breeding." In Michael Tyler, ed., *The Gastric Brooding Frog,* 129–135. London: Croom Helm.

Valentine, James W. 1977. "Metazoa." In A. Hallam, ed., *Patterns of Evolution as Illustrated by the Fossil Record.* New York: Elsevier.

Valentine, James W. 1985. "The Evolution of Complex Animals." In Laurie R. Godfrey, ed., *What Darwin Began: Modern Darwinism and Non-Darwinist Perspectives*, 258–272. Boston: Allyn and Bacon.

Valentine, James W. 1989. "How Good Is the Fossil Record? Clues from the California Plestocene." *Paleobiology* 15 (Spring):83–94.

Valentine, James W., and Douglas H. ERwin. 1987. "Interpreting Great Developmental Experiments." in Rudolf A. Raff, and Elizabeth E. Raff, eds., *Development as an Evolutionary Process*, 71–107. New York: Alan Liss.

Vaughan, Terry A. 1986. *Mammalogy*. 3d ed. Philadelphia: W. B. Saunders.

Voelker, William. 1986. *The Natural History of Living Mammals*. Medford N.J.: Plexus.

Waddington, C. H. 1975. *The Evolution of an Evolutionist*. Ithaca, N.Y.: Cornell University Press.

Walcott, Charles. 1989. "Show Me the Way You Go Home." *Natural History* (November):40–46.

Waldrop, M. Mitchell. 1990. "Spontaneous Order, Evolution, and Life." *Science*, March 30, 1543–1545.

Wall, Stephen V., and Russell P. Balda. 1983. "Remembrance of Seeds Stashed." *Natural History* 92 (September):60–65.

Wallace, Alfred R. 1881. *Natural Selection and Tropical Nature*. London: Macmillan.

Wallace, Alfred R. 1901. *Darwinism: An Exposition of the Theory of Natural Selection*. 3d ed. London: Macmillan.

Weber, Bruce H., et al. 1989. "Evolution in Thermodynamic Perspective: An Ecological Approach." In *Biology and Philosophy*, 4:373–405.

Weisburg, S. 1987. "Fungus Feel Their Way to Feast." *Science News*, April 4, 214.

Wellnhofer, Peter. 1990. "Archaeopteryx." *Scientific American* 262 (May):70–77.

Wells, M. J. 1962. *Brain and Behavior in Cephalopods*. Stanford: Stanford University Press.

Wells, M. J. 1978. *Octopus: Physiology and Behavior of an Advanced Invertebrate*. London: Chapman and Hall.

Welty, Joel C. 1982. *The Life of the Birds*. Philadelphia: Saunders.

Went, F. W. 1971. "Parallel Evolution." *Taxon* 20 (May):197–226.

Wesson, Robert. 1967. *The Imperial Order*. Berkeley: University of California Press.

Wesson, Robert. 1978. *State Systems: International Pluralism, Politics and Culture*. New York: Free Press.

Wesson, Robert. 1979. "Wrong Number." *Natural History* 88 (March):9–22.

West-Eberhard, Mary Jane. 1988. "Phenotypic Plasticity and 'Genetic' Theories of Insect Sociality." In Gary Greenberg and Ethel Tobach, eds., *Evo-*

lution of Social Behavior and Integrative Levels, 123–133. Hillsdale, N.J.: Laurence Erlbaum.

Whalley, P. 1987. "Insect Evolution during the Extinction of the Dinosauria." *Entomologia Generalis* 13 (1/2):119–124.

Wheeler, A. 1974. *Fishes of the World*. New York: Macmillan.

White, Randall. 1989. "Visual Thinking in the Ice Age." *Scientific American* 261 (July):92–99.

Whitley, Gilbert, and Joyce Allen. 1958. *The Seahorse and Its Relatives*. Newton, Mass.: Bromford.

Whyte, Lancelot L. 1965. *Internal Selection*. New York: Braziller.

Wicken, Jeff S. 1988. "Thermodynamics, Evolution, and Emergence." In Bruce H. Weber, David J. Depew, and James D. Smith, *Entropy, Information, and Evolution*, 139–169. Cambridge, Mass.: The MIT Press.

Wickler, Wolfgang. 1968. *Mimicry in Plants and Animals*. New York: World University Library.

Wiepkema, P. A. 1961. "An Ethological Analysis of the Reproductive Behavior of the Bitterling (*Rhodeus amarus*)." *Archives Néerlandaises de Zoologie* 14 (2):103–199.

Wilford, John N. 1988. "Human Teeth, Small Already, Continue to Shrink." *New York Times*, August 30, III, 1:3.

Wilkinson, Gerald S. 1990. "Food Sharing in Vampire Bats." *Scientific American* 262 (February):76–82.

Williams, George C. 1966. *Adaptation and Natural Selection: A Critique of Some Current Evolutionary Thought*. Princeton: Princeton University Press.

Williamson, P. G. 1981. "Palaeontological Documentation of Speciation in Cenozoic Molluscs from Turkana Basin." *Nature* 293 (October 8):437–443.

Williamson, R. G. 1987. "Selection or constraint." In K. S. W. Campbell and M. F. Day, eds., *Rates of Evolution*. Boston: Allen and Unwin.

Wills, Christopher. 1989. *The Wisdom of the Genes*. New York: Basic Books.

Wilson, Allan C. 1975. "Evolutionary Importance of Gene Regulation." *Stadler Genetics Symposium* 7.

Wilson, Allan C. 1985. "The Molecular Basis of Evolution." *Scientific American* 253 (October):164–173.

Wilson, Brayton F. 1984. *The Growing Tree*. Amherst: University of Massachusetts Press.

Wilson, D. S. 1977. "How Nepotistic Is the Brain Worm?" *Behavioral Ecology and Sociobiology* 2:421–425.

Wilson, Edward O. 1971. *The Insect Societies*. Cambridge: Harvard University Press.

Wilson, Edward O. 1978. *On Human Nature*. Cambridge: Harvard University Press.

Wilson, Edward O. 1980. *Sociobiology*. Abridged ed. Cambridge: Harvard University Press.

Wilson, Edward O. 1989. "Threats to Biodiversity." *Scientific American* 261 (September):108–116.

Wilson, Peter J. 1983. *Man, the Promising Primate: The Conditions of Human Evolution.* 2d ed. New Haven: Yale University Press.

Winstatt, William C., and Jeffrey C. Schank. 1988. "Two Constraints on the Evolution of Complex Adaptation and the Means of Their Avoidance." In *Evolutionary Progress,* ed. Matthew H. Nitecki, 231–273. Chicago: University of Chicago Press.

Wittenberg, James F. 1981. *Animal Social Behavior.* North Scituate, Mass: Duxbury Press.

Wöhrmann, K. 1990. "Genetic Variation." In *Population Biology: Ecological and Evolutionary Viewpoints,* ed. K. Wöhrmann and S. K. Jain, 7–26. Berlin: Springer Verlag.

Würsig, Bernard. 1989. "Cetaceans." *Science* 244 (June 30):1550–1557.

Wyles: Jeff S., and Joseph G. Kimbal, and Allan C. Wilson. 1983. "Birds, Behavior, and Anatomical Evolution." *Proceedings of the National Academy of Sciences USA* 80 (13–14):4394–4397.

Wynne-Edwards, V. C. 1984. "Intergroup Selection in the Evolution of Social Systems." In Robert N. Brandon and Richard M. Burian, eds., *Genes, Organisms, and Populations,* 42–51. Cambridge, Mass.: The MIT Press.

Young, J. Z. 1981. *The Life of the Vertebrates.* Oxford: Clarendon Press.

Zahavi, A. 1974. "Communal Nesting of the Arabian Babbler." *Ibis* 116:84–87.

Zeuner, Frederick E. 1963. *A History of Domesticated Animals.* New York: Harper & Row.

Zimmerman, Elwood C. 1960. "Possible Evidence of Rapid Evolution in Hawaiian Moths." *Evolution* 14:137–138.

Index